ADP-RIBOSYLATION REACTIONS: FROM BACTERIAL PATHOGENESIS TO CANCER

ADP-Ribosylation Reactions: From Bacterial Pathogenesis to Cancer

Edited by

RAFAEL ALVAREZ-GONZALEZ

Department of Molecular Biology and Immunology
University of North Texas Health Science Center
2500 Camp Bowie Boulevard
Fort Worth, TX 76107, USA

Reprinted from *Molecular and Cellular Biochemistry*, Volume 193 (1999)

Kluwer Academic Publishers

Dordrecht / Boston / London

Library of Congress Cataloging-in-Publication Data

ADP-ribosylation reactions: from bacterial pathogenesis to cancer /
edited by Rafael Alvarez-Gonzalez
p. cm. -- (Developments in molecular and cellular
biochemistry)
Papers from the 12th International Symposium on ADP-Ribosylation
Reactions, held in Cancun, Mexico, May 10–14, 1997.
ISBN 0-7923-8235-8 (alk. paper)
1. ADP-ribosylation--Congresses. 2. I. Alvarez-Gonzalez, Rafael,
II. International Symposium on ADP-Ribosylation Reactions (12th :
1995 : Cancun, Mexico) III. Series.
[DNLM: 1. Adenosine Diphosphate Ribose--metabolism congresses.
2. DNA Repair--physiology congresses. 3. Poly Adenosine Diphosphate
Ribose--metabolism congresses. W1DE998D 1998]
QP625.A29A364 1998
572'.7921--dc21
DNLM/DLC 98-29792
for Library of Congress CIP

ISBN 0-7923-8235-8

Published by Kluwer Academic Publishers,
P.O. Box 17, 3300 AA Dordrecht, The Netherlands.

Sold and distributed in North, Central and Latin America
by Kluwer Academic Publishers,
101 Philip Drive, Norwell, MA 02061, U.S.A.

In all other countries, sold and distributed by
Kluwer Academic Publishers,
P.O. Box 322, 3300 AH Dordrecht, The Netherlands.

Printed on acid-free paper

Printed in the Netherlands

Molecular and Cellular Biochemistry:

An International Journal for Chemical Biology in Health and Disease

CONTENTS VOLUME 193, Nos. 1 & 2, March 1999

ADP-RIBOSYLATION REACTIONS: FROM BACTERIAL PATHOGENESIS TO CANCER
Dr. Rafael Alvarez-Gonzalez

Dedication 1

Preface 3

F.R. Althaus, H.E. Kleczkowska, M. Malanga, C.R. Müntener, J.M. Pleschke, M. Ebner and B. Auer: Poly ADP-ribosylation: A DNA break signal mechanism 5–11
G. Pacheco-Rodriguez and R. Alvarez-Gonzalez: Measurement of poly(ADP-ribose) glycohydrolase activity by high resolution polyacrylamide gel electrophoresis: Specific inhibition by histones and nuclear matrix proteins 13–18
R. Alvarez-Gonzalez, T.A. Watkins, P.K. Gill, J.L. Reed and H. Mendoza-Alvarez: Regulatory mechanisms of poly(ADP-ribose) polymerase 19–22
S. Chatterjee, S.J. Berger and N.A. Berger: Poly(ADP-ribose) polymerase: A guardian of the genome that facilitates DNA repair by protecting against DNA recombination 23–30
J. Tatsumi-Miyajima, J.-H. Küpper, H. Takebe and A. Bürkle: *Trans*-dominant inhibition of poly(ADP-ribosyl)ation potentiates alkylation-induced shuttle-vector mutagenesis in Chinese hamster cells 31–35
D.A. Bobak: Clostridial toxins: Molecular probes of Rho-dependent signaling and apoptosis 37–42
M.G. Silletta, A. Colanzi, R. Weigert, M. Di Girolamo, I. Santone, G. Fiucci, A. Mironov, M.A. De Matteis, A. Luini and D. Corda: Role of brefeldin A-dependent ADP-ribosylation in the control of intracellular membrane transport 43–51
C. Trucco, V. Rolli, F. Javier Oliver, E. Flatter, M. Masson, F. Dantzer, C. Niedergang, B. Dutrillaux, J. Ménissier-de Murcia and G. de Murcia: A dual approach in the study of poly (ADP-ribose) polymerase: *In vitro* random mutagenesis and generation of deficient mice 53–60
R. Bortell, T. Kanaitsuka, L.A. Stevens, J. Moss, J.P. Mordes, A.A. Rossini and D.L. Greiner: The RT6 (Art2) family of ADP-ribosyltransferases in rat and mouse 61–68
E.L. Jacobson, W.M. Shieh and A.C. Huang: Mapping the role of NAD metabolism in prevention and treatment of carcinogenesis 69–74
J.-C. Amé, E.L. Jacobson and M.K. Jacobson: Molecular heterogeneity and regulation of poly(ADP-ribose) glycohydrolase 75–81
A.C. Boyonoski, L.M. Gallacher, M.M. ApSimon, R.M. Jacobs, G.M. Shah, G.G. Piorier and J.B. Kirkland: Niacin defiency increases the sensitivity of rats to the short and long term effects of ethylnitrosourea treatment 83–87
H.C. Lee, C. Munshi and R. Graeff: Structures and activities of cyclic ADP-ribose, NAADP and their metabolic enzymes 89–98
N. Raffaelli, M. Emanuelli, F.M. Pisani, A. Amici, T. Lorenzi, S. Ruggieri and G. Magni: Identification of the archaeal NMN adenylyltransferase gene 99–102
M. Miwa, S. Hanai, P. Poltronieri, M. Uchida and K. Uchida: Functional analysis of poly(ADP-ribose) polymerase in *Drosophila melanogaster* 103–108
J. Moss, E. Balducci, E. Cavanaugh, H.J. Kim, P. Konczalik, E.A. Lesma, I.J. Okazaki, M. Park, M. Shoemaker, L.A. Stevens and A. Zolkiewska: Characterization of NAD:arginine ADP-ribosyltransferases 109–113
H. Okamoto: The CD38-cyclic ADP-ribose signaling system in insulin secretion 115–118
R.W. Pero, B. Axelsson, D. Siemann, D. Chaplin and G. Dougherty: Newly discovered anti-inflammatory properties of the benzamides and nicotinamides 119–125
E.B. Affar, R.G. Shah and G.G. Poirier: Poly(ADP-ribose) turnover in quail myoblast cells: Relation between the polymer level and its catabolism by glycohydrolase 127–135
C.M. Simbulan-Rosenthal, D.S. Rosenthal, S. Iyer, H. Boulares and M.E. Smulson: Involvement of PARP and poly(ADP-ribosyl)ation in the early stages of apoptosis and DNA replication 137–148
M. Masutani, T. Nozaki, E. Nishiyama, T. Shimokawa, Y. Tachi, H. Suzuki, H. Nakagama, K. Wakabayashi and T. Sugimura: Function of poly(ADP-ribose) polymerase in response to DNA damage: Gene-disruption study in mice 149–152
J. Moss and M. Vaughan: Activation of toxin ADP-ribosyltransferases by eukaryotic ADP-ribosylation factors 153–157

Index to Volume 193 159–161

Molecular and Cellular Biochemistry **193:** 1, 1999.

Dedication

This special issue of *Molecular and Cellular Biochemistry* is dedicated to the 110 registered participants (representing 16 countries) of the '12th International Symposium on ADP-ribosylation Reactions; From Bacterial Pathogenesis to Cancer' that was held in Cancun, Mexico May 10–14, 1997. This issue is also dedicated to the Sociedad Mexicana de Bioquimica which celebrated its 40th Anniversary in 1997.

The photograph shows 91 of the International Symposium participants: *Upper line (from left to right)*- Cale Halbleib, Arnold Huang, Dr. James Slama, Dr. Katherine Wall, Dr. J.S. Miller, Dr. Todagashi Nozaki, Dr. Masanao Miwa, Dr. Yoshifurni Adachi, Dr. Val Adams, Dr. Hiroshi Okamoto, Dr. Kunihiro Ueda, Dr. Peter Zahradka, Dr. J.C. Arne, Dr. James B. Kirkland, Dr. Francis Schubert, Dr. Deborah Sardinha, Dr. Jianke Li, Gerald Schmid, Dr. John MacDermot, Dr. Mathias Ziegler, Joachim Griesenbeck, Dr. Shiao Li Oei, Mr. Scovassi, Dr. Girish Shah, Heinreich Berghammer, Dr. Anders Olsson, Dr. S Shall, and Dr. Gilbert deMurcia: *Middle line (from left to right)*- Dr. Zdenek Hostomsky, Dr. Minnie O'Farrell, Dr. S. Hostomska, Victor Suarez, Dr. Naoko Morinaga, Dr. Martha Vaughan, Dr. Ian Okasaki, Dr. Joel Moss, Dr. Enrico Balducci, Dr. Walter Patton, Dr Masatoshi Noda, David Koh, Dr. Melissa Shieh, Dr. Chinh Q. Vu, Donna Coyle, Dr. Maria diGirolamo, Dr. Pin Lan Li, Dr. Daniela Corda, Dr. Lina Ghibelli, Dr. Paola Caiafa, Cedric Muntener, Dr Piera Quesada, Jutta Pleschke, Dr. Barbara Saxty, Dr. Ana I. Scovassi, Dr. Maria D'Erme, Dr. Bernhard Auer, Sascha Benecke, Ralph Meyer, Dr. Zhao-Q. Wang, Dr. Yezhou Sheng, and Dr. Alexander Burkle; *Lower line (from left to right)*- Cecile LePage, Dr. J. Wietzerbin, Dr. Sunitha Kumari, Dr. Marietta Tuena de Gomez-Puyou, Dr. Satadel Chatterjee, Dr. Sosamma Berger, Dr. Nathan A. Berger, Dr. Mark Smulson, Sheila Srinivasan, Dr. Karen Maegley, Dr. Tracey Ruscetti, Dr. Elaine L. Jacobson, Dr. Myron K. Jacobson, Dr. Ronald Pero, Dr. Henry Meler, Dr. Marina Malanga, Dr. M. Rosario Faraonne-Menella, Dr. Marcia Puchi, Dr. Marla Imschenetzky, Dr. Armando Gomez-Puyou, Dr. Mitzuko Masutani, Dr. Nicola Curtin, Dr. Guy G. Poirier, and Dr. Mark Strohm; *Centre (from left to right)*- Dr. Rafael Alvarez-Gonzalez, Hilda Mendoza-Alvarez, and Dr. Giulio Magni.

Molecular and Cellular Biochemistry **193:** 3, 1999.

Preface

This special issue of *Molecular and Cellular Biochemistry* contains twenty two selected research papers and reviews from a total of one hundred and ten presentations registered at the '12th International Symposium on ADP-ribosylation Reactions: From Bacterial pathogenesis to Cancer' held in Cancun, Mexico from May 10–14, 1997. The Symposium was hosted by the Sociedad Mexicana de Bioquimica and was sponsored by the University of North Texas Health Science Center, Fort Worth, TX, USA. We hope that this special issue of *Molecular and Cellular Biochemistry* provides a state of the art source of information for basic scientists and clinicians who are interested in the molecular, biochemical, and cellular aspects of protein-(ADP-ribose) transfer reactions in human health and disease.

Enzymatically catalyzed ADP-ribose transfer reactions are ubiquitously found in biological systems from bacteria to human cells. The papers of this focused issue cover several examples of the diversity of functions fulfilled by (ADP-ribosyl)ation in a wide variety of living organisms. New developments and recent breakthroughs in the molecular and cellular aspects of: (i) cyclic ADP-ribose and calcium-dependent signal transduction pathways. (ii) the protein-mono (ADP-ribosyl)ation pathways and mechanisms of bacterial pathogenesis; (iii) niacin nutrition and cancer; as well as (iv) the role poly(ADP-ribosyl)ated DNA binding proteins in eucaryotic gene expression, DNA replication, DNA-repair, and programmed cell death (apoptosis) are highlighted in this issue.

Some of the examples include the discovery of: (i) the role of cyclic ADP-ribose, a cyclic nucleotide synthesized from βNAD^+ by an ADP-ribose cyclase activity associated with CD38 and/or NAD glycohydrolases, as an intracellular calcium mobilizing agent in signal transduction, insulin secretion, and cardiac contraction. In addition, this issue also highlights (ii) the cloning and sequencing of procaryotic and eucaryotic mono(ADP-ribosyl)transferases as well as a brief description of the physiological role of mono(ADP-ribosyl)transferases in either apoptosis or immune cell function. Another topic rapidly advancing in the area of protein-mono(ADP-ribosyl)ation is the participation of GTP-binding ADP-ribosylation factors (ARF's) in the regulation and modulation of secretory pathways. Major developments in the area of protein-poly(ADP-ribosyl)ation pertain to the status of (iii) niacin nutrition and cancer and the potential of tumor chemotherapy with newer and more potent inhibitors of ADP-ribose polymer synthesis. Recent advances In the areas of (iv) ADP-ribose polymer metabolism including: the x-ray crystal lographic and catalytic characterization of poly(ADP-ribose) polymerase (PARP), the enzyme that synthesizes protein-bound poly(ADP-ribose); the cloning, sequencing and characterization of mammalian of poly(ADP-ribose) glycohydrolase (PARG), the enzyme that degrades ADP-ribose polymers; the development of two additional transgenic mice that lack PARP (knockout mice) in France and Japan; the development of shuttle-vector mutagenesis techniques that evaluate the role of PARP in the repair of DNA damage; and the identification of DNA-binding proteins that heterodimerize with PARP or PARP-bound polymers, such as p53, in both normal cells and apoptotic cells following DNA-damage, are also emphasized in this issue.

The guest editor would like to express its gratitude to all the contributors to this volume for their sustained interest in the field of ADP-ribosylation reactions and for their participation in the '12th International Symposium on ADP-ribosylation Reactions'. Special thanks are extended to Oxigene, Inc. and Augoron pharmaceuticals, Inc. who sponsored this international event as well as to Dr. Myron K. Jacobson, Dr. Elaine L. Jacobson, and Dr. Joel Moss who provided invaluable help in putting together the scientific program for the Symposium as well as their support and guidance during the organization of the Symposium. The organizer is also grateful to Hilda Mendoza-Alvarez for her assistance with the multiple organizational details of the Symposium both in the USA and Mexico. Finally, all the authors of the papers published in this special issue of *Molecular and Cellular Biochemistry* would also like to express their gratitude to the Editor-in-Chief of this journal, Dr. Naranjan S. Dhalla, for publishing this collection of papers as a special issue. We also thank Dr. Peter Zahradka, Assistant Editor, for his help in finishing this task in a timely fashion.

RAFAEL ALVAREZ-GONZALEZ
Department of Molecular Biology and Immunology
University of North Texas Health Science Center
3500 Camp Bowie Boulevard
Fort Worth, TX
USA

Molecular and Cellular Biochemistry **193:** 5–11, 1999.

Poly ADP-ribosylation: A DNA break signal mechanism

Felix R. Althaus,[1] Hanna E. Kleczkowska,[1] Maria Malanga,[1] Cedric R. Müntener,[1] Jutta M. Pleschke,[1] Maria Ebner[2] and Bernhard Auer[2]

[1]*University of Zürich-Tierspital, Institute of Pharmacology and Toxicology, Zürich, Switzerland and* [2]*University of Innsbruck, Institute of Biochemistry, Innsbruck, Austria*

Abstract

Recent evidence obtained with transgenic knockout mice suggests that the enzyme poly(ADP-ribose)polymerase (PARP) does not play a direct role in DNA break processing [1, 2]. Nevertheless, inactivation of the catalytic or the DNA nick-binding functions of PARP affects cellular responses to genotoxins at the level of cell survival, sister chromatid exchanges and apoptosis [2, 3]. In the present report, we conceptualize the idea that PARP is part of a DNA break signal mechanism [4, 5]. *In vitro* screening studies revealed the existence of a protein family containing a polymer-binding motif of about 22 amino acids. This motif is present in p53 protein as well as in MARCKS, a protein involved in the regulation of the actin cytoskeleton. Biochemical analyses showed that these sequences are directly targeted by PARP-associated polymers *in vitro,* and this alters several molecular functions of p53- and MARCKS protein. PARP-deficient knockout mice from transgenic mice were found to exhibit several phenotypic features compatible with altered DNA damage signaling, such as downregulation and lack of responsiveness of p53 protein to genotoxins, and morphological changes compatible with MARCKS-related cytoskeletal dysfunction. The knockout phenotype could be rescued by stable expression of the PARP gene. - We propose that PARP-associated polymers may recruit signal proteins to sites of DNA breakage and reprogram their functions. (Mol Cell Biochem **193**: 5–11, 1999)

Key words: poly(ADP-ribose)polymerase, PARP-knockout mice, poly(ADP-ribose)-binding proteins, p53 protein, MARCKS protein

Introduction

We have previously shown that the enzymes poly(ADP-ribose) polymerase (PARP) and poly(ADP-ribose)glycohydrolase (PARG) may cooperate to shuttle histones off and onto DNA breakage sites [6]. *In vitro*, the shuttle mechanism establishes local accessibility for enzymes, which otherwise could not access histone-DNA complexes. We have previously speculated that this mechanism could play a role in DNA strand break processing, since *in vivo* DNA repair sites appear preferentially in unfolded regions of chromatin when probed with chemical or enzymatic probes [7]. However, with the availability of PARP-deficient cells from PARP-gene knockout mice, it became clear that PARP does not directly participate in DNA break processing but rather affects events downstream of DNA repair sites [1, 2]. This prompted us to reconsider the biological function of the shuttle mechanism.

In recent years it became clear that PARP is not the only enzyme capable of binding to DNA breaks. The tumor suppressor protein p53 is a likely competitor at DNA breaks [8] and so is the enzyme DNA-dependent protein kinase (DNA-PK; [9]). We noted that these proteins play a role in downstream signaling of DNA breaks although the complete mechanisms of action still remain to be elucidated. By analogy, PARP and recruitment of proteins to DNA breaks by PARP-associated polymers could play a role in DNA damage signaling. In fact, the PARP-PARG system exhibits all major hallmarks of a signal transduction mechanism (Fig. 1).

Address for offprints: F.R. Althaus, University of Zürich-Tierspital, Institute of Pharmacology and Toxicology, Winterthurerstrasse 260, CH-8057 Zürich, Switzerland

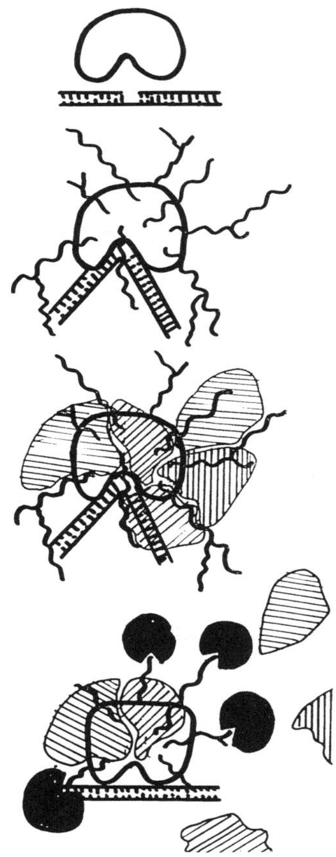

Fig. 1: Schematic view of the four major steps of DNA break signaling via the PARP-PARG system. *Sensor step,* PARP recognizes DNA strand break and binds to it. *Amplification step,* PARP binding stabilizes a V-shaped DNA conformation (for review see [10]), the catalytic activities increase 500-fold to produce PARP-bound polymers (this involves the formation of a catalytic PARP-PARP dimer [27, 28], which is ignored in the scheme). *Effector step,* PARP-associated polymers recruit proteins at the breakage site, the selection is specific for proteins containing a polymer-binding sequence motif. Some of the domain functions of polymer-bound proteins are temporarily suspended/altered ('reprogramming'). *Termination step,* PARG terminates the damage signal by degrading the polymers and thereby releasing polymer-bound proteins.

Sensor function

PARP is equipped with two zinc-fingers in the DNA binding domain which specifically recognize DNA single and double strand breaks [10]. The binding to the break triggers a 500-fold stimulation of the four catalytic activities required for ADP-ribose polymer synthesis. The formation of PARP-bound polymers - adding up to 350 kD of molecular mass to the 113 kD PARP-protein - causes a dramatic change in the microenvironment of the DNA break, equivalent to the amplification step in signal transduction. PARP-associated polymers represent a high density cluster of electronegative charges at the DNA breakage site [6]. The presumed effector function consists in the temporary recruitment of proteins from the vicinity of the DNA breaks, thereby disrupting their association with DNA or other proteins. This could lead to a temporary reprogramming of DNA damage singal pathways which would be terminated through the action of PARG. Thus, the PARP-PARG system could play a role in coupling DNA break processing with events downstream of DNA repair, such as cell cycling, control of genomic stability and cell death/survival programs [5].

In order to test some of the predictions of this model, we focused on two aspects. First, can we demonstrate direct interactions of signal proteins with PARP-associated polymers? Is there a polymer-binding sequence motif in these proteins which would allow us to define a family of polymer-binding proteins? Secondly, does the absence of PARP-gene functions in knockout-cells lead to signal disturbances? Can we restore these functions by stably reexpressing the PARP-gene in knockout cells? - Below, we review the evidence obtained so far supporting a role of poly ADP-ribosylation in DNA damage signaling.

Materials and methods

Polypeptides and oligonucleotides

Human and murine p53 proteins were immunopurified from Sf9 insect cells infected with recombinant baculoviruses as previously described [11] and were a generous gift of Dr. U. Hübscher (Department Biochemistry, University of Zürich-Irchel). Anti-p53 antibody Pab421 was from Calbiochem, Cambridge, Mass, USA, and anti-PARP antibody was produced in chicken. Custom-designed synthetic oligopeptides were obtained from Anawa Biomedical Services & Products, Wangen, Switzerland. Synthetic oligonucleotides were purchased from Microsynth (Switzerland). The following oligonucleotides were used: 5′-GACGAATGCGCCGC AAGAAGCCCCCATAGCGTTTGT-3′(36mer, N2) [4], 5′-CTAGAGGACATGCCCGGGCATGTCCT-3′ (26mer) and its complementary strand 5′AGGACATGCCCGGGCATGT CCTCTAG-3′, containing the palindromic consensus p53 target sequence [11, 12]. Single-stranded oligonucleotides were 5′-end-labeled with [^{32}P]γ-ATP and annealed to the complementary strand. (ADP-ribose) polymers were synthesized with purified PARP and isolated by affinity chromatography on dihydroxyboronate resin. They had an average chain length of 16 residues, with - 6% of total ADP-ribose moieties being incorporated in branched polymeric chains [13].

Cells from transgenic knockout mice

Embryonic fibroblasts from transgenic knock-out mice lacking a functional PARP gene were established in culture according to a standard 3T3-protocol [1]. 3T3-like fibroblasts negative for PARP expression were transformed with a linearized construct (starting from pREP 7, Invitrogen, Carlsbad CA, USA), containing human PARP cDNA [14] under the control of the RSV LTR promoter and a hygromycin resistance cassette by electroporation. After selection for stable transformants with 0.2 mg/ml hygromycin several clones were obtained expressing approximately the same protein levels of PARP as wild type 3T3 fibroblasts. After growth to confluency, cell lysates were processed for SDS-polyacrylamide gel electrophoresis and Western blots were performed with p53-antibody PAb421 from Calbiochem (Cambridge, MA, USA) as recommended by the supplier.

Poly(ADP-ribose) blot analyses

The polypeptides were blotted onto nitrocellulose membrane (Schleicher & Schuell, 0.2 µm pore size) using a Bio-Rad dot blot apparatus. Proteins on nitrocellulose were gold stained by using the Protogold staining kit (British BioCell International), as recommended by the manufacturer. Poly(ADP-ribose) blot analyses were performed as previously described [15]. Briefly, the nitrocellulose membrane was rinsed with three changes of Tris-buffered saline (TBS: 10 mM Tris, 0.15 M NaCl, pH 7.4) containing 0.05% (v/v) Tween 20 (TBST). Polymers of [^{32}P]ADP-ribose (0.5 µCi/nmol ADP-ribose; 0.5 nmol of total ADP-ribose) synthesized *in vitro* using purified PARP were diluted to 10 ml, with TBST and added to the nitrocellulose. After incubation for 1 h at room temperature, the membrane was extensively washed with TBST, dried and subjected to autoradiography.

DNA binding assays

Binding of p53 to ss- and ds-DNA was assayed by electrophoretic mobility shift [16]. Incubation of p53 with dsDNA oligonucleotide was carried out in 20 mM Hepes buffer, pH 7.9, containing 25 mM KCl, 2 mM $MgCl_2$, 0.1 mM EDTA, 0.025% NP-40, 0.1 mg/ml BSA, 10% glycerol, 0.2 ng ^{32}P-labeled probe DNA, 10–40 ng p53 protein, in a final volume of 20 µl. Where indicated, sonicated herring sperm DNA (SIGMA) was used as unspecific competitor (1–3 µg/ml). The reaction mixture was incubated at room temperature for 45 min and subsequently loaded onto a native 4% polyacrylamide gel containing 0.5×TBE buffer (45 mM Tris/borate, pH 8.3, 1 mM EDTA) and 0.05% NP-40. Samples were electrophoresed in 0.5 × TBE, at 200 V, for 90 min, at 4°C, and analyzed by autoradiography [16].

The binding of p53 to ssDNA was assayed by incubating the p53 protein (20–40 ng) in 40 mM Tris-HCl, pH 7.5, 1.5 mM dithiothreitol, 100 mM NaCl, 1 mM EDTA and 20 mg/ml BSA. The reaction was started by the addition of 0.3 ng (2 gl) of radiolabeled ssDNA. After incubation at 37°C for 30 min, reaction products were analyzed by electrophoresis in a non denaturing 5% polyacrylamide gel in 0.5 × TGE buffer (25 mM Tris/0.19 M glycine, pH 8.5, 1 mM EDTA), followed by autoradiography of the dried gel.

Effect of poly(ADP-ribose) on p53-DNA interactions

The effect of poly(ADP-ribose) on the DNA binding activities of p53 was assayed by preincubating the p53 protein (40 ng) in the absence or presence of the indicated amount of [^{3}H]poly(ADP-ribose) (160 dpm/pmol), for 15 min at room temperature in the binding buffer. [^{3}H]-labeling was used for a precise quantification of the ADP-ribose polymers. The reaction was started by the addition of 0.1–0.3 ng (1–2 µl) of radiolabeled DNA and the samples were analysed by electrophoretic mobility shift assay.

Results

Identification of a poly(ADP-ribose)- binding motif in nuclear proteins

We have previously shown that PARP-associated (or free) ADP-ribose polymers may specifically target histone tails for binding [15]. This binding involved very strong non-covalent interactions [17] as demonstrated by the resistance of the polymer-histone complexes to treatments with 0.2 M HCl, 0.2 M H_2SO_4, 1 M acetic acid, 1 M NaCl, 4 M urea, 0.2% Triton X-100 or 1% SDS. Based on this information, we searched for a putative polymer-binding sequence motif in the tail domain of histone H3 using synthetic oligopeptides of 20–26 amino acids in conjunction with the polymer-blot technique reported earlier [15]. The results helped identify some rules for a putative polymer-binding sequence motif which was used in our searches for further protein candidates from sequence databases. Testing these proteins as well as mutated versions of the polymer-binding sequence domains, we could further narrow down the essential elements for the polymer-binding motif. These included the presence of at least 3 hydrophobic amino acids spaced at 6 and 7 positions apart, and a combination of flanking and intervening clusters of positively charged amino acids (L.M. Pleschke, H.E. Kleczkowska & F.R. Althaus, manuscript in preparation). These motifs were found to be present in the tumor suppressor protein p53, which plays a prominent role in DNA damage signaling.

ADP-ribose polymers bind to specific domains of tumor suppressor protein p53

The ability of p53 to bind to ADP-ribose polymers was tested using the polymer blot assay as described previously.

Figure 2 shows a strong signal for poly(ADP-ribose)-binding to as little as 0.75 pmol of murine or human p53 protein. This binding was comparable to that of histone H1, which has a high affinity for poly(ADP-ribose) as previously reported [17]. No binding was observed with proteinase K or DNase I (Fig. 2), even if 400-times larger quantities were tested. Also, the polymer blot can be performed with automodified PARP which does not reduce the binding affinity of polymers (data not shown).

Biochemical identification of the polymer-binding sites in murine p53 was achieved with a set of synthetic oligopeptides containing partially overlapping sequences of murine p53 protein. These peptides were framed to contain a complete polymer-binding sequence as predicted from our sequence analyses. Three peptides of 23–26 amino acids tested positive in the polymer blot assay, while intervening and flanking sequences were negative. As shown schematically in Fig. 3, the peptides containing aa positions 153–178 and 231–253 were located in the sequence-specific DNA-binding domain of p53, while a third binding site (aa 326–348) coincided with the domain responsible for tetramerization of the protein.

We next determined whether poly(ADP-ribose)-binding could interfere with the DNA-binding functions of p53. First, sequence-specific binding was tested in an electrophoretic mobility shift assay using a 26 bp oligonucleotide containing the p53-target sequence as a probe [18]. The results showed that ADP-ribose polymers prevented sequence-specific DNA-binding of p53 in a dosedependent manner. This phenomenon was also observed when sonicated herring sperm DNA [5-fold excess over poly(ADP-ribose)] or a 26 nt single stranded DNA devoid of a p53-binding motif (125-fold excess) was present in the reaction mixture. Additional control experiments confirmed that heat-denatured p53 did not lead to complexes with the DNA probe or ADP-ribose polymers [18].

Apart from the sequence-specific binding function located in the core domain, p53 binds to single and double stranded DNA ends [8, 19]. This function maps to the C-terminal domain of p53 protein and has been correlated to the ability of p53 to promote DNA renaturation and strand transfer. A single-stranded radiolabeled oligonucleotide (36 nt) was used in an electrophoretic mobility shift assay and the action of poly(ADP-ribose) on p53-binding to ssDNA ends was determined and the autoradiographs of the gels were scanned for quantification. The results showed a dose-dependent inhibition of p53-ssDNA-complex formation. Interestingly, maximal inhibition of p53 binding was observed at poly(ADP-ribose) concentrations ranging from 40–100 mM (i.e. 0.8–2 pmoles in the reaction mixture), where sequence specific DNA binding was only moderately affected. Also, a comparison of the inhibition curves suggested different types of polymer interactions with the p53 core domains as compared to the C-terminal oligomerization domain [18].

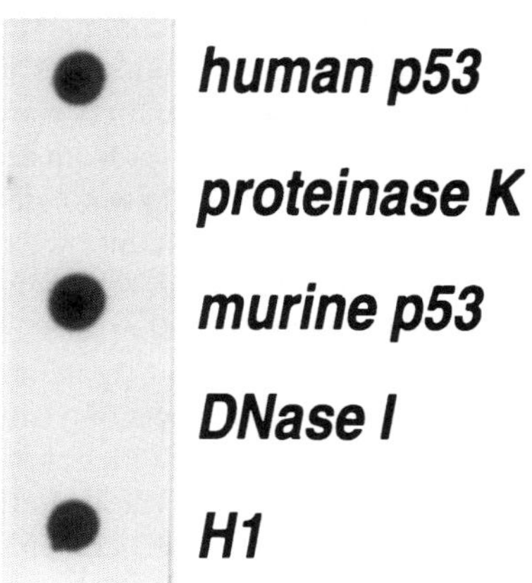

Fig. 2: Poly(ADP-ribose)-binding to murine and human p53. –40 ng of the indicated proteins were dot blotted onto nitrocellulose membrane and incubated with radiolabeled ADP-ribose polymers as described under Materials and methods. Protein-bound polymers were visualized by autoradiography. Histone H1, positive blot control; proteinase K and DNase I, negative controls.

ADP-ribose polymers bind to other (signal) proteins

The presence of a polymer-binding motif in MARCKS protein was identified by screening sequence databases with the putative motif and testing the purified protein in the polymer blot assay. MARCKS (*M*yristoylated-*A*lanine-*R*ich-*C*-*K*inase-*S*ubstrate) and MRP (*M*ARCKS-*R*elated-*P*rotein) belong to a protein family regulating rearrangernents of the actin cytoskeleton of cells [20, 21]. Polymer-blot analyses confirmed the presence of a 22 amino acid polymer-binding site in the effector domain of MARCKS and MRP. This domain is responsible for actin- and calmodulin-binding and is the target of phosphorylation by protein kinase C. Preliminary experiments indicate that polymer-binding may affect both calmodulin-binding and phosphorylation by protein kinase C *in vitro* (A. Schmitz, G. Vergères, J.M. Pleschke & F.R. Althaus, manuscript in preparation), suggesting that MARCKS/MRP functions could be further regulated by interactions with PARP-associated polymers.

Using the screening methods and biochemical testing, we have been able to identify a small family of proteins which share a common polymer-binding domain of 22–26 amino acids (Table 1). Among these are: p21 (WAF1/CIP1), DNA-PK catalytic and Ku70/86 subunits, XPA, and mismatch repair protein MSH6 (GTBP). All these proteins play a role in the expression of DNA damage, either at the level of cell cycle regulation (p21), damage recognition (XPA, MSH6), or downstream signaling (DNA-PK). - Further studies are

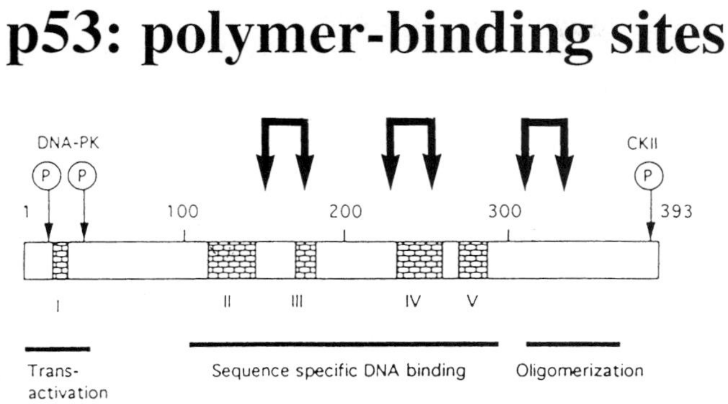

Fig. 3: Map of the p53 domain structure showing the sites of ADP-ribose polymer binding.Oligopeptides covering different regions of the p53 protein were tested in the polymer blot assay as described under Materials and methods. Arrows connected by bars indicate the sites of polymer-binding.

now aimed at determining the functional consequences of polymer-binding to these proteins and their association with modified PARP *in vivo*.

Searching for evidence of disturbed signal functions in PARP-knockout and PARP-knockout-rescue cells

Using 3T3-cells derived from PARP-deficient knockout mice, we found that p53 protein was almost entirely absent when probed by the Western blot technique. By contrast, PARP-proficient cells from control mice expressed very high levels of p53 protein. This suggested that the absence of a functional PARP-gene played a role in the regulation of constitutive p53 expression. That this phenomenon was causally related to PARP-gene function and not attributable to the genetic background of the transgenic animals was demonstrated following the stable transfection of the knockout cells with a functional PARP gene, since p53 levels recovered to normal levels in the knockout-rescue cells. Likewise, upregulation of p53 protein in response to inducers of apoptosis was deficient in knockout cells, but not in knockout-rescue cells (C.R. Müntener, J.M. Pleschke, H.E. Kleczkowska, M. Malanga, M. Ebner, B. Auer & F.R. Althaus, paper submitted).

Another phenotypic manifestation of PARP-deficiency in knockout cells - which was rescuable by transfection with a functional PARP-gene - was related to cytoskeletal disturbances. The knockout cells assumed a multinucleated, giant phenotype, spreading over a surface area 30–50 times larger than that of control cells, when plated at very low cells densities (e.g. ~ 200 cells/60 min plate). This finding was compatible with a disruption of actin regulation by MARCKS protein. However, the mechanistic basis of this phenomenon requires further investigations. Interestingly, it has been reported that MARCKS phosphorylation is completely inhibited under conditions where poly(ADP-ribose) synthesis is stimulated by treatment of 3T3-cells with an alkylating agent [22], suggesting a regulatory interdependence between MARCKS phosphorylation and poly ADP-ribosylation. The establishment of a functional linkage between PARP and MARCKS requires further investigations.

Table 1. Poly(ADP-ribose)-binding proteins so far identified*

p53
p21 (CDN1)
DNA-PKcs
Ku70/86
XPA
MSH6(MBP)
MARCKS
MRP
Histones

*based on polymer blot assay as described in Materials and methods

Discussion

Two lines of evidence presented here converge on the concept that PARP plays a role in DNA damage signaling. First, PARP has a DNA break sensor function [10]. Binding to DNA breaks triggers a dramatic response equivalent to amplification step in signal transduction pathways, i.e. PARP converts into a protein modified with a negatively charged polymer cluster and a drastic increase in the molecular mass of the modified protein. Our *in vitro* results demonstrate that the polymers may recruit proteins at DNA breakage sites and reprogram some of their domain functions. Among these proteins are p53 and MARCKS protein which play a role both in the regulation of constitutive as well as stress-induced functions of mammalian cells. The interaction with these proteins is exceptionally strong and involves a specific sequence-motif for binding. Secondly, evidence for a signal function of PARP *in vivo* could be derived from knockout and knockout-rescue cells. Deletion of the PARP-gene led to severe disturbances in the constitutive as well as stress-induced regulation of p53 protein. We speculate that PARP-associated ADP-ribose polymers could protect p53 protein from proteolytic degradation, which has been shown by others to be an important prerequisite for p53 upregulation following cellular exposure to DNA damaging agents [23]. Thus, PARP-associated polymers could be natural trans-activators of p53 protein. Preliminary evidence was also obtained for regulation of MARCKS protein function by PARP, although the mechanisms involved remain to be elucidated. MARCKS is a particularly interesting candidate for PARP regulation, since it controls important rearrangements of actin cytoskeletal functions and has been implicated in secretion and membrane trafficking processes, cellular motility, mitosis and transformation [20, 21].

The concept of PARP being part of a DNA damage signal mechanism also reconciles apparently controversial results from the literature. For example, many results obtained with different levels of inhibition of the catalytic activity of PARP (using competitive and non-competitive PARP inhibitors) were not consistent in themselves and most likely reflected different levels of disturbed signal functions in different cell systems [24]. Another confounding factor may be the extremely high doses of DNA damaging agents that have been used in the past to challenge the PARP system [24]. The high levels of DNA damage produced by these agents may not elicit the physiological damage signaling role of the PARP system and this may explain the discrepancy of results obtained with low or very low doses of PARP stimulators. In fact, PARP activation in human cells has an extremely low threshold, it is stimulated by nanomolar concentrations of alkylating agents that do not elicit a measurable DNA repair response but increase clonogenic survival of cells [25]. These findings further corroborate the signal function of PARP. In addition, different results obtained with inhibitor studies and molecular genetic approaches to disrupt PARP functions may simply reflect different targets of manipulation, e.g. catalytic functions vs. DNA-break binding of PARP. In the present concept, the DNA break-binding and activation of catalytic activities represent two consecutive steps of PARP-signaling whose disruption may produce different phenotypic responses in cells under genotoxic stress. Finally, as has been pointed out already [2–5], DNA strand break processing is apparently not a direct target of the PARP regulatory function but rather an indirect consequence of DNA damage signaling to keep DNA repair processes in tune with cell cycle events and downstream pathways leading to cellular recovery or death.

A major prediction and experimental challenge of the 'PARP-DNA break signal concept' is of course the direct demonstration of signal proteins associating with PARP-bound ADP-ribose polymers *in vivo*. Preliminary evidence in support of such interactions is the demonstration of a colocalization of the PARP protein with p53 in living cells as well as the reciprocal coprecipitation of PARP-p53 complexes from these cells by antibodies raised against either of the two partners [26]. These observations are compatible with our *in vitro* results suggesting the occurrence of PARP-p53-complexes *in vivo*.

Acknowledgements

We acknowledge the financial support from the Swiss National Foundation for Scientific Research (to F.R.A.) as well as from the Austrian Science Foundation and the Austrian Industrial Research Promotion Fund (to B.A.).

References

1. Wang ZQ, Auer B, Stingl L, Berghammer H, Haidacher D, Schweiger M, Wagner EF: Mice lacking ADPRT and poly(ADP-ribosyl)ation develop normally but are susceptible to skin disease. Genes Dev 9: 509–520, 1997
2. Ménissier-de Murcia J, Niedergang C, Trucco C, Ricoul M, Dutrillaux B, Marks M, Oliver FJ, Masson M, Dierich A, LeMeur M, Walztinger C, Chambon P, de Murcia G: Requirement of poly(ADP-ribose) polymerase in recovery from DNA damage in mice and in cells. Proc Natl Acad Sci USA 94: 7303–7307, 1997
3. Schreiber V, Hunting D, Trucco C, Gowans B, Grunwald D, de Murcia G, Ménissier-de Murcia J: Dominant-negative mutant of human poly(ADP-ribose)polymerase affects cell recovery, apoptosis, and sister chromatid exchange following DNA damage. Proc Natl Acad Sci USA 92: 4753–4757, 1995
4. Lindahl T, Satoh MS, Poirier GG, Klungland A: Posttranslational modification of poly(ADP-ribose)polymerase induced by DNA strand breaks. Trends Biochem Sci 20: 405–411, 1995
5. Althaus FR: Role of poly(ADP-ribose)polymerase in base excision repair. In: J.D. Hickson (ed) Base Excision Repair of DNA Damage. Springer/Landes Bioscience, Austin TX, 1997, pp. 169–181
6. Althaus FR: Poly ADP-ribosylation: A histone shuttle mechanism in DNA excision repair. J Cell Sci 102: 663–670, 1992
7. Mathis G, Althaus FR: Uncoupling of DNA excision repair and nucleosomal unfolding in poly (ADP-ribose)-depleted mammalian cells. Carcinogenesis 13: 135–138, 1990
8. Reed M, Woelker B, Wang P, Anderson MA, Tegtmeyer P: The C-terminal domain of p53 recognizes DNA damaged by ionizing radiation. Proc Natl Acad Sci USA 92: 9455–9459, 1995
9. Jackson SP, Jeggo PA: DNA double-strand break repair and V(D)J recombination: Involvement of DNA-PK. Trends Biochem Sci 20: 412–415, 1995
10. De Murcia G, Ménissier-de Murcia J: Poly(ADP-ribose)polymerase: A molecular nick sensor. Trends Biochem Sci 19: 172–176, 1994
11. Bargonetti J, Friedman PN, Kern SE, Vogelstein B, Prives C: Wild-type but not mutant p53 immunopurified proteins bind to sequences adjacent to the SV40 origin of replication. Cell 65: 1083–1091, 1991
12. Niewolik D, Vojtesek B, Kovarik J: p53-derived from human tumour cell lines and containing distinct point mutations can be activated to bind its consensus target sequence. Oncogene 10: 881–890, 1995
13. Malanga M, Bachmann S, Panzeter PL, Zweifel B, Althaus FR: Poly(ADP-ribose) quantification at the fentomol level in mammalian cells. Anal Biochem 228: 245–251, 1995
14. Kaiser P, Auer B, Schweiger M: Inhibition of cell proliferation in *Saccharomyces cerevisiae* by expression of human NAD-ADP-ribosyltransferase requires the DNA binding domain ('zinc finger'). Mol Gen Genet 232: 231–239, 1992
15. Panzeter PL, Zweifel B, Malanga M, Waser SH, Richard MC, Althaus FR: Targeting of histone tails by poly(ADP-ribose). J Biol Chem 268: 17662–17664, 1993
16. Jayaraman L, Prives C: Activation of p53 sequence-specific DNA binding by short single strands of DNA requires the p53 C-terminus. Cell 81: 1021–1029, 1995
17. Panzeter P, Realini C, Althaus FR: Noncovalent interactions of poly(adenosine diphosphate ribose) with histones. Biochemistry 31: 1379–1385, 1992

18. Malanga M, Pleschke JM, Kleczowska HE, Althaus FR: Poly(ADP-ribose) binds to specific domains of p53 and alters its DNA binding functions. J Biol Chem 273: 11839–11843, 1998
19. Bakalkin G, Selivanova G, Yakovleva T, Kiseleva E, Kashuba E, Magnusson KP, Szekely L, Klein G, Terenius L, Wiman KG: p53 binds single-stranded DNA ends through the C-terminal domain and internal DNA segments via the middle domain. Nucleic Acids Res 23: 362–369, 1995
20. Aderem AA: The MARCKS brothers: A family of protein kinase C substrates. Cell 71: 713–716, 1992
21. Hartwig JH, Thelen M, Rosen A, Janmey PA, Naim AC, Aderem AA: MARCKS is an actin filament crosslinking protein regulated by protein kinase C and calcium-calmodulin. Nature 356: 618–622, 1992
22. Shin I, Kam Y, Ha KS, Kang K, Joe CO: Inhibition of the phosphorylation of a myristoylated alanine-rich C kinase substrate by methyl methanesulfonate in cultured MH 3T3 cells. Mutat Res 351: 163–171, 1996
23. Li X, Coffino P: Identification of a region of p53 that confers lability. J Biol Chem 271: 4447–4451, 1996
24. Althaus FR, Richter C: ADP-ribosylation of proteins. Mol Biol Biochem Biophys 37: 1–125, 1987
25. Kleczkowska HE, Althaus FR: Response of human keratinocytes to extremely low concentrations of N-methyl-N′-nitro-N-nitrosoguanidine. Mutat Res 367: 151–159, 1996
26. Wesierska-Gadek J, Bugajaska-Schretter A, Cemi C: ADP-ribosylation of p53 tumor suppressor protein: Mutant but not wild-type p53 is modified. J Cell Biochem 62: 90–101, 1996
27. Mendoza-Alvarez H, Alvarez-Gonzalez R: Poly(ADP-ribose) polymerase is a catalytic dimer and the automodification reaction is intermolecular. J Biol Chem 268: 22575–22580, 1993
28. Panzeter P L, Althaus F R: DNA strand break-mediated partitioning of poly(ADP-ribose)polymerase function. Biochemistry 33: 9600–9605, 1994

Molecular and Cellular Biochemistry **193:** 13–18, 1999.

Measurement of poly(ADP-ribose) glycohydrolase activity by high resolution polyacrylamide gel electrophoresis: Specific inhibition by histones and nuclear matrix proteins

Gustavo Pacheco-Rodriguez and Rafael Alvarez-Gonzalez

Department of Molecular Biology and Immunology, University of North Texas Health Science Center at Fort Worth, Texas, USA

Abstract

We have developed a novel enzyme assay that allows the simultaneous determination of noncovalent interactions of poly(ADP-ribose) with nuclear proteins as well as poly(ADP-ribose) glycohydrolase (PARG) activity by high resolution polyacrylamide gel electrophoresis. ADP-ribose chains between 2 and 70 residues in size were enzymatically synthesized with pure poly(ADP-ribose) polymerase (PARP) and were purified by affinity chromatography on a boronate resin following alkaline release from protein. This preparation of polymers of ADP-ribose was used as the enzyme substrate for purified PARG. We also obtained the nuclear matrix fraction from rat liver nuclei and measured the enzyme activity of purified PARG in the presence or absence of either histone proteins or nuclear matrix proteins. Both resulted in a marked inhibition of PARG activity as determined by the decrease in the formation of monomeric ADP-ribose. The inhibition of PARG was presumably due to the non-covalent interactions of these proteins with free ADP-ribose polymers. Thus, the presence of histone and nuclear matrix proteins should be taken into consideration when measuring PARG activity. (Mol Cell Biochem **193**: 13–18, 1999)

Key words: histone(s), nuclear matrix, poly(ADP-ribose), glycohydrolase, enzyme assay

Introduction

Poly(ADP-ribosyl)ation of proteins is a reversible post-translational modification reaction elicited by DNA strand breaks [1]. The synthesis and degradation of poly(ADP-ribose) are catalyzed by poly(ADP-ribose) polymerase (PARP) [E.C.2.4.2.30], a DNA-dependent enzyme, and poly(ADP-ribose) glycohydrolase (PARG), respectively [2]. This pathway plays a role in modulating chromatin structure and function [3]. ADP-ribose polymers are covalently linked to PARP, which is the main protein acceptor, both *in vitro* [4] and *in vivo* [5]. Other protein targets for covalent poly(ADP-ribosyl)ation in the eucaryotic cell nucleus include histone [6] and nuclear matrix proteins [7]. Experiments with eucaryotic cells in metaphase show that PARP and other nuclear matrix proteins, such as the lamins, are poly(ADP-ribosyl)ated [8] in a cell cycle dependent manner [9]. Not surprisingly, it has also been shown that a significant fraction of basal poly(ADP-ribose) is tightly associated with the nuclear matrix [10]. Since histone proteins bind non-covalently to poly(ADP-ribose) [11], it is likely that the association of poly(ADP-ribose) with the nuclear matrix is mediated by both covalent and non-covalent interactions. Should noncovalent interactions of poly(ADP-ribose) molecules with chromatin proteins occur *in vivo,* it may result in the protection of ADP-ribose polymers from the degradative function of PARG. Here, we have evaluated the effects of nuclear matrix and individual histone proteins on the activity of PARG. In order to achieve this objective, we have developed a highly sensitive and reproducible assay to

Address for offprints: R. Alvarez-Gonzalez, Department of Molecular Biology and Immunology, University of North Texas Health Science Center, 3500 Camp Bowie Boulevard, Fort Worth, Texas 76107, USA

simultaneously determine non-covalent interactions of chromatin proteins with free ADP-ribose polymers and the enzymatic activity of PARG by high resolution PAGE.

Materials and methods

Materials

β-NAD$^+$, and ADP-ribose were purchased from Sigma Chemical Co. (St. Louis, MO, USA). Histone H1, H2A, H2B, H3 and H4 were obtained from Boehringer, Mannheim (Indianapolis, IN, USA). [Adenylate-^{32}P]NAD (250 Ci/mmol) was from ICN Biomedicals (Irvine, CA). Electrophoresis reagents were purchased from Bio-Rad (Richmond, CA, USA). The Partisil 10-SAX column was supplied by Whatman Inc. (Bridewell Place, NJ, USA).

Purification of PARP

Poly(ADP-ribose)polymerase was purified from calf thymus by the method of Zahradka and Ebisuzaki [12].

Synthesis and purification of protein-free poly(ADP-ribose)

Protein-free [^{32}P]Poly(ADP-ribose) was synthesized in a 500 µl reaction mixture containing 75 nM PARP, 100 mM Tris-HCl, pH 8.0, 10 mM DTT, 10 mM $MgCl_2$, 20 µg/ml of active calf thymus DNA and 10 µM [^{32}P]NAD$^+$. The reaction mixture was incubated for 10 min at 37°C. Next, the 20% trichloroacetic acid precipitable material was obtained. Subsequently, this material was incubated with 0.2 N NaOH containing 20 mM EDTA for 2 h at 60°C to release polymers from protein. Finally, protein-free [^{32}P]poly(ADP-ribose) was purified on dihydroxyboronyl-Bio Rex as described by Alvarez-Gonzalez *et al.* [13].

Isolation of nuclei

Nuclei were isolated from rat liver by the method of Blobel and Potter [14].

Isolation of the nuclear matrix

Rat liver nuclei were fractionated as described by Alvarez-Gonzalez and Ringer [15]. Briefly, nuclei were incubated at 37°C for 45 min to allow the action of endogenous nucleases. Next, digested nuclei were centrifuged (780 × g) for 15 min at 4°C. The pellet was resuspended in 10 mM Tris-HCl, pH 7.4 (Buffer A) containing 200 µM $MgCl_2$. Next, three extractions were carried out with buffer A, 200 µM $MgCl_2$ to extract chromatin. The bulk amount of chromatin was eliminated with Buffer A, 200 µM $MgCl_2$, 2 M NaCl. Finally, the nuclear matrix was obtained by treatment of the remaining subnuclear structures with 1% Triton X-100 and 5 mM $MgCl_2$. After washing the pellet twice in buffer A, 5 mM $MgCl_2$, to remove Triton X-100, nuclear matrices were resuspended in buffer A containing 5 mM $MgCl_2$. Subnuclear fractions obtained were kept at –70°C until used.

Purification of PARG

Poly(ADP-ribose) glycohydrolase was partially purified from calf thymus by a 30–60% ammonium sulfate cut. Next, the precipitated material was passed through Sephadex G-25 and the eluate was applied to a DNA-cellulose column. Fractions containing PARG activity were pooled and used in this study as partially purified PARG.

Enzyme assay for PARG

The enzyme assay was carried out in a 25 or 50 µl reaction mixture containing 100 mM Tris-HCl, pH 8.0, 10 mM DTT, 30 nM [^{32}P]poly(ADP-ribose) and 6 µg/ml of purified PARG. The reaction mixture was incubated for 5 min at 37°C. The reaction was then stopped with an equal volume of electrophoresis loading buffer and the samples were placed on ice. To evaluate the effect of either nuclear matrix or histone proteins, the same assay was carried out in the presence of the indicated amounts of each protein mixture. The products generated were separated by high resolution polyacrylamide gel electrophoresis [16]. Finally, the products were analyzed by spot densitometry using the IS-1000 version 2.00 digital imaging system from Alpha Innotech Corp. (CA, USA).

Chromatographic identification of the reaction product generated by incubating [^{32}P]poly(ADP-ribose) with purified PARG

[^{32}P]poly(ADP-ribose) was quantitatively hydrolyzed with purified PARG. [^{32}P]poly(ADP-ribose) (30 nM) was treated with 6 µg/ml of PARG in a 50 µl reaction mixture for 120 min. The enzyme incubation mixture contained 50 mM KH_2PO_4, and 10 mM $MgCl_2$. Hydrolytic products were subsequently separated by HPLC on a Whatman Partisil 10-SAX column (250 mm × 4.6 mm i.d.) preceded by a guard column (50 mm × 1.5 mm i.d.). Initial chromatographic conditions were 125 mM KH_2PO_4, pH 4.7 (buffer B). Next, a linear gradient was established using 125 mM KH_2PO_4, pH 4.7 containing 0.5 KCl (buffer C).

Results

Enzyme assay for PARG

A highly sensitive and reproducible method to measure poly(ADP-ribose) glycohydrolase assay by high resolution polyacrylamide gel electrophoresis was developed. First, [^{32}P]poly(ADP-ribose) was synthesized with homogeneously pure poly(ADP-ribose) polymerase from calf thymus and [^{32}P] NAD^+. Next, the polymer was chemically detached from protein with 0.2 N KOH, 20 mM EDTA and subsequently purified by affinity chromatography on dihydroxyboronyl-Bio Rex. This boronate does not bind AMP [13], the alkaline product of free monomeric ADP-ribose. Therefore, our substrate preparation for PARG was completely devoid of monomeric ADP-ribose (Fig. 1A, lane 1). This was particularly important since monomeric ADP-ribose is the main degradation product of ADP-ribose chains generated by PARG [2]. The electrophoretic distribution of these polymers

B

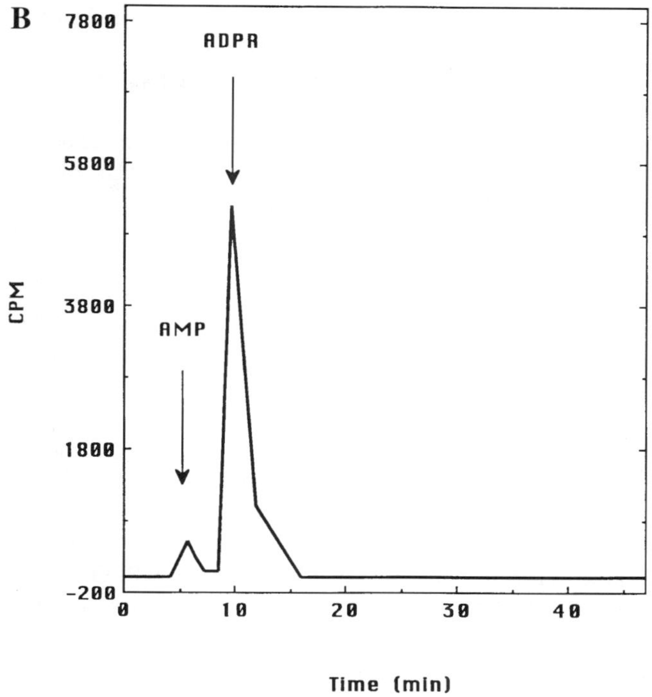

C

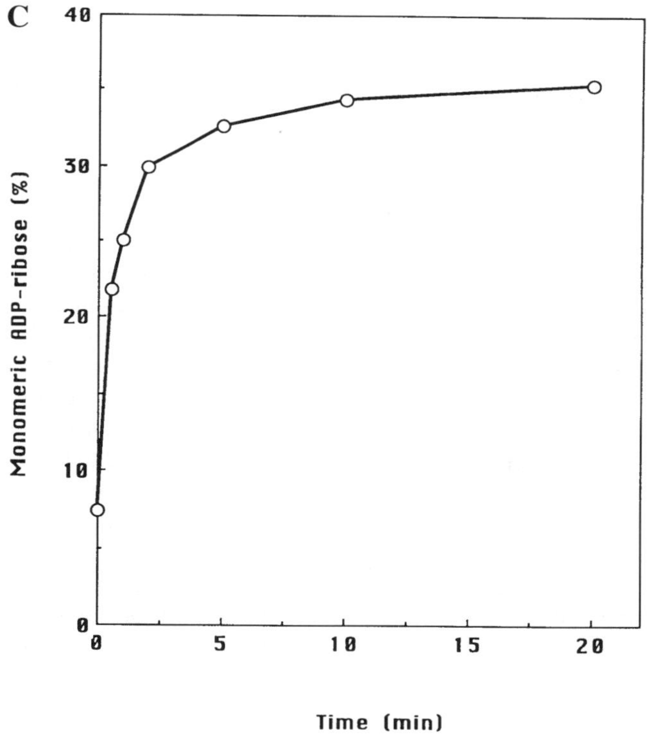

A

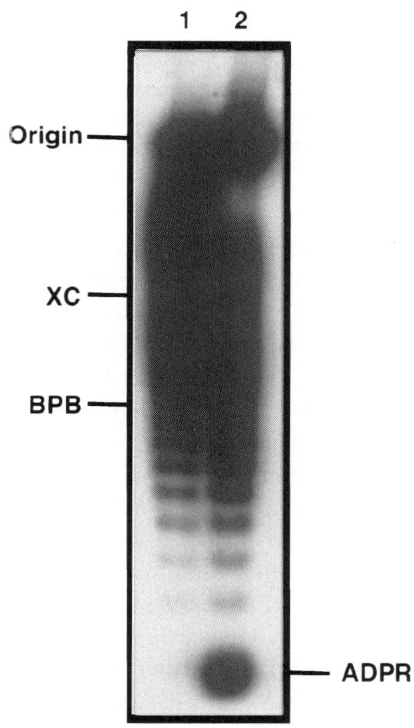

Fig. 1. (A) Measurement of PARG activity by high resolution polyacrylamide gel electrophoresis. Lane 1, control, [^{32}P] (ADP-ribose)$_{2-70}$ (1.5 nmoles). Lane 2, polymers (1.5 nmoles) following incubation with rat liver nuclei (25 μg of protein). The reaction was carried out for 10 min at 37°C. Bromophenol blue (BPB) and xylene cyanol (XC) co-migrate with ADP-ribose chains of 8 and 20 residues, respectively; (B) Chromatographic identification of ADP-ribose, the hydrolytic product generated by PARG. HPLC separation was carried out as described under Materials and methods; (C) Time-dependent formation of monomeric ADP-ribose in incubations of 50 μl containing 30 nM [^{32}P] poly(ADP-ribose) and 6 μg/ml of PARG.

showed chains of 2–70 residues in size. As expected, Fig. 1A, lane 2 shows that incubation of protein-free [^{32}P]poly-(ADP-ribose) with rat liver nuclei generated [^{32}P] radiolabeled monomeric ADP-ribose as a result of the activity of endogenous PARG. Figure 1A, lane 2 also shows that nuclear proteins interacted non-covalently with free poly(ADP-ribose) because the amount of radiolabeled material detected at the origin of the gel increased significantly following incubation with the nuclear extract.

Characterization of the enzymatic products generated by PARG

Hydrolytic products of PARG activity were obtained from an incubation of [^{32}P] (ADP-ribose)$_{2-70}$ and purified PARG and subsequently separated by SAX-HPLC (Fig. 1B). Two [^{32}P] radiolabeled compounds were observed. These peaks co-eluted with authentic ADP-ribose and AMP. Once we confirmed that the main nucleotide generated was monomeric ADP-ribose, we proceeded to measure its formation as a function of time by high resolution PAGE. The kinetics of PARG activity with (ADP-ribose)$_{2-70}$ as a substrate is shown in Fig. 1C.

Specific inhibition of PARG activity by histone and nuclear matrix proteins

Poly(ADP-ribose) interacts non-covalently with nuclear proteins as detected by the mobility shift of poly(ADP-ribose) molecules of 20–30 residues in size to the origin of the gel (Fig. 1A, lane 2 and Fig. 2, lane 2). This observation confirms previous reports in the literature regarding the strong ionic interactions of free ADP-ribose chains with histone proteins [11]. Figure 3A shows the effect of each individual histone protein on the activity of purified PARG. When the concentration of each histone protein was equi-molar with that of free ADP-ribose polymers, histone H2B proved to be the weakest inhibitor of poly(ADP-ribose) degradation. Surprisingly, when free polymers were pre-incubated with increasing amounts of the nuclear matrix extract, a decrease in the formation of monomeric ADP-ribose was also observed (Fig. 3B) following densitometric scanning of the X-ray films. In summary, both histone and nuclear matrix proteins inhibit PARG activity, presumably via their ionic interactions with free ADP-ribose polymers.

Discussion

PARG has a very high affinity for long ADP-ribose chains [17]. However, ionic interactions of histone proteins with free

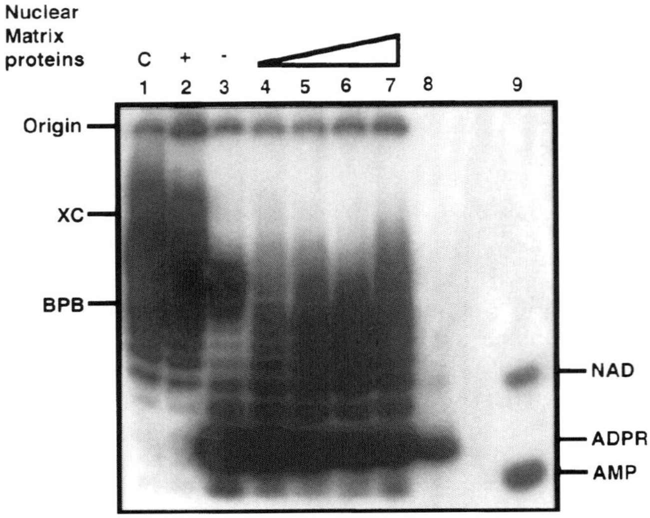

Fig. 2. Size distribution of ADP-ribose polymers following incubation with 150 ng of PARG in the presence of increasing amounts of nuclear matrix proteins. A 25 μl reaction mixture containing 30 nM [^{32}P] poly-(ADP-ribose) and increasing amounts of nuclear matrix protein was pre-incubated for 5 min at 37°C. Next, PARG (6 μg/ml) was added and the reaction was carried out for 5 min at 37°C. The radiolabeled products were resolved by high resolution polyacrylamide gel electrophoresis (20% gels). Lane 1, control, [^{32}P] (ADP-ribose)$_{2-70}$; lane 2, incubation of the substrate with nuclear matrix (336 μg of protein/ml) in the absence of PARG; lane 3, incubation of the (ADP-ribose)$_{2-70}$ with PARG only; lanes 4–7 correspond to incubations of [^{32}P] poly(ADP-ribose) with PARG in the presence of 168, 336, 672, 1344 μg of protein/ml of nuclear matrix, respectively; lane 8, [^{32}P] (ADP-ribose); lane 9, [^{32}P] NAD$^+$, and [^{32}P] AMP. Bromophenol blue (BPB) and xylene cyanol (XC) co-migrate with (ADP-ribose)$_8$ and (ADP-ribose)$_{20}$, respectively.

poly(ADP-ribose) appears to strongly inhibit PARG activity [11]. Here, we observe that a nuclear matrix protein extract also inhibits PARG in a concentration-dependent manner. Due to the fact that the nuclear matrix is devoid of histone proteins, we conclude that some of the nuclear matrix proteins also interact ionically with free poly(ADP-ribose) and inhibit PARG. Our conclusion is compatible with the previously reported non-covalent interactions of poly(ADP-ribose) with histone proteins as examined by: (i) a phenol/partitioning assay [11]; (ii) a poly(ADP-ribose)-western blotting approach [18], and a photoaffinity labeling procedure [19]. In our poly(ADP-ribose) preparations, the lack of monomeric ADP-ribose allowed us to simultaneously monitor PARG activity as well as non-covalent interactions of long polymers with chromatin protein(s). Nozaki *et al.,* [18] showed that the polypeptides of a HeLa cell nuclear extract strongly interacted non-covalently with ADP-ribose chains larger that 20 residues.

Although histone proteins were the most prominent proteins identified, it should be noted that some of the most efficient polypeptides that blot radiolabeled poly(ADP-ribose) electrophoretically migrate in the molecular weight range of

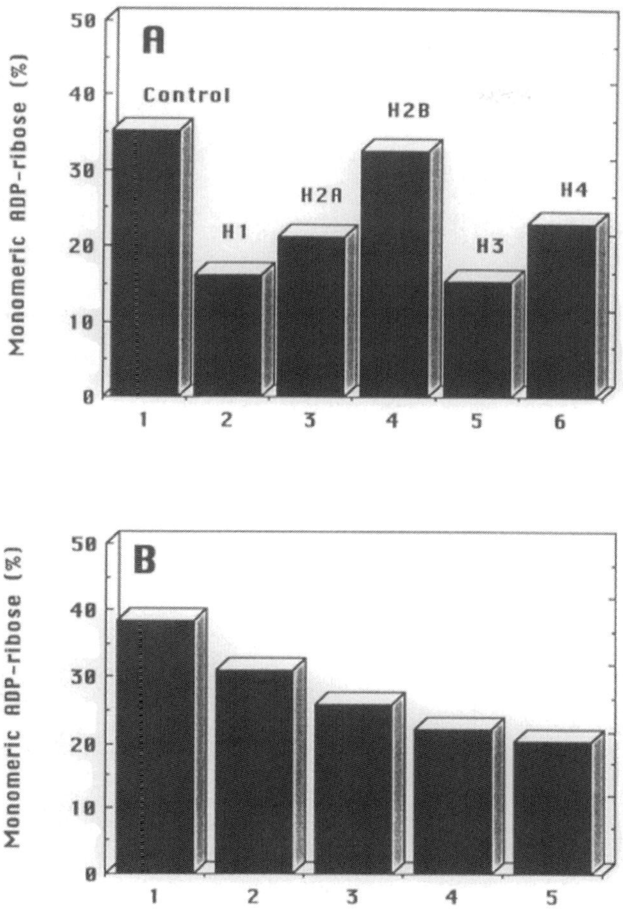

Fig. 3. Inhibition of purified PARG by either histone or increasing amounts of nuclear matrix proteins. (A) Inhibition of PARG by histone proteins. Incubations of 50 µl contained 1.5 nmoles of poly(ADP-ribose) and an equal amount of the indicated histone. The enzyme assays were carried out as described under Materials and methods. Bar 1, control, (monomeric ADP-ribose formed in the absence of histone proteins. Bars 2–6 show the percent of monomeric ADP-ribose product in the presence of different histone proteins; (B) Inhibition of PARG by a nuclear matrix extract. The incubation conditions of the reaction are described in legend to Fig. 2. The hydrolytic products were resolved by high resolution PAGE on a 20% gel, and the percent of ADP-ribose formed was quantified by densitometric analysis. Bar 1, control (monomeric ADP-ribose) formed in the absence of nuclear matrix; Bars 2–5 show the percent of monomeric ADP-ribose produced in the presence of 168, 336, 672, 1344 µg of protein/ml, respectively.

the nuclear matrix lamins. Our conclusion is also consistent with the notion that the nuclear matrix contains a large proportion of cationic proteins [20]. In fact, ADP-ribose polymers associated with the nuclear matrix requires high ionic strength buffers for extraction [10].

The specific inhibition of PARG by histone and nuclear matrix proteins also supports the hypothesis that non-covalent interactions of poly(ADP-ribose) with cationic proteins play an important regulatory role in poly(ADP-ribose) catabolism. Given the fact that PARP and enzyme-bound polymers of ADP-ribose are strongly associated with the nuclear matrix [15], it is possible that PARG activity may be associated with the nuclear matrix fraction as well. Studies in this laboratory that address this question are currently under way.

Acknowledgement

This project was partially supported by grants GM45451 from NIH and 9678-014 from the Texas Advanced Research Program.

References

1. Benjamin RC, Gill DM: ADP-ribosylation in mammalian cell ghosts. Dependence of poly(ADP-ribose) synthesis on strand breakage in DNA. J Biol Chem 255: 10493–10510, 1980
2. Miwa M, Sugimura T: Splitting of the ribose-ribose linkage of poly(adenosine diphosphate-ribose) by a calf thymus extract. J Biol Chem 246: 6362–6364, 1971
3. De Murcia GM, Huletsky A, Poirier GG: Modulation of chromatin structure by poly(ADP-ribosyl)ation. Biochem Cell Biol 66: 626–635, 1988
4. Ogata N, Ueda K, Kawaichi M, Hayaishi O: Poly(ADP-ribose) synthetase, a main acceptor of poly(ADP-ribose) in isolated nuclei. J Biol Chem 256: 4135–4137, 1981
5. Adamietz P: Poly(ADP-ribose) synthase is the major endogenous nonhistone acceptor for poly (ADP-ribose) in alkylated rat hepatoma cells. Eur J Biochem 169: 365–372, 1987
6. Adamietz P, Rudolph A: ADP-ribosylation of nuclear proteins *in vivo*. Identification of histone H2B as a major acceptor for mono- and poly(ADP-ribose) in dimethyl sulfate-treated hepatoma AH 7974 cells. J Biol Chem 259: 6841–6846, 1984
7. Adolph KW, Song MKH: Variations in ADP-ribosylation of nuclear scaffold proteins during the HeLa cell cycle. Biochem Biophys Res Commun 126: 840–847, 1985
8. Song HMK, Adolph KW: ADP-ribosylation of nonhistone proteins during the HeLa cell cycle. Biochem Biophys Res Commun 115: 938–945, 1983
9. Adolph KW: ADP-ribosylation of nuclear proteins labeled with [^{3}H]adenosine: Changes during the HeLa cycle. Biochem Biophys Acta 909: 222–230, 1987
10. Cardenas-Corona ME, Jacobson EL, Jacobson MK: Endogenous polymers of ADP-ribose are associated with the nuclear matrix. J Biol Chem 262: 14863–14866, 1987
11. Panzeter PL, Realini CA, Althaus FR: Noncovalent interactions of poly(adenosine diphosphate ribose) with histones. Biochemistry 31: 1379–1385, 1992
12. Zahradka P, Ebisuzaki K: Poly(ADP-ribose) polymerase is a zinc metalloenzyme. Eur J Biochem 142: 503–509, 1984
13. Alvarez-Gonzalez R, Juarez-Salinas H, Jacobson EL, Jacobson MK: Evaluation of immobilized boronates for studies of adenine and pyridine nucleotide metabolism. Anal Biochem 135: 69–77, 1983
14. Blobel G, Potter VR: Nuclei from rat liver: Isolation method that combines purity with high yield. Science 154: 1662–1665, 1966
15. Alvarez-Gonzalez R, Ringer DP: Nuclear matrix associated poly(ADP-ribose) metabolism in regenerating rat liver. FEBS Lett 236: 362–366, 1988

16. Alvarez-Gonzalez R, Jacobson MK: Characterization of polymers of adenosine diphosphate ribose generated *in vitro* and *in vivo*. Biochemistry 26: 3218–3224, 1987
17. Hatakeyama K, Nemoto Y, Ueda K, Hayaishi O: Purification and characterization of poly(ADP-ribose) glycohydrolase. Different modes of action on large and small poly(ADP-ribose). J Biol Chem 261: 14902–14911, 1986
18. Nozaki T, Masutani M, Akagawa T, Sugimura T, Esumi H: Non-covalent interaction between poly(ADP-ribose) and cellular proteins: an application of a poly(ADP-ribose)-Western blotting method to detect poly(ADP-ribose) binding on protein-blotted filter. Biochem Biophys Res Commun 198: 45–51, 1994
19. Strohm MS, Jacobson MK: Photoaffinity labelling of core histones and histone H1 using photoactive ADP-ribose polymers. FASEB J 10: A699, 1996
20. Berezney R, Mortillaro MJ, Ma H, Wei X, Samarabandu J: The nuclear matrix: A structural milieu for genomic function. Int Rev Cytol 162A: 2–66, 1995

Molecular and Cellular Biochemistry **193**: 19–22, 1999.

Regulatory mechanisms of poly(ADP-ribose) polymerase

Rafael Alvarez-Gonzalez, Trent A. Watkins, Paramjit K. Gill, Jason L. Reed and Hilda Mendoza-Alvarez

The Department of Molecular Biology & Immunology, University of North Texas Health Science Center at Fort Worth, Texas, USA

Abstract

Here, we describe the latest developments on the mechanistic characterization of poly(ADP-ribose) polymerase (PARP) [EC 2.4.2.30], a DNA-dependent enzyme that catalyzes the synthesis of protein-bound ADP-ribose polymers in eucaryotic chromatin. A detailed kinetic analysis of the automodification reaction of PARP in the presence of nicked dsDNA indicates that protein-poly(ADP-ribosyl)ation probably occurs via a sequential mechanism since enzyme-bound ADP-ribose chains are not reaction intermediates. The multiple enzymatic activities catalyzed by PARP (initiation, elongation, branching and self-modification) are the subject of a very complex regulatory mechanism that may involve allosterism. For instance, while the NAD^+ concentration determines the average ADP-ribose polymer size (polymerization reaction), the frequency of DNA strand breaks determines the total number of ADP-ribose chains synthesized (initiation reaction). A general discussion of some of the mechanisms that regulate these multiple catalytic activities of PARP is presented below. (Mol Cell Biochem **193**: 19–22, 1999)

Key words: poly(ADP-ribose)polymerase, kinetics, allosterism, regulation

Introduction

The synthesis of protein-bound ADP-ribose polymers in eucaryotic chromatin is catalyzed by poly(ADP-ribose) polymerase (PARP) [E.C. 2.4.2.30], a homodimeric enzyme of 1014 amino acids (113 kDa)/subunit that utilizes βNAD^+ as the ADP-ribose donor and DNA-binding proteins (heterodimeric), including itself, as ADP-ribose polymer covalent acceptors. The ADP-ribose polymerizing activity of this enzyme (elongation and branching) is substantially increased in the presence of nicked or broken dsDNA [1, 2]. As a result, PARP covalently modifies a number of nucleic acid-binding proteins with a very strong polyanion [3–11].

At present, the biochemical signal(s) (regulation) that determine the acceptor specificity of PARP as well as the magnitude of poly(ADP-ribosyl)ation (number of ADP-ribose chains/acceptor molecule and the chain length of the polymer synthesized) are not known in detail. Below, we review the structural properties of PARP as they pertain to the catalytic function(s) of this enzyme.

Primary structure of PARP

Three distinct structural domains of PARP have been identified following limited proteolysis of pure enzyme with papain and/or α-chymotrypsin [12–14]. These domains correspond to: (i) a 46 kDa DNA-binding amino-terminal domain; (ii) a 22 kDa centrally located automodification domain; and (iii) a 54 kDa carboxy-terminal catalytic domain. The DNA-binding domain contains a duplicated sequence (amino acid residues 2–97 and 106–207) in which 35 amino acids are highly conserved. These peptides show the typical 'zinc-finger' structure of the form $CXaa_2$-C-Xaa_{28-30}-H-Xaa_2-C that is found in several DNA-binding proteins. In fact, zinc fingers 1 and 11 allow the enzyme to specifically bind double [15] and single strand-breaks [16] on DNA, respectively.

The automodification domain of PARP contains protein-protein binding motifs, e.g., peptide sequences involved in both homo- and heterodimerization [18] recognition. In this catalytic dimerization process, protein-protein interactions appear to stabilize homodimeric and heterodimeric (PARP)-DNA

Address for offprints: R. Alvarez Gonzalez, Department of Molecular Biology and Immunology, University of North Texas Health Science Center, 3500 Camp Bowie Boulevard, Fort Worth, TX 76107-2690, USA

complexes [19–22] which in turn facilitate intermolecular protein-poly(ADP-ribosyl)ation. This domain also contains 15 highly conserved Glu residues [17]. These Glu's include some of the auto-poly(ADP-ribosyl)ation sites.

The carboxy-terminal region of PARP, the 54 kDa NAD-binding domain, contains Glu 988, a highly conserved amino acid residue that is directly involved in catalysis of the ADP-ribose transfer reaction carried out by procaryotic [23, 24] and eucaryotic [23–26] enzymes.

Catalytic activities associated with PARP

The enzymatic automodification reaction of PARP as well as the heterologous poly(ADP-ribosyl)ation process involves three chemically distinct reactions. First, PARP is responsible for (i) the Glu-specific binding of one ADP-ribose unit from NAD^+ to a protein acceptor amino acid side chain, which usually involves mono-ester bond formation (initiation reaction) [27]. Next, (ii) PARP proceeds to the characteristic (2'-1") ribose-ribose glycosidic bond formation of the ADP-ribose chain elongation reaction [28–30]. Finally, (iii) the (2"-1"') ribose-ribose bonding between ADP-ribose units (branching reaction) [26, 31] takes place at the ratio of one every forty ADP-ribose units polymerized [32]. While the initiation reaction occurs via a quasi-distributive mechanism [20, 33], the ADP-ribose polymerization (elongation and branching reactions) take place via a highly processive mechanism [34].

ADP-ribose chain initiation

We have recently developed a highly specific assay to characterize the mono(ADP-ribosyl)transferase activity displayed by PARP [35] independently of ADP-ribose polymerization (elongation and branching combined). In that study we found that the initiation reaction of ADP-ribose polymer synthesis requires nicked dsDNA for enzyme activity. We also observed that the rates of the reaction increased with the square of the enzyme concentration. Therefore, the mono(ADP- ribosyl) transferase activity of PARP involves enzymatic dimerization and it probably occurs intermolecularly. Careful stoichiometric analysis of substrates and products involved in catalysis indicated that a maximum of four amino acid residues/mol of PARP were mono(3′-dADP-ribosyl)ated (8 moles of 3′-dADP-ribose/PARP dimer).

The kinetic constants that we determined for this reaction, although obtained with 3′-dNAD as a substrate [35], were consistent with the notion that the auto-mono(3′-dADP-ribosyl)ation reaction of PARP proceeds at a rate which is 200-fold slower than ADP-ribose polymerization. In addition, PARP displayed 50-fold higher efficiency (kcat/km) at polymerizing protein-bound ADP-ribose with NAD^+ as a substrate than at mono(ADP-ribosyl)ating proteins with 3′-$dNAD^+$ as a substrate. Finally, we also observed [35] that the rates of mono(ADP-ribosyl)ation (number of chains) increased as a function of the concentration of nicked dsDNA strongly suggesting that co-enzymic dsDNA is the main determining factor of the extent of protein-poly(ADP-ribosyl)ation.

ADP-ribose chain elongation

The elongation of ADP-ribose chains has been shown to occur via a protein-distal elongation mechanism [28–30]. In this model, the 2′-hydroxyl group of the non-reducing end of the growing ADP-ribose chain (polymer tail) is utilized as the acceptor for the covalent binding of the newly added ADP-ribose unit. This mechanism of ADP-ribose polymerization predicts that the automodification reaction of PARP is bimolecular or intermolecular in nature and that rates of auto-poly(ADP-ribosyl)ation should increase with the square of the enzyme concentration. In agreement with this hypothesis, we have observed that rates of automodification increase with second order kinetics as a function of the enzyme concentration [20]. In fact, sigmoidal kinetic behavior was also observed in the auto-mono(ADP-ribosyl)ation of PARP with 3′-$dNAD^+$ as a substrate [35]. Thus, our results are consistent with the conclusion that PARP performs the ADP-ribose chain initiation and polymerization reactions as a catalytic dimer [20].

As indicated above, PARP catalyzes the initiation, elongation, and branching reactions of ADP-ribose polymer synthesis. For every ADP-ribose chain initiated [protein-mono(ADP-ribosyl)ation], PARP catalyzes over 200 elongation cycles [36, 37]. Therefore, ADP-ribose chain elongation represents the main catalytic function of this protein.

We have also performed experiments to determine the molecular, kinetic, chemical, and regulatory mechanisms of ADP-ribose chain elongation. In this series of experiments, we have observed that the average size of the polymer is strictly dependent on the concentration of NAD^+ utilized in the assay [20]. Therefore, it appears that ADP-ribose polymerization may be allosterically regulated by the substrate itself. Indeed, we have observed that ADP-ribose polymer synthesis shows sigmoidal saturation kinetics as a function of the concentration of NAD^+.

ADP-ribose polymer branching

For every ADP-ribose polymer synthesized (>200 units), PARP catalyzes over 5–7 branching events [31–33]. At present, very little is known about the biological significance of the branched structure of this chromatin-bound polynucleotide. Our mechanistic understanding of how PARP catalyzes this reaction is limited to a recent report by the group of deMurcia and collaborators [26] who identified some of the amino acids involved in polymer branching by polymerase chain reaction random mutagenesis of the poly(ADP-ribose) catalytic domain in *Escherichia coli.* Their results indicate spatial proximity of the amino acid residues involved in chain elongation (Glu988) and branching (Tyr986) and suggest that the catalytic site(s) for polymer elongation and branching may overlap stereochemically. Other amino acid residues that affected the branching reaction to some extent when mutated include Arg847, Glu923, and Gly972.

It should be pointed out that the chemical role of each of these amino acid residues in catalysis as well as the protein-protein interactions involved in determining the acceptor specificity of protein-poly(ADP-ribosyl)ation remains to be clarified.

Further mechanistic studies are required to determine the chemistry involved in each of the catalytic functions performed by PARP.

Acknowledgements

This project was supported by a grant from the National Institutes of Health (GM45451) to RAG.

References

1. Benjamin RC, Gill DM: Poly(ADP-ribose) Synthesis *in vitro* programmed by damaged DNA. A comparison of DNA molecules containing different types of strand breaks. J Biol Chem 255: 10502–10508, 1980
2. Menissier-deMurcia J, Molinete M, Gradwohl G, Simonin F, deMurcia G: Zinc-binding domain of poly(ADP-ribose) polymerase participates in the recognition of single-stranded breaks on DNA. J Mol Biol 210: 229–2337, 1989
3. Huletsky A, Niedergang C, Frechette A, Aubin R, Gaudreau A, Poirier GG: Sequential ADP-ribosylation pattern of nucleosomal histones. Eur J Biochem 146: 277–285, 1985
4. Boulikas T: At least 60 ADP-ribosylated variant histones are present in nuclei from dimethylsulphate-treated and untreated cells. EMBO J 7: 57–67, 1988
5. Boulikas T: Poly(ADP-ribosylated) histones in chromatin replication. J Biol Chem 265: 14638–14647, 1990
6. Yoshihara K, Itaya A, Tanaka Y, Ohashi Y, Ito K, Teraoka H, Tsukuda K, Matsukage A, Kamiya T: Inhibition of DNA polymerase α, DNA polymerase β, terminal deoxynucleotidyltransferase and DNA ligase II by poly(ADP-ribosyl)ation reaction *in vitro.* Biochem Biophys Res Commun 128: 61–67, 1985
7. Ohashi Y: Effect of ionic strength on chain elongation in ADP ribosylation of various nucleases. J Biochem 99: 971–979, 1986
8. Yoshihara K, Hashida T, Tanaka Y, Oghushi H: Enzyme-bound early product of purified poly(ADP-ribose) polymerase. Biochem Biophys Res Commun 78: 1281–1288, 1977
9. Kawaichi M, Ueda K, Hayaishi O: Multiple auto-poly(ADP-ribosyl)ation of rat liver poly(ADP-ribose) synthetase: Mode of modification and properties of automodified synthetase. J Biol Chem 256: 9483–9489, 1981
10. Adamietz P: Poly(ADP-ribose) synthetase is the major endogenous non-histone acceptor for poly(ADP-ribose) in alkylated rat hepatoma cells. Eur J Biochem 169: 365–372, 1987
11. Kreimeyer A, Wieickens K, Adamietz P, Hilz H: DNA-repair associated ADP-ribosylation *in vivo*. Modification of histone H1 differs from that of the principal acceptor proteins. J Biol Chem 259: 890–896, 1984
12. Nishikimi N, Ogasawara K, Kameshita I, Taniguchi T, Shizuta Y: Poly(ADP-ribose) synthetase. The DNA binding domain and the auto modification domain. J Biol Chem 257: 6102–6105, 1982
13. Kameshita I, Matsuda Z, Taniguchi T, Shizuta Y: Poly(ADP-ribose) synthetase. Separation and identification of three proteolytic fragments as the substrate binding domain, the DNA-binding domain, and the automodification domain. J Biol Chem 259: 4770–4776, 1984
14. Kameshita I, Matsuda M, Nishikimi N, Ushiro H, Shizuta Y: Reconstitution and poly(ADP-ribosyl)ation of proteolytically fragmented poly(ADP-ribose) synthetase. J Biol Chem 261: 3863–3868, 1986
15. Ikejima M, Noguchi S, Yamashita R, Ogura T, Sugimura T, Gill DM, Miwa M: The zinc fingers of human poly(ADP-ribose) polymerase are differentially required for the recognition of DNA breaks and nicks and the consequent enzyme activation. J Biol Chem 265: 21907–21913, 1991.
16. Gradwohl G, Menessier de Murcia G, Molinete M, Simonin F, Koken M, Hoeijmakers JHJ, deMurcia G: The second zinc finger domain of poly(ADP-ribose) polymerase determines specificity for single-stranded breaks in DNA. Proc Natl Acad Sci USA 87: 2990–2994, 1990
17. Uchida K, Hanai S, Ishikawa K-I, Ozawa Y-I, Uchida M, Sugimura T, Miwa M: Cloning of cDNA encoding *Drosophila* poly(ADP-ribose) polymerase: Leucine zipper in the automodification domain. Proc Natl Acad Sci USA 90: 3481–3485, 1993
18. Buki KG, Bauer PI, Hakam A, Kun E: Identification of domains of poly(ADP-ribose) polymerase for protein binding and self-association. J Biol Chem 270: 3370–3377, 1995
19. Bauer PI, Buki KG, Hakam A, Kun E: Macromolecular association of ADP-ribosyltransferase and its correlation with enzymic activity. Biochem J 270: 17–26, 1990
20. Mendoza-Alvarez H, Alvarez-Gonzalez R: Poly(ADP-ribose) polymerase is a catalytic dimer and the automodification reaction is intermolecular. J Biol Chem 268: 17575–17580, 1993
21. Panzeter P, Althaus FR: DNA strand break-mediated partitioning of poly(ADP-ribose) polymerase function. Biochem 33: 9600–9605, 1994
22. Le Cam E, Fack F, Menissier-de Murcia J, Cognet JAH, Barbin A, Sarantoglou V, Revet B, Delain E, de Murcia G: Conformational analysis of a 139 base-pair DNA fragment containing a single-stranded break and its interaction with human poly(ADP-ribose) polymerase. J Mol Biol 235: 1062–1071, 1994
23. Domenighini M, Rappuoli R: Three conserved consensus sequences identify the NA-binding site of ADP-ribosylating enzymes, expressed by bacteria, T-even bacteriophages, and eucaryotes. Mol Microbiol 21: 667–674, 1996
24. Marsischky GT, Wilson BA, Collier RJ: Role of glutamic acid 988 of human poly(ADP-ribose) polymerase in polymer formation. Evidence for active site similarities to the ADP-ribosylating toxins. J Biol Chem 270: 3247–3254, 1995
25. Takada T, Iida K, Moss J: Conservation of a common motif in enzymes catalyzing ADP-ribose transfer. J Biol Chem 279: 541–544, 1995
26. Rolli V, O'Farrell M, Menissier-deMurcia J, deMurcia G: Random mutagenesis of the poly(ADP-ribose) catalytic domain reveals amino acids involved in polymer branching. Biochem 36: 12147–12154, 1997

27. Kawaichi M, Ueda K, Hayaishi O: Initiation of poly(ADP-ribosyl)-histone synthesis by poly(ADP-ribose) synthetase. J Biol Chem 255: 816–819, 1980
28. Ueda K, Kawaichi M, Okayama H, Hayaishi O: Poly(ADP-ribosyl)ation of nuclear proteins: Enzymatic elongation of chemically synthesized ADP-ribose histone adducts. J Biol Chem 254: 679–687, 1979
29. Taniguchi T: Reaction mechanism of poly(ADP-ribose) synthetase. Biochem Biophys Res Commun 147: 1008–1012, 1987
30 Alvarez-Gonzalez R: 3′-deoxyNAD$^+$ as a substrate for poly(ADP-ribose) polymerase and the reaction mechanism of poly(ADP-ribose) elongation. J Biol Chem 263: 17690–17696, 1988
31. Miwa M, Saikawa N, Yamaizumi Z, Nishimura S, Sugimura T: Structure of poly(adenosine diphosphate ribose): Identification of 2'-(1"-ribosyl-2"- (or 3"-)(1'"-ribosyl)adenosine-5',5",5'"-tris-(phosphate) as a branch linkage. Proc Natl Acad Sci USA 76: 595–599, 1979
32. Keith G, Desgres J, deMurcia G: Use of two-dimensional thin-layer chromatography for the components study of poly(adenosine diphosphate ribose). Anal Biochem 191: 309–313, 1990
33. Kawaichi M, Ueda K, Hayaishi O: Multiple auto-poly(ADP-ribosyl)ation of rat liver poly(ADP-ribose) synthetase: Mode of modification and properties of automodified synthetase. J Biol Chem 256: 9483–9489, 1981
34. Naegeli H, Loetscher P, Althaus FR: Poly ADP-ribosylation of proteins: Processivity of a post-translational modification. J Biol Chem 264: 14382–14385, 1989
35. Mendoza-Alvarez H, Alvarez-Gonzalez R: Characterization of the mono(ADP-ribosyl)transferase activity associated with poly(ADP-ribose) polymerase. (Submitted), 1998
36. Alvarez-Gonzalez R, Jacobson MK: Characterization of polymers of adenosine diphosphate ribose generated *in vitro* and *in vivo*. Biochemistry 26: 3218–3224, 1987
37. Kiehibauch CC, Aboul-Ela N, Jacobson EL, Ringer DP, Jacobson M K: High resolution fractionation and characterization of ADP-ribose polymers. Anal Biochem 208: 26–34, 1993

Molecular and Cellular Biochemistry **193**: 23–30, 1999.

Poly(ADP-ribose) polymerase: A guardian of the genome that facilitates DNA repair by protecting against DNA recombination

Satadal Chatterjee, Sosamma J. Berger and Nathan A. Berger
Case Western Reserve University School of Medicine and Cancer Research Center, Cleveland, Ohio, USA

Abstract

We have studied the clonogenic survival response to X-rays and MNNG of V79 Chinese hamster cells and two derivative cell lines, ADPRT54 and ADPRT351, deficient in poly(ADP-ribose) polymerase (PARP) activity. Under conditions of exponential growth, both PARP-deficient cell lines are hypersensitive to X-rays and MNNG compared to their parental V79 cells. In contrast, under growth-arrested, confluent conditions, V79 and PARP-deficient cells become similarly sensitive to X-rays and MNNG suggesting that PARP may be involved in the repair of X-ray or MNNG-induced DNA damage in logarithmically growing cells but not in growth-arrested confluent cells. This suggestion, however, creates a dilemma as to how PARP can be involved in DNA repair in only selected growth phases while it is functionally active in all growth phases. To explain these paradoxical results and resolve this dilemma we propose a hypothesis based on the consistent observation that inhibition of PARP results in a significant increase in sister chromatid exchange (SCEs). Thus, we propose that PARP is a guardian of the genome that protects against DNA recombination. We have extended this theme to provide an explanation for our results and the studies done by many others. (Mol Cell Biochem **193**: 23–30, 1999)

Key words: PARP, interference with PARP activity, DNA damage and repair, DNA recombination, DNA replication and growth arrest

Introduction

Poly(ADP-ribose) polymerase (PARP) is a nuclear enzyme activated specifically by DNA strand breaks caused directly or indirectly by a variety of agents [1–3]. The enzyme binds to the DNA strand breaks and hydrolyzes its sole substrate NAD to nicotinamide and adenosine diphosphate ribose (ADPR) units. The same enzyme covalently links ADPR units to proteins and then links successive ADPR units to form extended and, sometimes, branched chains of poly(ADP-ribose) [4]. The enzyme itself is the major acceptor for polymer synthesis, however, polymers can also be associated with other acceptor proteins [8]. This process is important in facilitating DNA repair following low to moderate levels of DNA damage. However, high levels of DNA damage can activate PARP to a great enough extent so that utilization of the substrate NAD becomes the predominant biological effect. Thus, high levels of PARP activation can cause rapid NAD and ATP depletion leading to cell death before there is an opportunity to repair the DNA damage. This ATP depletion is a suicide mechanism that virtually eliminates the possibility of survival of highly damaged cells [1, 9, 10].

Recent evidence suggests that PARP/NAD may be involved in signal transduction pathways that are responsible for controlling the expression of p53, a tumor suppressor gene product, and GRP78, a stress protein, initially discovered to be regulated by glucose levels [11–13]. We have previously shown that deficiency of PARP/NAD is associated with the downregulation of p53 and upregulation of GRP78 [11–13]. As a consequence of these associations, apoptosis is prevented and there is a significant reduction of clonogenic cytotoxicity induced by topoisomerase II inhibitors such as etoposide [11–13].

A variety of cytogenetic studies aimed at measuring sister chromatid exchange (SCE) frequency following inhibition of PARP activity indicate a role for PARP in prevention of DNA

Address for offprints: N.A. Berger, School of Medicine, Case Western Reserve University, 10900 Euclid Avenue, Cleveland, Ohio 44106-4915, USA

recombination. Cell lines treated with various inhibitors of PARP show a dose-dependent increase in SCE frequency [14–19]. For example, 3-AB, a potent PARP inhibitor which is usually not cytotoxic and does not cause chromosomal aberrations or mutations is a potent inducer of SCE [15]. Cell lines deficient in PARP or its substrate NAD also show a nearly 10 fold increase in spontaneous SCE frequency [20, 21]. PARP-DBD, a cell line that overexpresses PARP DNA binding domain, is functionally deficient in PARP activity because of a competition for DNA breaks between overexpressed DBD and resident PARP molecules. These cells also exhibit an increase in spontaneous SCE frequency compared to that of control cells [22]. Cells derived from PARP knockout mice also show a significant increase in SCE [23]. Taken together, these results indicate that the inhibition of PARP activity causes a substantial increase in baseline SCE frequency thus suggesting that normal PARP activity may function to prevent DNA recombination. Although numerous functions and activities have been ascribed to PARP in various cellular processes [2], the most consistent observation is that interference with PARP activity results in a substantial increase in SCE or recombination frequency.

Further support for the involvement of PARP in DNA recombination has emerged from a variety of cytogenetic and molecular genetic studies. When a stably transformed cell line, obtained by transfection of NIH 3T3 cells with human c-Ha-Ras, was treated with benzamide there was a deletion of the transfected oncogene resulting in reversion of the transformed phenotype. However, endogenous mouse c-Ha-ras was not lost during culture in the presence of benzamide suggesting that PARP may be involved in DNA recombination and that its inhibition results in the rejection of randomly integrated DNA [24]. Shall's group has demonstrated that inhibition of PARP results in the reduction of random integration of transfected DNA molecules into the genome [25]. Another study has shown that treatment of mouse fibroblasts with the PARP-inhibitor, 3-methoxybenzamide, inhibits random integration of transfected DNA and increases the rate of intrachromosomal homologous recombination by nearly 4-fold [26, 27]. Both gene conversions as well as single crossovers are increased to a similar extent suggesting a mechanistic association between these two types of intrachromosomal rearrangements in mammalian cells and their regulation by PARP activity.

Cytogenetic effects of DNA damaging agents can be measured in metaphase chromosomes as SCE and/or chromosomal aberrations. Treatment of cells with PARP inhibitors and alkylating agents synergistically increases SCE frequencies [28–30]. The degree of synergism is dependent on the concentrations of both the inhibitor and the alkylating agent. PARP-DBD cells, deficient in PARP activity, exhibit increased DNA recombination as demonstrated by higher frequencies of MNNG-induced SCE compared to the parental HeLa cells [22].

Inhibitors of PARP by themselves do not cause DNA strand breaks, however, their presence in association with alkylating agents or ionizing radiation significantly increases the strand break frequency compared to that produced by DNA damaging agents alone [1, 31–34]. In addition, in most cases, a delay in rejoining of strand breaks is observed in the presence of these inhibitors [1, 33]. Although inhibitors of PARP are usually not cytotoxic by themselves, they significantly potentiate the X-ray and alkylating agent-induced cell killing to varying degrees in different cell types. The degree of potentiation depends on the types and concentrations of alkylating agents and inhibitors as well as on the status of cell growth [1, 32–37]. Jacobsons' group reported that 3-methoxybenzamide increased DNA strand breaks following MNNG in both quiescent and dividing cells, however, it had no effect on MNNG-induced cytotoxicity in growth-arrested cells but was very co-cytotoxic with MNNG in dividing cells [35, 36]. These studies suggest that the substantial increase in alkylating agent induced DNA strand breaks observed in the presence of PARP inhibitors does not directly contribute to the increased cytotoxicity, or in other words, inhibitor-induced increase in DNA strand breaks produced by alkylating agents is dissociated from the inhibitor-induced potentiation of cytotoxicity inflicted by alkylating agents. Furthermore, although PARP inhibitors effectively potentiate MNNG-induced cytotoxicity in exponentially growing cells, they fail to produce such effects in growth arrested cells suggesting a role of PARP in DNA repair in logarithmically growing cells but not in growth arrested cells.

Using PARP-deficient cell lines, we have shown that (i) PARP deficiency results in increased sensitivity to X-ray and MNNG in proliferating cells but not in growth arrested cells, and (ii) PARP-deficient and normal cells exhibit similar rates of DNA strand break repair [38]. These observations suggest that PARP may facilitate cell survival following DNA damage in proliferating cells but not in growth arrested cells.

The above findings can be summarized as follows. (i) Any interference with PARP activity results in increased SCE/recombination; (ii) The synergistic increase in DNA strand breaks, observed in the presence of a PARP inhibitor and an alkylating agent, does not contribute to increased cytotoxicity; (iii) Combination of PARP inhibitors and DNA damaging agents can synergistically increase SCE frequency; (iv) PARP inhibition results in potentiation of cytotoxic effects of DNA damaging agents in proliferating cells but not in growth arrested cells.

Based on these results and the experiments described below, we propose the hypothesis that the major role of PARP is to protect against DNA recombination. This hypothesis explains the observations outlined above, resolves the indicated paradoxes and fits well with a wide variety of studies examining the interaction between PARP inhibitors and DNA damaging agents.

Materials and methods

Cell lines

Description and maintenance of the cell lines was described previously [20]. Briefly, the mutant cell lines ADPRT54 and ADPRT351 have 5 and 11% PARP activity compared to their parental V79 Chinese hamster cell line [20]. All cell lines were maintained in α-modified minimum essential medium (GIBCO-BRL, Gaithersburg, MD, USA) buffered to pH 7.2 with 25 mM N-2-hydroxyethyl-piperazine-N′-2-ethane-sulfonic acid (HEPES) and supplemented with 100 units/ml penicillin, 100 μg/ml streptomycin, and 5% heat inactivated fetal calf serum. Plateau phase cultures were obtained by growing cells to confluence with daily medium changes. To confirm the status of growth arrested confluent cells DNA synthesis rate was measured by labeling cells with ^{3}H-thymidine and counting radioactivity in TCA insoluble fractions. Under the conditions of arrested growth DNA synthesis rate was 1–4% compared to that of exponentially growing cells.

X-ray and MNNG exposure

For X-irradiation, cells were irradiated in a Gammacell 1000 (Nordion International Inc., Kanata, Ontario, Canada) at a dose rate of 8.07 Gy/min. The cells were exposed to various doses of X-rays at room temperature and the fraction of surviving cells was determined by clonogenic survival assays as described previously [39–42]. MNNG was obtained from Sigma Chemical Co. (St. Louis, MO, USA). MNNG stock solution, 10 mg/ml, in 100% DMSO was prepared fresh before each experiment and subsequently diluted in 0.9% NaCl. All drug treatments were carried out in growth medium at 37°C.

Results

Table 1 summarizes the IC_{50} values from previously described experiments [38] showing that under exponential conditions of growth, the PARP-deficient cell lines, ADPRT351 and ADPRT54 are 4.67 and 3.09 times more sensitive to MNNG than their parental V79 cells. However, PARP-deficient and V79 cells show no difference in sensitivity to MNNG when the cells are grown under growth arrested confluent conditions. Similar results were obtained when the cells were exposed to X-rays. On the basis of IC_{50} values, logarithmically growing ADPRT351 and ADPRT54 are 3.99 and 3.29 times more sensitive to X-rays compared to V79 cells. These differences in sensitivity are virtually abolished when cells are treated under growth arrested confluent conditions. Despite different levels of survival in exponentially growing cells, after DNA damage in normal and PARP-deficient cells, we have previously shown that the extent of X-ray and MNNG-induced DNA strand breaks and their rate of repair are similar in these cell lines suggesting that enhanced cytotoxicity observed in PARP deficient cell lines is not caused by impaired DNA strand break repair.

Discussion

Our results demonstrate that PARP deficiency results in increased sensitivity to X-ray and MNNG in proliferating cells but not in growth arrested cells. These observations are paradoxical since they suggest that PARP facilitates cell survival following DNA damage in proliferating cells but not in growth arrested cells. This paradox creates a dilemma as to how PARP can be involved in the DNA repair process in certain growth phases but not others when the enzyme is functionally active throughout the cell cycle [2]. In order to explain these paradoxical results and resolve the dilemma we hypothesize that (i) PARP protects against homologous and non-homologous DNA recombination as demonstrated by increased SCE and genetic rearrangements when it is interfered with, and (ii) failure of PARP to protect against DNA recombination could account for the increased cytotoxicity

Table 1.

Agent	IC_{50} (cycling cells)			IC_{50} (growth arrested cells)		
	V79	ADPRT54	ADPRT351	V79	ADPRT54	ADPRT351
X-ray (Gy)	3.35 (1.00)	1.02 (3.29)	0.84 (3.99)	1.61 (1.00)	1.31 (1.29)	1.48 (1.09)
MNNG (μg/ml)	0.40 (1.00)	0.13 (3.09)	0.09 (4.67)	0.26 (1.00)	0.26 (1.00)	0.26 (1.00)

IC_{50} values are defined as the dose of the agent necessary to reduce the clonogenic survivals to 50% and are expressed in the units listed following each agent. MNNG treatment was for 1 h. Relative sensitivity of a given cell line was determined by dividing IC_{50} for V79 by IC_{50} for that cell line and is shown in the parenthesis.

associated with PARP interference. Thus, the increased cytotoxicity that occurs when PARP activity is interfered with in combination with DNA damage is due to an increase in aberrant or non-homologous DNA recombination rather than with any attenuation in resealing of DNA strand breaks. Since DNA recombination is a cell proliferation associated event, the cytotoxic effects of interference with PARP are manifest in growing but not growth arrested cells.

The concepts outlined above are illustrated in Fig. 1 and elaborated in the following paragraphs. PARP binds to DNA single- and double-strand breaks that arise directly as a consequence of DNA damage (e.g. ionizing radiation), or indirectly following activation of the enzymatic excision repair process initiated by DNA damage. Consequently, the enzyme becomes activated and undergoes the auto-poly-(ADP-ribosyl)ation reaction [2, 3, 8, 34, 43–45]. The polymers attached to the DNA-bound enzyme are more acidic than DNA (two negative charges per monomer unit compared to only one per nucleotide of DNA) and thus electrostatic repulsion dissociates the automodified PARP from the DNA [2, 3, 8, 34, 43–45]. This dissociated polymer-bound PARP is inactive, however, rapid action of poly(ADP-ribose) glycohydrolase (PARG) releases PARP from the bound polymers thus enabling the enzyme to bind to DNA breaks and regain activity [2, 3, 8, 34, 43–45]. Thus, with the cooperation of PARG, PARP shuttles between its DNA-bound active state, and automodified, inactive state. This relationship between PARP and PARG provides a mechanism for cyclical shielding and uncovering of DNA strand breaks.

Under normal growth conditions, semiconservative DNA replication consists of three steps that include: (i) DNA strand separation; (ii) discontinuous DNA synthesis; and (iii) ligation of newly synthesized DNA fragments. In addition, cells spontaneously undergo some recombination during the normal DNA replication process, as demonstrated by the low but constant level of background SCE in normal untreated cells [14–21, 28]. These recombination events may function as a bypass mechanism for spontaneously damaged DNA/inaccessible DNA ends and/or they may function to close gaps between fragments that are not acted upon by the normal ligation process.

Under normal circumstances, recombination occurs infrequently despite the multitude of strand ends created during discontinuous DNA synthesis. We propose that the normal function of PARP is to bind reversibly to these DNA strand ends whether generated during replication and/or repair processes and transiently shield them from the actions of other enzymes that act on DNA strand ends. Thus, binding of PARP to DNA strand ends could shield against binding by a variety of proteins including DNA polymerases, ligases, endonucleases, proteins involved in DNA recombination and others. As alluded to earlier, the association-dissociation cycle of PARP from DNA strand ends would allow other enzymes to compete for strand ends and complete their respective processes. Assuming that all proteins working on DNA strand ends are present in cells in limited numbers, the overall effect of the reversible binding of PARP would be to modulate the rate of processes working on DNA strand ends and prevent access to enzymes such as endonucleases and proteins involved in recombination while providing sufficient time for access by enzymes such as polymerases and ligases. However, interruption or interference with any of the steps involved in the normal function of PARP such as failure to bind to DNA ends (enzyme deficiency) or failure to dissociate from DNA ends (treatment of cells with PARP inhibitors or overexpression of the PARP-DNA-binding domain) will facilitate the recombination process and nuclease action leading to the appearance of SCE and both homologous and non-homologous rearrangements.

In general, SCE represent homologous recombination events and thus are not cytotoxic. However, as the frequency of recombination increases, it is likely that there is a proportional increase in imperfect (nonhomologous) recombination. The latter produce DNA rearrangements and deletions that are potentially cytotoxic. Thus, when replicating cells are subjected to DNA damage the number of DNA strand ends may exceed the protective capacity of PARP and lead to an increased frequency of cytotoxic, nonhomologous recombination events whereas when growth arrested cells are subjected to DNA damage the absence of DNA strand ends associated with the replication process results in a lower overall frequency of strand breaks which do not exceed PARP protective capacity. At low levels of recombination, it is likely that most SCE will be homologous in nature and, therefore, the SCE will not be associated with an increase in cytotoxicity. However, at higher levels of recombination the process could lead to cytotoxicity by creating lethal genomic deletions and/or rearrangements [46].

Exposure of cells to DNA damaging agents leads to an increase in DNA strand breaks. These may be produced directly or indirectly as part of the excision repair process [47] and predispose to increased recombination events which can lead to mutation and/or cell death. When replicating cells are exposed to DNA damaging agents, the situation becomes more complicated. In addition to the DNA strand ends associated with the normal replication process described above, there is an increase in recombination processes, as demonstrated by the dose-dependent increase in SCE formation that occurs in response to treatment with DNA damaging agents. We suggest that this increase in SCE occurs because the number of DNA strand ends generated by the combination of semiconservative replication and the excision repair process exceeds the capacity of PARP to completely protect against recombination events.

Replicating cells with impaired PARP activity show a 10–20 fold increase in SCE compared to normal, untreated cells

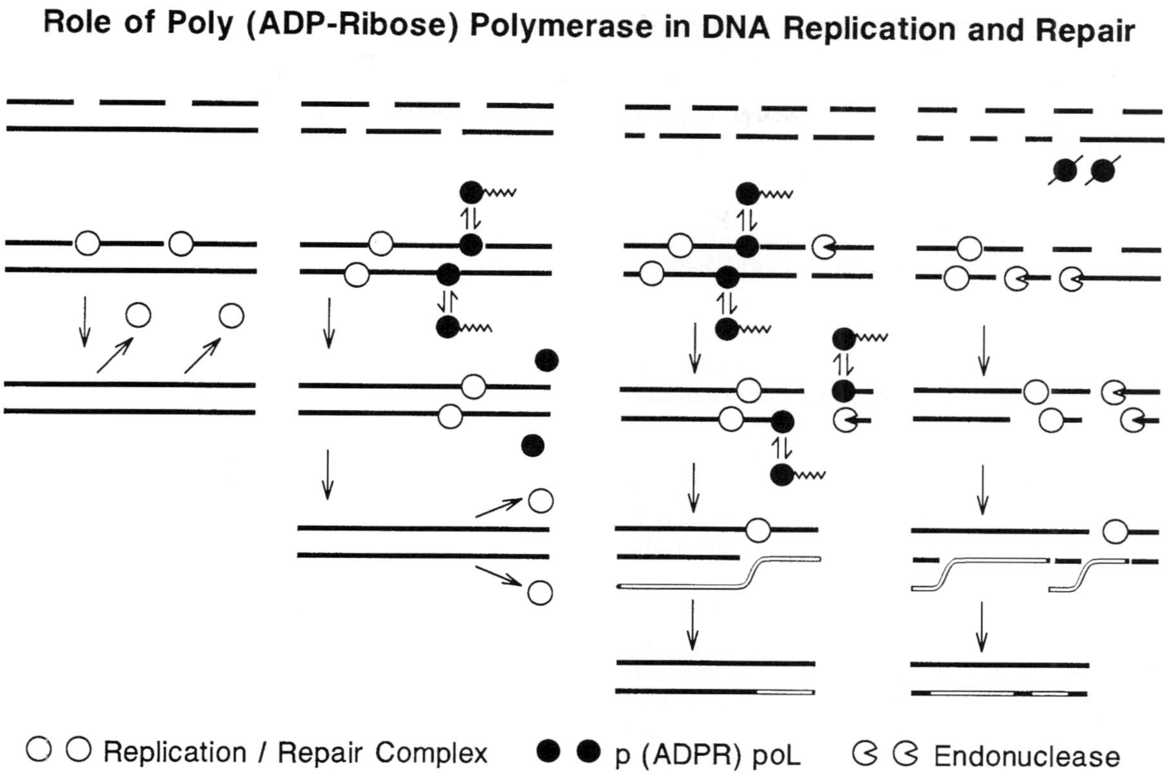

Fig. 1. Model of poly(ADP-ribose) polymerase role in DNA replication and repair. Open circles indicate DNA replication/repair complex, closed circles represent poly(ADP-ribose) polymerase, jagged lines represent poly(ADP-ribose), and notched circles represent endonuclease. We define DNA replication/repair complex as a multi-protein complex responsible for DNA replication/repair. This model assumes a limited number of molecules of DNA replication/repair complex and poly(ADP-ribose) polymerase. Although we have indicated two molecules of each enzyme system, we do not mean to imply that there are equal numbers of each enzyme system present, rather we suggest that each enzyme system is present in limited quantities. The first panel on the left shows DNA with a limited number of strand ends. These could be the result of direct breaks, the excision repair process or replication intermediates resulting from replicative DNA synthesis. In this situation, there are sufficient number of DNA replication/repair complexes to directly seal all DNA strand disruptions and restore DNA integrity. In the second column, there are an increased number of DNA strand ends that exceed the available number of DNA replication/repair complexes. In this situation, all DNA replication/repair complexes are engaged. The DNA strand ends that exceed the capacity of DNA replication/repair complexes are protected by poly(ADP-ribose) polymerase which is activated by binding DNA strand ends to synthesize poly(ADP-ribose). Polymer synthesis causes the enzyme to dissociate from the DNA. Subsequent degradation of poly(ADP-ribose) by poly(ADP-ribose) glycohydrolase allows the enzyme to return, bind another DNA strand end and reinitiate polymer synthesis. Poly(ADP-ribose) polymerase cycles on and off of DNA strand ends until DNA replication/repair complexes are available to join the disruption, and restore DNA integrity. The third column illustrates the situation that occurs when the number of DNA strand ends increase to the point that they are in excess of the number of DNA replication/repair complexes and poly(ADP-ribose) polymerase molecules. This could occur following DNA damage and excision repair in replicating cells where strand ends are already increased due to the presence of replication intermediates. Under these conditions, endonucleases begin to act on some of the free ends and the recombination process is used to repair other sections. This process results in DNA deletions and recombination. The latter process appears as sister chromatid exchanges which usually represent homologous recombination. However, as the frequency of SCE and DNA deletions increase, there is an increased probability for the occurrence of chromosomal aberrations, non-homologous recombination events, gene deletions, and cytotoxicity. The fourth column illustrates the situation with increased DNA strand breaks and inactive poly(ADP-ribose) polymerase. The latter condition could be due to mutations which inactivate the enzyme, the presence of enzyme inhibitors or the absence of sufficient NAD to provide substrate for enzyme activity. In this situation, the number of free strand ends, far exceeds the capacity to be joined by the existing number of DNA replication/repair complexes but poly(ADP-ribose) polymerase is not available to protect the excess free ends. As a result of the lack of protection, DNA strand ends are subject to more extensive endonuclease action resulting in DNA deletions. DNA strand ends and gaps are ultimately rejoined by ligation and recombination resulting in chromosomal aberrations, SCE, increased nonhomologous recombination, gene deletion, and increased cytotoxicity. The latter condition explains the observations in poly(ADP-ribose) polymerase deficient cell lines treated with DNA damaging agents during logarithmic growth. When the increased strand ends occur during the replication process and represent replication intermediates, the absence of poly(ADP-ribose) polymerase could result in more strand ends available for recombination and SCE accounting for the increased frequency of SCE in poly(ADP-ribose) polymerase deficient cell lines. In contrast, the diagram in the left column explains the situation in poly(ADP-ribose) polymerase deficient cell lines which are treated with DNA damaging agents during growth arrest. These cells have a limited number of DNA strand breaks relative to DNA replication/repair complexes, since during growth arrest, there are no ends due to replication intermediates. In this case, poly(ADP-ribose) polymerase is not required to protect DNA strand ends and the DNA integrity is restored even though the cells are poly(ADP-ribose) polymerase deficient.

[14–21, 28]. This can be explained by an insufficient amount of available functional PARP molecules relative to the number of DNA strand ends produced by discontinuous DNA synthesis. When these cells are further stressed by increasing doses of DNA damaging agents, the number of recombinational events is further increased and the chance of nonhomologus recombination also increases. Thus, cells with impaired PARP activity will be more susceptible to the induction of recombination and will undergo both homologous and nonhomologous recombination at much lower doses of DNA damaging agents compared to cells with normal PARP activity. Consequently, treatment of replicating cells with a given dose of DNA damaging agent will result in enhanced cell killing in cells with impaired PARP activity relative to the cells with normal poly(ADP-ribose) synthetic activity. Since only replicating cells can undergo recombination, impaired PARP activity will result in increased cytotoxicity in replicating cells treated with DNA damaging agents, but not in growth arrested cells.

Our hypothesis explains the observation made by Boorstein and Pardee who showed that the lethality of 3-aminobenzamide to a population of asynchronously cycling cells treated with the alkylating agent, methyl methane sulphonate (MMS), was maximum at S-phase and this lethality could be prevented by treating cells with the DNA synthesis inhibitor aphidicolin [37]. This study concluded that DNA synthesis is required to reach a step in DNA repair at which 3-AB causes lethality. Thus, these studies suggest that PARP inhibitors are mostly effective in potentiating cytotoxicity of alkylating agents in replicating cells, however, this effectiveness is significantly reduced in cells that are not cycling. Our hypothesis provides an explanation for the observations by Jacobsons' group, that PARP inhibitors potentiate the effect of MNNG in proliferating cells but not in quiescent cells [35, 36]. Likewise, our proposal explains the observation by Wiericke and Morgan who demonstrated that 3-AB potentiates X-ray-induced chromosomal aberrations maximally at S-phase while no potentiation is observed when cells are irradiated in G_0 [48]. Since chromosomal aberrations often lead to cell death these results can be interpreted to suggest that PARP inhibitors potentiate cytotoxic effects of X-ray in S-phase cells but not in G_0 cells. In addition, our hypothesis explains the observations, made by several laboratories, that inhibition of PARP activity by various approaches results in potentiation of cytotoxicity by alkylating agents. For example, Smulson's group has shown that a derivative of HeLa cell line (PADRP-as [7]), that is 90% depleted of PARP because of the expression of a PARP-antisense construct, exhibits increased susceptibility to nitrogen mustard and methyl methane sulfonate relative to the control cells based on clonogenic survival assays [49–51]. The groups of Burkle and deMurcia have shown that a HeLa-derived cell line, deficient in PARP activity because of the overexpresion of the PARP-DNA binding domain, is hypersensitive to MNNG [22, 52, 53].

Using an *in vitro* DNA repair assay system, Lindahl's group in association with Poirier have shown that rejoining of DNA single strand breaks from γ-rays, bleomycin, or neocarzinostatin exposure is catalyzed inefficiently in cell extracts containing PARP, but no NAD [54–56]. However, addition of NAD greatly increases the efficiency of end rejoining. This efficient rejoining is blocked by 3-AB. Furthermore, selective depletion of PARP from cell extracts improves the repair of DNA exposed to a variety of DNA-damaging agents by removing the NAD-dependent mode of repair. These results agree with our previous observations that PARP-deficient cells with normal NAD levels are equally efficient in rejoining DNA strand breaks when compared with cells that are normal in PARP-NAD metabolism [38]. Furthermore, these observations are in agreement with our suggestion that any interference with the on-off cycle of PARP will promote homologous and/or non-homologous recombination. Our hypothesis is also in agreement with the suggestion by Shall's group that poly-(ADPR) synthesis facilitates the DNA ligation process [57]. However, it does not support direct ADP-ribosylation and subsequent activation of ligase.

The last two decades have provided us with ample evidence that interference with PARP activity results in the potentiation of cytotoxicity of a variety of DNA damaging agents with particular emphasis on alkylating agents [1, 32–38]. Although initially these observations were derived from the studies with inhibitors of PARP, whose specificities were later questioned [32, 58–61], all the subsequent studies with various model systems such as PARP-deficient cell lines, NAD-deficient cell lines, cell lines with overexpressed PARP-antisense construct, and cell lines with overexpressed PARP-DNA-binding domain substantiated the earlier findings [22, 38–40, 42, 49–53]. Our hypothesis that PARP functions to prevent DNA recombination and that interference with PARP activity results in increased DNA recombination events provides a mechanistic framework to understand these results. The foundation for our proposal is built on the highly consistent observation that inhibition of PARP activity leads to a substantial increase in SCE frequency. While our proposal provides an initial step towards understanding PARP activity and explaining the consequences of PARP inhibition in mediating cytotoxic effects of DNA damaging agents, clearly, critical evaluation of this hypothesis with carefully planned studies will further improve our understanding of the protective role of PARP.

Acknowledgement

These studies were supported in part by National Cancer Institute Grants R29CA65920, P01CA51183, P30CA43703, and U01CA63200.

References

1. Berger NA: Poly(ADP-ribose) in the cellular response to DNA damage. Radiat Res 101: 4–15, 1985
2. Ueda K, Hayaishi O: ADP-ribosylation. Annu Rev Biochem 54: 73–100, 1985
3. Althaus FR, Richter C: ADP-ribosylation of proteins. Enzymology and biological significance. Mol Biol Biochem Biophys 37: 1–237, 1987
4. Alvarez-Gonzalez R, Jacobson MK: Characterization of polymers of adenosine diphosphate ribose generated *in vitro* and *in vivo*. Biochemistry 26: 3218–3224, 1987
5. Alvarez-Gonzalez R, Althaus FR: Poly(ADP-ribose) catabolism in mammalian cells exposed to DNA-damaging agents. Mutat Res 218: 67–74, 1989
6. Alvarez-Gonzalez R, Pacheco-Rodriguez G, Mendoza-Alvarez H: Enzymology of ADP-ribose polymer synthesis. Mol Cell Biochem 138: 33–37, 1994
7. Mendoza-Alvarez H, Alvarez-Gonzalez R: Poly(ADP-ribose) polymerase is a catalytic dimer and the automodification reaction is intermolecular. J Biol Chem 268: 22575–22580, 1993
8. Lautier D, Lagueux J, Thibodeau J, Menard L, Poirier GG: Molecular and biochemical features of poly (ADP-ribose) metabolism. Mol Cell Biochem 122: 171–193, 1993
9. Berger NA, Berger SJ: Metabolic consequences of DNA damage: The role of poly (ADP-ribose) polymerase as mediator of the suicide response. Basic Life Sci 38: 357–363, 1986
10. Berger NA, Sims JL, Catino DM, Berger SJ: Poly(ADP-ribose) polymerase mediates the suicide response to massive DNA damage: Studies in normal and DNA-repair defective cells. Princess Takamatsu Symp 13: 219–226, 1983
11. Chatterjee S, Cheng MF, Berger RB, Berger SJ, Berger NA: Effect of inhibitors of poly(ADP-ribose) polymerase on the induction of GRP78 and subsequent development of resistance to etoposide. Cancer Res 55: 868–873, 1995
12. Chatterjee S, Cheng MF, Berger SJ, Berger NA: Induction of M(r) 78,000 glucose-regulated stress protein in poly(adenosine diphosphate-ribose) polymerase- and nicotinamide adenine dinucleotide-deficient V79 cell lines and its relation to resistance to the topoisomerase II inhibitor etoposide. Cancer Res 54: 4405–4411, 1994
13. Whitacre CM, Hashimoto H, Tsai ML, Chatterjee S, Berger SJ, Berger NA: Involvement of NAD-poly(ADP-ribose) metabolism in p53 regulation and its consequences. Cancer Res 55: 3697–3701, 1995
14. Hori T: High incidence of sister chromatid exchanges and chromatid interchanges in the conditions of lowered activity of poly(ADP-ribose)polymerase. Biochem Biophys Res Commun 102: 38–45, 1981
15. Morgan WF, Bodycote J, Doida Y, Fero ML, Hahn P, Kapp LN: Spontaneous and 3-aminobenzamide-induced sister-chromatid exchange frequencies estimated by ring chromosome analysis [published erratum appears in Mutagenesis 1988 May; 3(3): 2861.] Mutagenesis 1: 453–459, 1986
16. Morgan WT, Wolff S: Induction of sister chromatid exchange by 3-aminobenzamide is independent of bromodeoxyuridine. Cytogenet Cell Genet 38: 34–38, 1984
17. Natarajan AT, Csukas I, van Zeeland AA: Contribution of incorporated 5-bromodeoxyuridine in DNA to the frequencies of sister-chromatid exchanges induced by inhibitors of poly-(ADP-ribose)polymerase. Mutat Res 84: 125–132, 1981
18. Oikawa AH, Tohda H, Kanai M, Miwa M, Sugimura T: Inhibitors of poly(adenosine diphosphate ribose) polymerase induce sister chromatid exchanges. Biochem Biophys Res Commun 97: 1311–1316, 1980
19. Utakoji T, Hosoda K, Umezawa K, Sawamura M, Matsushima T, Miwa M, Sugimura T: Induction of sister chromatid exchanges by nicotinamide in Chinese hamster lung fibroblasts and human lymphoblastoid cells. Biochem Biophys Res Commun 90: 1147–1152, 1979
20. Chatterjee S, Hirschler NV, Petzold SJ, Berger SJ, Berger NA: Mutant cells defective in poly(ADP-ribose) synthesis due to stable alterations in enzyme activity or substrate availability. Exp Cell Res 184: 1–15, 1989
21. Chatterjee S, Petzold SJ, Berger SJ, Berger NA: Strategy for selection of cell variants deficient in poly(ADP-ribose) polymerase. Exp Cell Res 172: 245–257, 1987
22. Schreffier V, Hunting D, Trucco C, Gowans B, Grunwald D, De Murcia G, De Murcia JM: A dominant-negative mutant of human poly(ADP-ribose) polymerase affects cell recovery, apoptosis, and sister chromatid exchange following DNA damage. Proc Natl Acad Sci USA 92: 4753–4757, 1995
23. Haidacher D, Ebner M, Berghammer H, Auer B: Poly(ADP-ribosyl) transferase (ADPRT) deficient cell lines: A model system to examine the biological role of ADPRT. The 12th International Symposium on ADP-ribosylation Reactions: From Bacterial Pathogenesis to Cancer, Abstract #13, 1997
24. Nakayasu M, Shima H, Aonuma S, Nakagama H, Nagao M, Sugimura T: Deletion of transfected oncogenes from NIH 3T3 transformants by inhibitors of poly(ADP-ribose) polymerase. Proc Natl Acad Sci USA 85: 9066–9070, 1988
25. Farzaneh F, Panayotou GN, Bowler LD, Hardas BD, Broom T, Walther C, Shall S: ADP-ribosylation is involved in the integration of foreign DNA into the mammalian cell genome. Nucleic Acids Res 16: 11319–11326, 1988
26. Waldman AS, Waldman BC: Stimulation of intrachromosomal homologous recombination in mammalian cells by an inhibitor of poly(ADP-ribosylation). Nucleic Acids Res 19: 5943–5947, 1991
27. Waldman BC, Waldman AS: Illegitimate and homologous recombination in mammalian cells: Differential sensitivity to an inhibitor of poly(ADP-ribosylation). Nucleic Acids Res 18: 5981–5988, 1990
28. Morgan MT, Cleaver JE: 3-Aminobenzamide synergistically increases sister-chromatid exchanges in cells exposed to methyl methanesulfonate but not to ultraviolet light. Mutat Res 104: 361–366, 1982
29. Park SD, Kim CG, Kim MG: Inhibitors of poly(ADP-ribose) polymerase enhance DNA strand breaks, excision repair, and sister chromatid exchanges induced by alkylating agents. Environ Mutagen 5: 515–525, 1983
30. Schwartz JL, Morgan WF, Weichselbaum RR: Different efficiencies of interaction between 3-aminobenzamide and various monofunctional alkylating agents in the induction of sister chromatid exchanges. Carcinogenesis 6: 699–704, 1985
31. Ben Hur E: Involvement of poly (ADP-ribose) in the radiation response of mammalian cells. Int J Rad Biol Rel Stud Phys Chem Med 46: 659–671, 1984
32. Cleaver JE, Morgan WF: Poly(ADP-ribose)polymerase: A perplexing participant in cellular responses to DNA breakage. Mutat Res 257: 148, 1991
33. Shall S: ADP-ribose in DNA repair: A new component of DNA excision repair. Adv Rad Biol 11: 1–69, 1984
34. Chatterjee S, Berger NA: Poly(ADP-ribose) polymerase in response to DNA damage. Chapter 22, In: J.A. Nickoloff, M. Hoekstra (eds). DNA Damage and Repair-Biochemistry, Genetics and Cell Biology, vol 2. Humana Press, 487–515, 1998
35. Jacobson EL, Meadows R, Measel J: Cell cycle perturbations following DNA damage in the presence of ADP-ribosylation inhibitors. Carcinogenesis 6: 711–714, 1985

36. Jacobson EL, Smith JY, Wielckens K, Hilz H, Jacobson MK: Cellular recovery of dividing and confluent C3H10T1/2 cells from N-methyl-N′-nitro-N-nitrosoguanidine in the presence of ADP-ribosylation inhibitors. Carcinogenesis 6: 715–718, 1985
37. Boorstein RJ, Pardee AB: 3-Aminobenzamide is lethal to MMS-damaged human fibroblasts primarily during S phase. J Cell Physiol 120: 345–353, 1984
38. Chatterjee S, Berger NA: Growth-phase-dependent response to DNA damage in poly(ADP-ribose) polymerase deficient cell lines: Basis for a new hypothesis describing the role of poly(ADP-ribose) polymerase in DNA replication and repair. Mol Cell Biochem 138: 61–69, 1994
39. Chatterjee S, Cheng MF, Berger SJ, Berger NA: Alkylating agent hypersensitivity in poly(adenosine diphosphate-ribose) polymerase deficient cell lines. Cancer Commun 3: 71–75, 1991
40. Chatterjee S, Cheng MF, Trivedi D, Petzold SJ, Berger NA: Camptothecin hypersensitivity in poly(adenosine diphosphate-ribose) polymerase-deficient cell lines. Cancer Commun 1: 389–394, 1989
41. Chatterjee S, Trivedi D, Petzold SJ, Berger NA: Mechanism of epipodophyllotoxin-induced cell death in poly(adenosine diphosphate-ribose) synthesis-deficient V79 Chinese hamster cell lines. Cancer Res 50: 2713–2718, 1990
42. Chatterjee S, Cheng MF, Berger NA: Hypersensitivity to clinically useful alkylating agents and radiation in poly(ADP-ribose) polymerase-deficient cell lines. Cancer Commun 2: 401–407, 1990
43. Ferro AM, Olivera BM: Poly(ADP-ribosylation) *in vitro*. Reaction parameters and enzyme mechanism. J Biol Chem 257: 7808–7813, 1982
44. Zahradka P, Ebisuzaki K: A shuttle mechanism for DNA-protein interactions. The regulation of poly(ADP-ribose) polymerase. Eur J Biochem 127: 579–585, 1982
45. Althaus FR, Hofferer L, Klewkowska HE, Malanga M, Naegeli H, Panzeter PL, Realini CA: Histone shuttling by poly ADP-ribosylation. Mol Cell Biochem 138: 53–59, 1994
46. Berger NA, Chatterjee S, Schmotzer JA, Helms SR: Etoposide (VP-16-213)-induced gene alterations: Potential contribution to cell death. Proc Natl Acad Sci 88: 8740–8743, 1991
47. Friedberg EC, Walker GC, Siede W: DNA repair and Mutagenesis. Am Soc Microbiol Washington DC, 1995
48. Wiencke JK, Morgan WF: Cell cycle-dependent potentiation of X-ray-induced chromosomal aberrations by 3-aminobenzamide. Biochem Biophys Res Commun 143: 372–376, 1987
49. Ding R, Pommier Y, Kang VH, Smulson M: Depletion of poly(ADP-ribose) polymerase by antisense RNA expression results in a delay in DNA strand break rejoining. J Biol Chem 267: 12804–12812, 1992
50. Ding R, Smulson M: Depletion of nuclear poly(ADP-ribose) polymerase by antisense RNA expression: Influences on genomic stability, chromatin organization, and carcinogen cytotoxicity. Cancer Res 54: 4627–4634, 1994
51. Stevnsner T, Ding R, Smulson M, Bohr VA: Inhibition of gene-specific repair of alkylation damage in cells depleted of poly(ADP-ribose) polymerase. Nucleic Acids Res 22: 4620–4624, 1994
52. Kupper JH, de Murcia G, Burkle A: Inhibition of poly(ADP-ribosyl)ation by overexpressing the poly(ADP-ribose) polymerase DNA-binding domain in mammalian cells. J Biol Chem 265: 18721–18724, 1990
53. Molinete M, Vermeulen W, Burkle A, de Murcia JM, Kupper JH, Hoeijrnakers JH, de Murcia G: Overproduction of the poly(ADP-ribose) polymerase DNA-binding domain blocks alkylation-induced DNA repair synthesis in mammalian cells. EMBO J 12: 2109–2117, 1993
54. Satoh MS, Lindahl T: Role of poly(ADP-ribose) formation in DNA repair. Nature 356: 356–358, 1992
55. Satoh MS, Poirier GG, Lindahl T: Dual function for poly(ADP-ribose) synthesis in response to DNA strand breakage. Biochemistry 33: 7099–8106, 1994
56. Satoh MS, Poirier GG, Lindahl T: NAD(+)-dependent repair of damaged DNA by human cell extracts. J Biol Chem 268: 5480–5487, 1993
57. Creissen D, Shall S: Regulation of DNA ligase activity by poly(ADP-ribose). Nature 296: 271–272, 1982
58. Hunting DJ, Gowans BJ, Henderson JF: Specificity of inhibitors of poly(ADP-ribose) synthesis. Effects of nucleotide metabolism in cultured cells. Mol Pharmacol 28: 200–206, 1985
59. Milam KM, Thomas GH, Cleaver: Disturbances in DNA precursor metabolism associated with exposure to an inhibitor of poly(ADP-ribose) synthetase. Exp Cell Res 165: 260–268, 1986
60. Cleaver JE, Morgan WF: Poly(ADP-ribose) synthesis is involved in the toxic effects of alkylating agents but does not regulate DNA repair. Mutation Res 150: 69–76, 1985
61. Cleaver JE, Milam KM, Morgan W: Do inhibitor studies demonstrate a role for poly(ADP-ribose) in DNA repair? Radiat Res 101: 16–28, 1985

Molecular and Cellular Biochemistry **193:** 31–35, 1999.

Trans-dominant inhibition of poly(ADP-ribosyl)ation potentiates alkylation-induced shuttle-vector mutagenesis in Chinese hamster cells

Junko Tatsumi-Miyajima,[1] Jan-Heiner Küpper,[2] Hiraku Takebe[1] and Alexander Bürkle[2]

[1]*Institute of Radiation Genetics, Faculty of Medicine, Kyoto University, Kyoto, Japan and* [2]*Abteilung Tumorvirologie, Deutsches Krebsforschungszentrum, Heidelberg, Germany*

Abstract

In most eukaryotic cells, the catalytic activation of poly(ADP-ribose) polymerase (PARP) represents one of the earliest cellular responses to the infliction of DNA damage. To study the biological function(s) of poly(ADP-ribosyl)ation, we have established stable transfectants (COM3 cells) of the SV40-transformed Chinese hamster cell line C060 which conditionally overexpress the PARP DNA-binding domain upon addition of dexamethasone. We could demonstrate that DNA-binding domain overexpression, which leads to *trans*-dominant inhibition of poly(ADP-ribosyl)ation, potentiates the cytotoxicity of alkylation treatment and of γ-radiation [21]. Likewise, carcinogen-induced gene amplification, viewed as a manifestation of genomic instability, was potentiated by the overexpression of the PARP DNA-binding domain [22]. Recently, we studied the effect of *trans*-dominant PARP inhibition on mutagenesis by employing a shuttle-vector assay in which mutagen-exposed plasmid pYZ289 is electroporated into COM3 cells. We could show that dexamethasone-induced overexpression of the PARP DNA-binding domain in COM3 cells potentiates the mutagenicity of the alkylating agent *N*-methyl-*N*-nitrosourea, while no effect of dexamethasone treatment on mutation frequency was recorded in control cells lacking the PARP DNA-binding domain transgene. Taken together, our results further substantiate the role of poly(ADP-ribosyl)ation in the maintenance of genomic integrity and stability under conditions of genotoxic stress. (Mol Cell Biochem **193**: 31–35, 1999)

Key words: poly(ADP-ribose) polymerase, mutagenesis, genomic instability, alkylating agents, shuttle vector

Introduction

Poly(ADP-ribosyl)ation of proteins is catalyzed by the nuclear enzyme poly(ADP-ribose) polymerase (PARP; EC 2.4.2.30) and represents an immediate response of most eukaryotic cells to DNA damage (for review see [1–3]). This damage responsiveness is due to the fact that PARP selectively binds to DNA strand breaks via its amino-terminal DNA-binding domain, resulting in the activation of its catalytic function.

Although the precise biological role of PARP is not known, this enzyme has been linked with several cellular responses to DNA damage, for instance DNA base excision repair [4–6], and cell death by necrosis [7–9] or apoptosis [10–13]. Furthermore, it was deduced from various experimental systems that stimulation of the poly(ADP-ribosyl)ation pathway in carcinogen-treated cells counteracts the phenomenon of genetic instability [14], assessed as sister-chromatid exchanges [15] or integration of foreign DNA [16–18]. Another form of genetic instability is the amplification of cellular DNA sequences, e.g. oncogenes or drug resistance genes, which play an important role in carcinogenesis. We have previously shown that the chemical ADP-ribosylation inhibitor 3-aminobenzamide potentiates Simian virus 40 (SV40) DNA amplification induced by the alkylating agent *N*-methyl-*N'*-nitro-*N*-nitrosoguanidine (MNNG) in the

Present address: J-H Küpper, Universitat Tübingen, Institut für Pathologie, Liebermeisterstr. 8, D-72076 Tübingen, Germany
Address for offprints: A. Bürkle, DKFZ, Abteilung F0100, Im Neuenheimer Feld 242, D-69120 Heidelberg, Germany

SV40-transformed hamster cell line C060 [19]. Likewise, MNNG-induced methotrexate resistance and amplification of the dihydrofolate reductase gene in CHO cells was potentiated by competitive ADP-ribosylation inhibitors [20]. Given the accumulating evidence of nonspecific inhibitor effects, we developed the stably transfected C060 subline COM3, in which inhibition of poly(ADP-ribosyl)ation is achieved by the dexamethasone (Dex)-inducible overexpression of the mere DNA-binding domain (DBD) of PARP, the latter acting as a dominant-negative mutant of this enzyme [21]. Induction of the DBD led to a 90% reduction of steady-state levels of poly(ADP-ribose) after γ-irradiation of the cells. We have shown that PARP DBD overexpression potentiates the cytotoxicity of the alkylating agent MNNG and of γ-radiation, while there is no significant influence on normal cell growth nor on the episomal replication of a supertransfected shuttle vector plasmid [21]. More recently, we found that carcinogen-induced gene amplification is also greatly potentiated by PARP DBD overexpression [22], which agrees with our earlier results [21]. Here, we show that the induction of mutations in a carcinogen-exposed shuttle vector plasmid is affected in the same way.

Materials and methods

Cells and induction treatment

The cell clone COM3 is a stable transfectant of the SV40-transformed hamster cell line C060 and expresses the DBD of human PARP upon induction with Dex [21]. COR3, another C060 transfectant, which lacks the PARP DBD gene but is matched for the content of glucocorticoid receptor, served as a control [21].

Shuttle vector mutagenesis

The shuttle vector plasmid pYZ289, comprising the *E. coli supF* gene as a mutational target gene as well as the mouse polyoma virus replicon [23], was resuspended in 1 × TE buffer and exposed to various concentrations of *N*-methyl-*N*-nitrosourea (MNU) for 1 h at 37°C, followed by ethanol precipitation and resuspension of the plasmid in PB S (137 mM NaCl, 2.7 mM KCl, 1.5 mM KH_2PO_4, 8.1 mM Na_2HPO_4). COM3 or COR3 cells that had been incubated or not with 50 nM Dex for 24 h, respectively, were transfected by electroporation exactly as described [21]. After electroporation, Dex-pretreated cells were incubated in Dex-free medium for 48 h, followed by cell lysis, Hirt extraction of low-molecular-weight DNA, and digestion of this DNA with restriction endonuclease *Dpn*I, to eliminate from further analysis those plasmid molecules that had failed to undergo eukaryotic replication. Mutation frequency was assayed by transformation of *E. coli* strain NMM7070 (*supF*⁻) and scoring of bacterial colonies for β-galactosidase activity [23]. Mutation frequency is expressed as the ratio of mutant (i.e. white) colonies / total (i.e. white plus blue) colonies.

Results and discussion

By using the overexpressed DBD of human PARP as a *trans*-dominant inhibitor of poly(ADP-ribosyl)ation in the stably transfected cell line COM3 [21], we set out to investigate the effect of this type of PARP inhibition on chemically induced mutagenesis. We chose shuttle-vector mutagenesis as an experimental system to address this question. A scheme of the experimental procedure is given in Fig. 1. Basically, naked plasmid DNA was exposed *in vitro* to the methylating carcinogen MNU, so as to generate premutagenic lesions on the target DNA (i.e., the bacterial

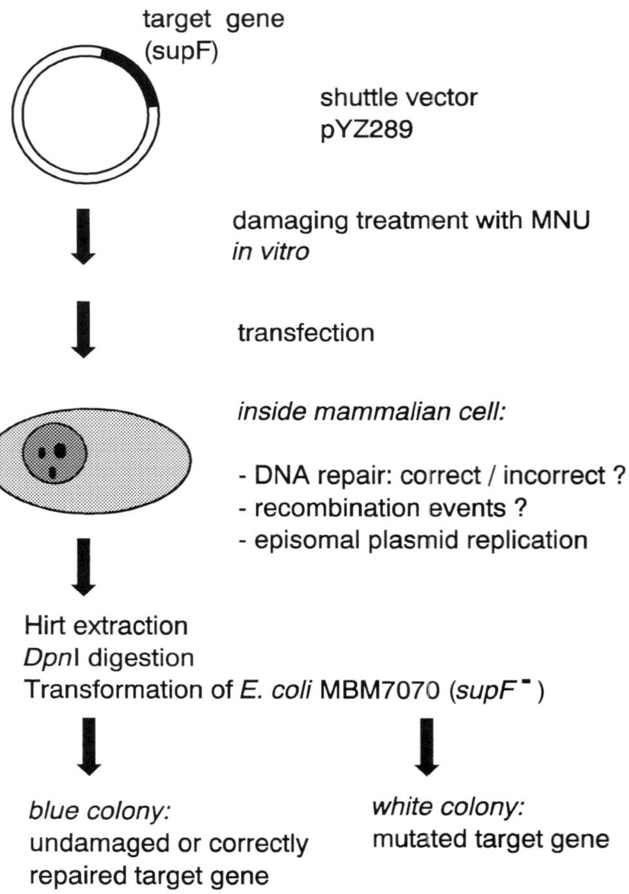

Fig. 1. Scheme of shuttle-vector mutagenesis assay. For details, see Materials and methods.

supF gene). We chose MNU since this compound strongly induces PARP activity, most likely due to the nicks in DNA that rapidly occur during base excision repair which is triggered by methylation damage. The plasmid DNA was then electroporated into COM3 or COR3 cells, respectively. COM3 cells carry stable insertions [i] of a Dex-inducible expression construct for the PARP DBD, and [ii] of a constitutively active expression construct for the human glucocorticoid receptor, allowing for high-level expression of DBD [21]. Induction of DBD expression by application of Dex 24 h before electroporation will therefore lead to *trans*-dominant inhibition of poly(ADP-ribosyl)ation by the time of introduction of the damaged exogenous DNA. To check for potential nonspecific effects of Dex in combination with glucocorticoid receptor overexpression, we used the cell line COR3 as a control. This cell line overexpresses human glucocorticoid receptor at the same level as COM3 but lacks the PARP DBD cDNA [21]. In the nuclei of successfully electroporated cells, the DNA lesions on the shuttle vector are exposed to the cellular DNA repair machinery. Removal of lesioned DNA sequences may happen by a variety of mechanisms, such as DNA base excision repair or recombinational events. In any case, the plasmid molecules can replicate episomally in the eukaryotic cell by virtue of the polyoma virus replicon they are carrying. Mutagenesis may result from DNA replication across the site of an unrepaired lesion, or from error-prone DNA repair, including recombinational repair. The mutation frequency is determined by extraction of plasmid DNA from the cells, selection of replicated molecules by *Dpn*I restriction endonuclease digestion, transformation of plasmid molecules into *E. coli* strain MBM7070 (*supF*$^-$), and measuring the relative frequency of β-galactosidase-negative colonies. Negativity for β-galactosidase is due to the lack of a functional *supF* gene on the shuttle-vector molecule, as a consequence of mutagenesis.

As shown in Fig. 2, the *in-vitro* treatment of the shuttle vector with increasing concentrations of MNU resulted in a decreasing yield of bacterial colonies (a measure of plasmid 'survival') if the damaged plasmid was directly transformed into *E. coli* (with omission of *Dpn*I digestion in this case), thus indicating the acquisition of a significant number of inactivating DNA lesions under the chosen conditions of MNU treatment.

When the same preparations of damaged plasmid were electroporated in COM3 or COR3 cells that had not been pretreated with Dex (Fig. 3, open symbols), there was a dose-dependent increase in the plasmid mutation frequency in both cell lines, as expected. Pretreatment of COM3 cells with 50 nM Dex (Fig. 3, filled square symbols), resulting in *trans*-dominant PARP inhibition [21], led to a significant potentiation of the mutation induction by MNU at all concentrations tested, but interestingly had no effect on mutation frequency in the absence of MNU treatment. Dex pretreatment

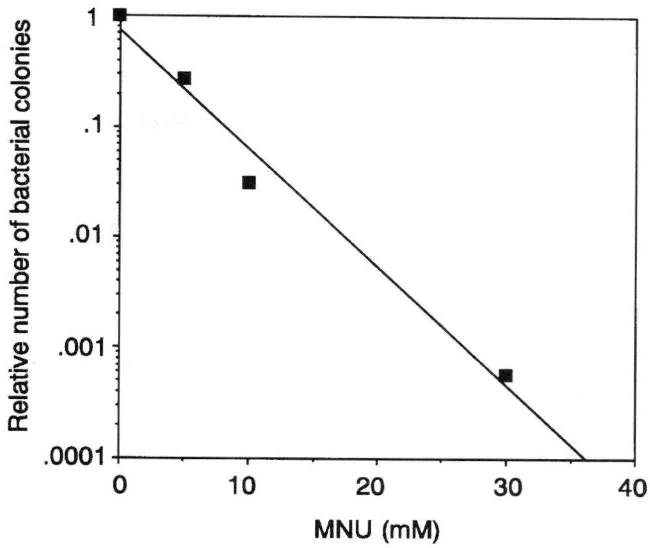

Fig. 2. Survival of MNU-treated shuttle vector directly transformed into *E. coli* MBM7070(*supF*$^-$). Plasmid pYZ289 was exposed to MNU *in vitro* at the concentrations indicated and then transformed into *E. coli*. Relative plasmid survival was deduced from the numbers of colonies obtained at each concentration divided by the value obtained from undamaged plasmid and is plotted versus MNU concentration.

of control COR3 cells (Fig. 3, filled circles) had no significant effect on the yield of MNU-induced mutations, thus confirming that the effect of Dex observed in COM3 cells was specifically due to PARP DBD overexpression.

Control experiments revealed that the yield of replicated plasmid recovered from electroporated COM3 cells was not affected by Dex pretreatment of the cells at all concentrations of MNU applied for the *in vitro* damaging of the plasmid (data not shown). Therefore, the differences we detected in the fraction of mutated plasmid molecules (Fig. 3) cannot be explained by any differential yield of plasmid.

It will be interesting to determine the so-called mutational spectra by sequencing the mutational target genes of mutated plasmids, so as to see whether or not the potentiation of MNU-induced shuttle vector mutagenesis we observed in COM3 cells overexpressing the PARP DBD is accompanied by qualitative differences in the mutations identified by this assay. Furthermore, it will be interesting to compare the shuttle vector data shown here with mutation rates of chromosomal genes, such as *HPRT*, in this cellular system.

Results of previous inhibitor studies on the role of poly (ADP-ribosyl)ation in mutagenesis have been conflicting [24–32]. This may have been due to differences in the experimental systems (including the mutational target genes) studied and/ or to problems with the specificity of ADP-ribosylation inhibitors used under the various conditions. Our present results reveal that *trans*-dominant inhibition of poly(ADP-

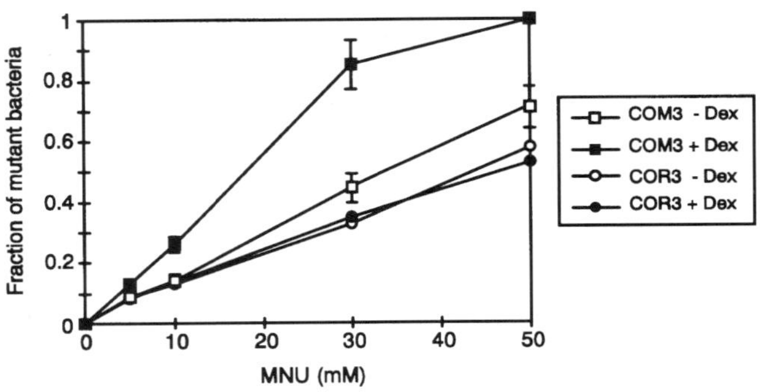

Fig. 3. Mutation frequency of MNU-treated shuttle vector passaged through COM3 or COR3 cells. COM3 cells, which overexpress the human PARP DNA-binding domain after induction with dexamethasone (Dex), and COR3 control cells were induced or not with Dex as indicated, followed by electroporation of plasmid pYZ289 that had been exposed to the indicated concentrations of MNU *in vitro*. Plasmid mutation frequencies were determined as described in Materials and methods and are plotted a as function of MNU concentration. Data for COM3 cells represent mean values ± 1 S.D. from 10 independent experiments. COR3 cells were assayed in 2 independent experiments, yielding almost identical results. Mean values are shown.

ribosyl)ation, as induced by overexpression of the PARP DBD, specifically potentiates the mutagenicity of MNU in a shuttle-vector system. Data presented here together with previous results on PARP inhibition and carcinogen-inducible gene amplification [19, 20, 22] further substantiate the role of poly(ADP-ribosyl)ation in the maintenance of genomic integrity and stability under conditions of genotoxic stress.

Acknowledgements

This work was supported by the Deutsche Forschungsgemeinschaft (grants Bu 698/2-3 and 446 JAP-113/74/0) and was facilitated by participating in the Concerted Action Programme 'Molecular gerontology: the identification of links between ageing and the onset of age-related diseases' (BMH1 CT94 1710) sponsored by the EU Commission (Biomed 2).

References

1. Althaus FR, Richter C: ADP-ribosylation of proteins. Enzymology and biological significance. Mol Biol Biochem Biophys 37: 1–237, 1987
2. de Murcia G, Ménissier de Murcia J: Poly (ADP-ribose) polymerase: A molecular nick sensor. Trends Biochem Sci 19: 172–176, 1994
3. Lindahl T, Satoh MS, Poirier GG, Klungland A: Post-translational modification of poly(ADP-ribose) polymerase induced by DNA strand breaks. Trends Biochem Sci 20: 405–411, 1995
4. Molinete M, Vermeulen W, Bürkle A, Ménissier-de Murcia J, Küpper J-H, Hoeijmakers JH, de Murcia G: Overproduction of the poly(ADP-ribose) polymerase DNA-binding domain blocks alkylation-induced DNA repair synthesis in mammalian cells. EMBO J 12: 2109–2117, 1993
5. Satoh MS, Lindahl T: Role of poly(ADP-ribose) formation in DNA repair. Nature 356: 356–358, 1992
6. Satoh MS, Poirier GG, Lindahl T: NAD(+)-dependent repair of damaged DNA by human cell extracts. J Biol Chem 268: 5480–5487, 1993
7. Berger NA: Poly(ADP-ribose) in the cellular response to DNA damage. Radiat Res 10: 14–15, 1985
8. Heller B, Wang Z-Q Wagner EF, Radons J, Bürkle A, Fehsel K, Burkart V, Kolb H: Inactivation of the poly(ADP-ribose) polymerase gene affects oxygen radical and nitric oxide toxicity in islet cells. J Biol Chem 270: 11176–11180, 1995
9. Zhang J, Dawson VL, Dawson TM, Snyder SH: Nitric oxide activation of poly(ADP-ribose) synthetase in neurotoxicity. Science 263: 687–689, 1994
10. Kaufmann SH, Desnoyers S, Ottaviano Y, Davidson NE, Poirier GG: Specific proteolytic cleavage of poly(ADP-ribose) polymerase: An early marker of chemotherapy-induced apoptosis. Cancer Res 53: 3976–3985, 1993
11. Tewari M, Quan LT, O'Rourke K, Desnoyers S, Zeng Z, Beidler DR, Poirier GG, Salvesen GS, Dixit M: Yama/CPP32 beta, a mammalian homolog of CED-3, is a CrmA-inhibitable protease that cleaves the death substrate poly(ADP-ribose) polymerase. Cell 81: 801–809, 1995
12. Nicholson DW, Ali A, Thornberry NA, Vaillancourt JP, Ding CK, Gallant M, Gareau Y, Griffin PR, Labelle M, Lazebnik YA, *et al.*: Identification and inhibition of the ICE/CED 3 protease necessary for mammalian apoptosis. Nature 376: 37–43, 1995
13. Negri C, Donzelli M, Bernardi R, Rossi L, Bürkle A, Scovassi AI: Multiparametric staining to identify apoptotic human cells. Exp Cell Res 234: 174–177, 1997
14. Ding R, Smulson M: Depletion of nuclear poly(ADP-ribose) polymerase by antisense RNA expression: influences on genetic stability, chromatin organization, and carcinogen cytotoxicity. Cancer Res 54: 4627–4634, 1994
15. Schreiber V, Hunting D, Trucco C, Gowans B, Grunwald D, de Murcia G, Ménissier-deMurcia J: A dominant-negative mutant of human poly(ADP-ribose) polymerase affects cell recovery, apoptosis, and sister chromatid exchange following DNA damage. Proc Natl Acad Sci USA 92: 4753–4757, 1995
16. Farzaneh F, Panayotou GN, Bowler LD, Hardas BD, Broom T, Walther C, Shall S: ADP-ribosylation is involved in the integration of foreign DNA into the mammalian cell genome. Nucl Acids Res 16: 11319–11326, 1988
17. Gäken JA, Tavassoli M, Gan S-U, Vallian S, Giddings I, Darling D, Galea-Lauri J, Thomas MG, Abedi H, Schreiber V, Ménissier-de Murcia J, Collins NWL, Shall S, Farzaneh F: Efficient retroviral infection of mammalian cells is blocked by inhibition of poly(ADP-ribose) polymerase activity. J Virol 70: 3992–4000, 1996
18. Petersen J, Dandri M, Bürkle A, Zhang L, Rogler CE: Increase in the frequency of hepadnavirus DNA integrations by oxidative DNA damage and inhibition of DNA repair. J Virol 71: 5455–5463, 1997
19. Bürkle A, Meyer T, Hilz H, zur Hausen H: Enhancement of *N*-methyl-*N'*-nitro-*N*-nitrosoguanidine-induced DNA amplification in a Simian virus 40-transformed Chinese hamster cell line by 3-aminobenzamide. Cancer Res 47: 3632–3636, 1987
20. Bürkle A, Heilbronn R, zur Hausen H: Potentiation of carcinogen-induced methotrexate resistance and dihydrofolate reductase gene amplification by inhibitors of poly(adenosine diphosphate-ribose) polymerase. Cancer Res 50: 5756–5760, 1990

21. Küpper J-H, Müller M, Jacobson MK, Tatsumi-Miyajima J, Coyle DL, Jacobson EL, Bürkle A: *Trans*-dominant inhibition of poly (ADP-ribosyl)ation sensitizes cells against γ-irradiation and *N*-methyl-*N'*-nitro-*N*-nitrosoguanidine but does not limit DNA replication of a polyomavirus replicon. Mol Cell Biol 15: 3154–3163, 1995
22. Küpper J-H, Müller M, Bürkle A: *Trans*-dominant inhibition of poly(ADP-ribosyl)ation potentiates carcinogen-induced gene amplification in SV40-transformed Chinese hamster cells. Cancer Res 56: 2715–2717, 1996
23. Moriwaki S, Yagi T, Nishigori C, Imamura S, Takebe H: Analysis of *N*-methyl-*N*-nitrosourea-induced mutations of a shuttle vector plasmid propagated in O^6-methylguanine-DNA methyltransferase-deficient cells in comparison with proficient cells. Cancer Res 51: 6219–6223, 1991
24. Natarajan AT, van Zeeland AA, Zwanenburg TS: Influence of inhibitors of poly(ADP-ribose) polymerase on DNA repair, chromosomal alterations, and mutations. Princess Takamatsu Symp 13: 227–242, 1983
25. Schwartz JL, Morgan WF, Brown-Lindquist P, Afzal V, Weichselbaum RR, Wolff S: Comutagenic effects of 3-aminobenzamide in Chinese hamster ovary cells. Cancer Res 45: 1556–1559, 1985
26. Kumar A, Kiefer J, Schneider E, Crompton NE: Enhanced cell killing, inhibition of recovery from potentially lethal damage and increased mutation frequency by 3-aminobenzamide in Chinese hamster V79 cells exposed to X-rays. Int J Radiat Biol Relat Stud Phys Chem Med 47: 103–112, 1985
27. Guo XC, Cleaver JE: Mutation frequencies from X-rays, ultraviolet light, and methyl methanesulfonate in Chinese hamster ovary cells incubated with 3-aminobenzamide. Mutagenesis 1: 237–239, 1986
28. Bhattacharjee SB, Bhattacharyya N: Influence of inhibitors of poly(ADP-ribose) polymerase activity on X-ray and ultraviolet light induced killing and mutation in Chinese hamster V79 cells. Carcinogenesis 7: 1267–1271, 1986
29. Bhattacharyya N, Bhattacharjee SB: The effect of nicotinamide on *N*-methyl-*N'*-nitro-*N*-nitrosoguanidine induced killing and mutation. Chem Biol Interact 60: 287–295, 1986
30. Nunbhakdi V, Jacobson EL: Effects of a poly(ADP-ribose) polymerase inhibitor on mutation frequencies in dividing and quiescent C3H10T1/2 cells. Mutat Res 180: 249–256, 1987
31. Ghosh R, Bhattacharjee SB: Influence of benzamide on killing and mutation of density-inhibited V79 cells by MNNG. Mutat Res 225: 137–141, 1989
32. Magnusson J, Ramel C: Inhibitor of poly(ADP-ribose)transferase potentiates the recombinogenic but not the mutagenic action of alkylating agents in somatic cells *in vivo* in *Drosophila melanogaster.* Mutagenesis 5: 511–514, 1990

Molecular and Cellular Biochemistry **193:** 37–42, 1999.

Clostridial toxins: Molecular probes of Rho-dependent signaling and apoptosis

David A. Bobak
Departments of Medicine and Microbiology, University of Virginia School of Medicine, Charlottesville, VA, USA

Abstract

The Rho family small GTPases are members of the Ras superfamily of small GTPases. Rho proteins were first determined to act as key regulators of many types of actin cytoskeletal-dependent cellular functions. Recent work by several investigators indicates that Rho GTPases are also critical modulators of several important intracellular and nuclear signal transduction pathways. Certain clostridial toxins and exoenzymes covalently modify, and thereby inactivate, specific types of Rho family GTPases. As such, these microbial enzymes have proven invaluable in helping to identify structural and functional attributes of Rho GTPases. (Mol Cell Biochem **193**: 37–42, 1999)

Key words: Rho, GTPase, toxins, *Clostridium*, signal transduction, apoptosis

Introduction

In the past 30 years or so, study into the molecular mechanisms of action of microbial toxins has not only shed light on the etiology of many important diseases, but has directly led to the elucidation of many important and basic biologic functions as well. There are several notable examples from this field of investigation: cholera and pertussis toxins and the discovery of heterotrimeric G-proteins, diptheria toxin and insight into the role of elongation factor II in protein synthesis, and molecular details of the neuroexocytotic process brought to light by the study of botulinal and clostridial neurotoxins. Recently, a variety of toxins and exoenzymes secreted by certain species of *Clostridium* bacteria have been discovered to covalently modify and inactivate several members of the Rho family of signal-transducing small GTPases [1–3]. Using these toxins and exoenzymes as molecular probes, several investigators have been able to determine that Rho GTPases are critical regulators of a wide variety of important cellular and nuclear signal transduction pathways.

Rho small GTPases

Rho family proteins are members of the Ras superfamily of signal-transducing small GTPases [4, 5]. In 1985, a gene was cloned from *Aplysia* and, on the basis of its sequence similarity to the *ras* oncogene, was named Rho (i.e. 'Ras homologue'). Subsequently, several forms of *rho* genes were isolated from yeast and human cDNA libraries. In addition, several rho-related genes have also been discovered. A listing of the currently known Rho and Rho-like proteins is outlined in Table 1. Of these several forms of Rho, the three forms known as Rho, Rac, and Cdc42 have been the most well-studied among the group [6–8].

The Rho family of proteins belongs to a larger superfamily of small GTPases related to the product of the *ras* oncogene. These GTPases are all of generally similar size and overall structure to Ras and are particularly homologous in the domains of proteins known to be involved in the binding and hydrolysis of GTP. Using structural and functional similarities, the Ras superfamily GTPases can be grouped into several different subfamilies: Ras/Ras-related proteins, Rho/

Table 1. Small GTPases of the Rho family.

Type	Mammalian	*Drosophila*	*Dictyostelium*	Yeast
Rho	RhoA–C	Rho1		Rho1
Rac	Rac1 & 2	DRac 1 & 2	Rac 1–3, & B	
Ccd42	Cdc42hs	DCdc42		Cdc42sc & sp
Misc.	TC10 Rho D, E, & G		Rac A, C, D, & E	Rho 2–4

Address for offprints: D.A. Bobak, Box 485-HSC, University of Virginia School of Medicine, Charlottesville, VA 22908, USA

Rho-related proteins, ARFs, Rabs, and a group of Misc. small GTPases [4, 5].

Regulation of Rho activity

As is generally true of most small GTPases, the activity state of Rho proteins is governed by a cycle of guanine nucleotide binding and modulated by complex interactions with a number of regulatory factors [4, 6–11]. Some of the key features of this activation/ deactivation cycle for Rho GTPases are shown in Fig. 1. Basal, or 'inactive,' Rho is found in the cytosolic phase, bound to GDP and complexed with a factor known as Rho guanine nucleotide dissociation inhibitor (GDI). RhoGDI is believed to sustain the 'off' phase of Rho by stabilizing the GDP-bound state of the Rho protein and masking Rho's COOH-terminal membrane targeting domain. In response to upstream activating signals, the exchange of GDP for GTP occurs on Rho, shifting the tertiary structure of the protein into an activated, or 'on,' state. In some instances, this exchange can be facilitated by another type of regulatory factor, known as a guanine nucleotide exchange factor (GEF). At, or about, the same time, RhoGDI dissociates from Rho and the fully activated GTPase then translocates to specific subcellular, membrane-associated compartments within the cell.

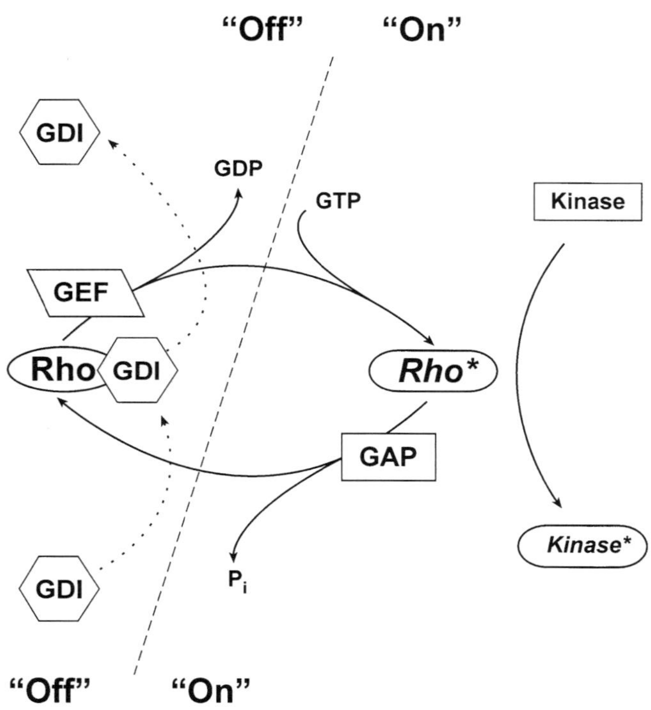

Fig. 1. Activation and regulation cycle of Rho GTPases. The cycle is described in the text.

Activated Rho-GTP recruits one of several known Rho effector molecules to Rho's subcellular location. Currently characterized Rho effectors include protein kinases, lipid kinases, phospholipases, putative adaptor proteins, and phosphatases [6, 7, 12–14]. The specificity of Rho signaling is believed to result from Rho's putative ability to recruit a specific effector, or complex of effectors, to a particular subcellular compartment in a time- and space-dependent manner. Termination of the Rho-dependent signal likely results in a return of the Rho protein to its basal, GDP-bound state. Some hydrolysis of GTP to GDP occurs intrinsically, but specialized factors that accelerate this reaction, known as GTPase accelerating proteins (GAPs), have been identified for Rho family GTPases [9]. Rho-GDI can assist in the release of Rho-GDP from the target membrane and likely helps complete the activation cycle. A cycle for Rho is illustrated here, but the same general cycle and compliment of regulatory factors have been described for the Rho-related proteins Rac and Cdc42 as well. Although the simple version of this cycle presented here appears intricate enough, the true complexity is much greater because many of the GEFs and GAPs for the Rho family GTPases probably have effector activities themselves. For example, several Rho GEFs are known to be oncogenes (e.g. Vav, Db1, Ost, Lbc) [10, 13]. The intricacies of the relationships between Rho GTPases and their associated factors and effectors helps to explain how Rho proteins can be implicated in the regulation of such a wide variety of cellular functions.

Ras was the first of the small GTPases to be crystalized and analyzed at the structural level [4, 5, 15]. Since that time, the three-dimensional structures of a number of wild type and mutant small GTPases, including members of the Rho family, have been solved and published. When compared with Ras at the amino acid sequence level, Rho small GTPases exhibit identities in the range of ~30 to ~ 60%, with most of the identity occurring in the domains involved in binding and hydrolysis of GTP. The overall topological structure of the small GTPases, though, appears to be highly conserved, and, from the results of numerous structural- and mutagenesis-based studies, several important functional domains have been identified. A cartoon outlining these particular domains for the Rho GTPases is shown in Fig. 2.

The Rho protein motifs responsible for the binding and hydrolysis of GTP are critical for activation and function of the molecule [2, 4, 6–8, 15, 16]. These regions include the amino acid sequences GXXXXGKS/T, DXXGQ, NKXD, and SXK (as are outlined in Fig. 2). Several well-characterized mutations have been described within these GTP-binding domains. In some cases, the mutated proteins bind GTP with high avidity and are resistant to GTP hydrolysis, resulting in a GTPase that is locked in the 'on' conformation. These activated forms of small GTPases are commonly described as constitutively active or dominant active forms. Conversely,

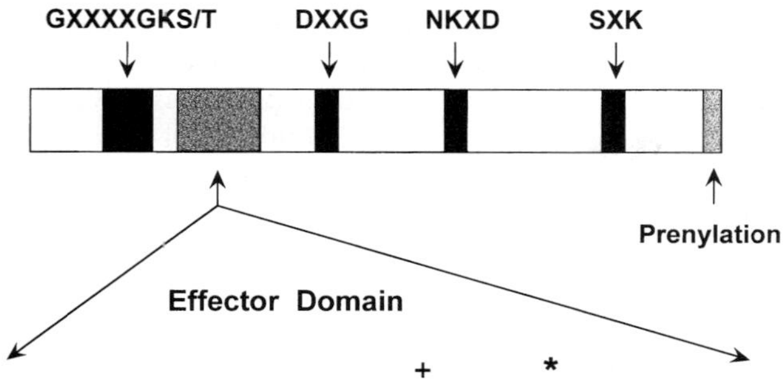


```
                        +      *
RhoA     29-QFPEVYVP T VFE N YVADIE-47
Rac1     27-AFPGEYIP T VFD N YSANVM-45
Cdc42    27-KFPSEYVP T VFD N YAVTVM-45
Ha-Ras   27-HFVDEYDP T IED S YRKQVV-45
```

Fig. 2. Structural domains within Rho GTPases. The domains involved in GTP-binding/hydrolysis, prenylation, and effector interaction are highlighted. *Clostridium botulinum* exoenzyme C3 ADP-ribosylates Rho at Asn[41] '*'), whereas *Clostridium difficile* toxins monoglucosylate Rho/Rac/Cdc42 at Thr[37] (or equivalent '+').

certain mutations result in a protein that binds GDP with high affinity and is resistant to nucleotide exchange. These 'unactivatible' forms of small GTPases are frequently referred to as being constitutively inactive or dominant negative. Both types of mutant small GTPases have proven valuable in determining the function of many types of small GTPases, including Rho family proteins.

In common with Ras proteins, Rho family GTPases contain a characteristic COOH-terminal sequence pattern - CAAX (C – a cysteine residue, A – an aliphatic amino acid residue and X – a serine or methionine residue for Ras proteins and a leucine residue for Rho family). The CAAX motif is a signal that triggers the onset of a series of important post-translational modifications consisting of prenylation of the cysteine residue, proteolytic cleavage of AAX, and carboxymethylation of the terminal cysteine [2, 4, 6–8, 15, 16]. For Ras proteins, the C-terminal prenyl modification is farnesyl whereas Rho family proteins are geranylgeranylated. Prenylation of small GTPases, including Rho family proteins, is generally believed to be required to direct the activated GTPase to its proper target subcellular location. A number of prenyltransferases have been currently been identified and their structural and functional attributes are being characterized in detail.

A most interesting region of the Rho family GTPases corresponds to the amino acid sequences from ~29 to ~47 (as numbered in RhoA). Based on analogies with earlier studies in Ras, this region has commonly been referred to as the effector domain (ED) of Rho. Although this nomenclature is still frequently used, it is now clear that there are other important domains within Rho that are likely to be equally important for directing the effector functions of the molecule [5, 6, 8, 16]. The Rho ED overlaps the switch I region, an area elucidated from the three dimensional structure that is known to be involved in regulating the GTP-binding pocket of Rho. This area is also believed to influence the nature of the interaction of Rho proteins with their regulatory and effector proteins. It is not surprising to learn, then, that the ED/switch I region also serves as the target for modification by a number of bacterial toxins [1–3].

Clostridial toxins inactivate Rho family GTPases

The mechanism of action of a number of clostridial toxins is now known to involve the covalent modification and inactivation of certain Rho family GTPases (Table 2). Rho-modifying clostridial toxins can be classified into one of two types based on the type of transferase reaction catalyzed: ADP-ribosylation or UDP-glucosylation/UDP-glucosaminlyation [1–3]. ADP-ribosylating toxins include those transferases produced by certain strains of *Clostridium botulinum* and *C. limosum*. Of note is the fact that two non-clostridal transferases, produced by *Staphylococcus aureus* and *Bacillus cereus*, ADP-ribosylate Rho proteins in a manner identical to that of the clostridial transferases. Exoenzyme C3 of *C. botulinum* is the most-well studied of all the Rho-modifying bacterial toxins/exoenzymes and members of this group of transferases are usually referred to as the C3-like exoenzymes. C3 ADP-ribosylates RhoA, B, and C at Asn[41], a modification that functionally inactivates the molecule (Fig. 2). Although the homologous Asn site (Asn[39]) is conserved within the sequences of Rac1, Rac2, and Cdc42, evidence to date indicates that these proteins are not efficiently ADP-ribosylated by C3-like exoenzymes [1–3, 16, 17]. C3, therefore, functions as a potent and specific probe of Rho function in cells.

C3 transferase, however, is relatively cell-impermeant due to the lack of specific cellular binding domains. Most

Table 2. Clostridal toxins that modify and inactivate Rho GTPases.

Exoenzymeltoxin	Enzymatic activity/substrate	Target protein
Clostridium botulinum exoenzyme C3	ADP-ribosylation NAD	RhoA–C
Clostridium limosum transferase	ADP-ribosylation NAD	RhoA–C
Clostridium difficile toxin A	UDP-glucosylation UDP-Glc	All Rho, Rac, & Cdc42
Clostridium difficile toxin B	UDP-glucosylation UDP-Glc	All Rho, Rac, & Cdc42
Clostridium sordellii hemorrhagic toxin	UDP-glucosylation UDP-Glc	All Rho, Rac, & Cdc42
Clostridium sordellii lethal toxin	UDP-glucosylation UDP-Glc	Rac, Cdc42, Ras, Rap, & Ral
Clostridium novyi α-toxin	UDP-glucosaminylation UDP-GlcNAc	All Rho, Rac, & Cdc42

investigators, therefore, have had to rely on techniques such as microinjection or permeabilization to achieve adequate cellular uptake of C3. We recently developed a transient gene expression system capable of rapid and efficient intracellular expression of C3 [18, 19]. The success of this approach promises to greatly expand the utility of C3-like transferases in elucidating the biochemical roles of Rho proteins in intact cells. For example, we have used this system to demonstrate that expression of C3 induces the formation of multinucleate giant cells and activates apoptosis in EL4 lymphocytes [19]. We also used this approach to determine that Rho is an important regulator of adhesion-dependent and -independent pathways of apoptosis in adherent cells [18].

The second group of Rho-modifying toxins is comprised of several different monoglucosylating toxins that use UDP-Glc or UDP-GlcNAc as substrates [2, 3]. The term large clostridial toxins has been used to refer to this particular group of toxins. Of the glucosyltransferase toxins, toxins A and B of *Clostridium difficile* are the prototypes. These two toxins differ from the others listed in Table 2 because they also act as the causative agents of a clinically important disease of humans known as antibiotic-associated colitis [20–22]. Pseudomembranous colitis, the most serious form of this disease, is a severe inflammatory diarrhea that usually occurs in the setting of antibiotic administration. When tested *in vivo* in experimental animals, toxin A is primarily enteropathogenic whereas toxin B is inactive in this regard. In cell lines, both toxins induce cellular retraction, rounding, and cytotoxicity, though toxin B is generally found to be the more potent cytotoxin *in vitro*. Because the morphological cytoskeletal changes induced by toxins A and B resembled some of the changes that had been observed for C3, certain investigators had speculated that toxins A and B could act via a C3-like effect on Rho.

Though the toxins were found to not be ADP-ribosyltransferases, Aktories and colleagues discovered that toxins A and B were, in fact, monoglucosyltransferases that used UDP-glucose as the substrate and Rho family GTPases as the target acceptor proteins [1–3]. Intriguingly, the site of glucosylation was determined to be at Thr^{37} of Rho A (Fig. 2). This amino acid residue is very close to Asn^{41}, the residue serving as the site of ADP-ribosylation by the C3-like exoenzymes (Fig. 2, Table 1). Studies to date suggest that these toxins A and B functionally act similarly to the C3-like exoenzymes and inactivate Rho GTPases. There are some important differences between toxins A or B and the C3-like exoenzymes, however [1–3, 20, 22, 22a]. For example, toxins A and B appear to efficiently glucosylate and inactivate Rho, Rac, and Cdc42. In addition, the toxins contain cell-binding domains and are efficiently taken up by a variety of cell types. Evidence suggests that the glucosyltransferase activity of the toxins is located within the amino terminal one-third of the molecules and that this transferase activity is essential, or at least very important, for the toxins' effects at the cellular level [1–3, 20, 22]. Toxins A and B are huge molecules (~300 kDa), however, and may well contain other functional domains that are unrelated to the modification of Rho, but necessary for cellular intoxication or the production of disease *in vivo*. In addition to toxins A and B, other large clostridial toxins have been identified and found to glucosylate or glucosaminylate certain Rho family GTPases (Table 2) [2, 3]. Notably, *C. sordellii* lethal toxin is also able to modify the Ras family GTPases Ras, Rap, and Ral. Detailed analysis of the actions and effects of C. *sordellii* and C. *novyi* toxins are just beginning.

Rho GTPases as regulators of signal transduction pathways controlling cell proliferation, transformation, and apoptosis

Rho proteins have traditionally been thought of as important regulators of a variety of cytoskeletal-dependent functions [16, 17, 23]. A series of reports by Hall and colleagues have suggested that, at least in some fibroblast cell lines, each mammalian Rho family member regulates a specific type of actinbased cytoskeletal structure [7, 23, 24]: Rho - focal adhesions and stress fibers, Rac - lamellipodia and ruffling, and Cdc42 - filopodia. Based on these findings, a hierarchical model of activity for Rho GTPases has evolved (outlined in Fig. 3A) [2, 7, 23, 24]. In the last few years, though, it has become apparent that Rho family members also help modulate other cytoskeletal-dependent cellular functions, including cell-cell adhesion, cell spreading, cell motility, and cytokinesis [6, 7, 16, 25]. Further, a large number of other, apparently actin-independent cellular functions and enzymes have been found to be activated or regulated by Rho GTPases [6–8, 13, 14], e.g., a variety of protein kinases, lipid kinases, and phospholipases; the phagocyte NADPH superoxide system; Ras-dependent malignant transformation; nuclear signaling through Jun kinase-, p38 kinase-, NFkB- and serum factor-dependent pathways; and apoptosis. When the large number of Rho family members, RhoGEFs, RhoGAPs, and Rho-associated kinases /phospholipases/ adaptors/ effectors are added to this diverse array of Rho-dependent functions, the adequacy of the current model seems somewhat limited.

In many situations, therefore, consideration of modified or alternate models for Rho function can be justified. Recently, Lim and colleagues found that in the instance of neuronal growth cone remodeling, Rho family GTPases are not activated in a manner that is explainable by the classically held model of linear hierarchy but by one in which Rho GTPases appear to function in a parallel and possibly competitive manner [7, 25, 27]; others have made similar

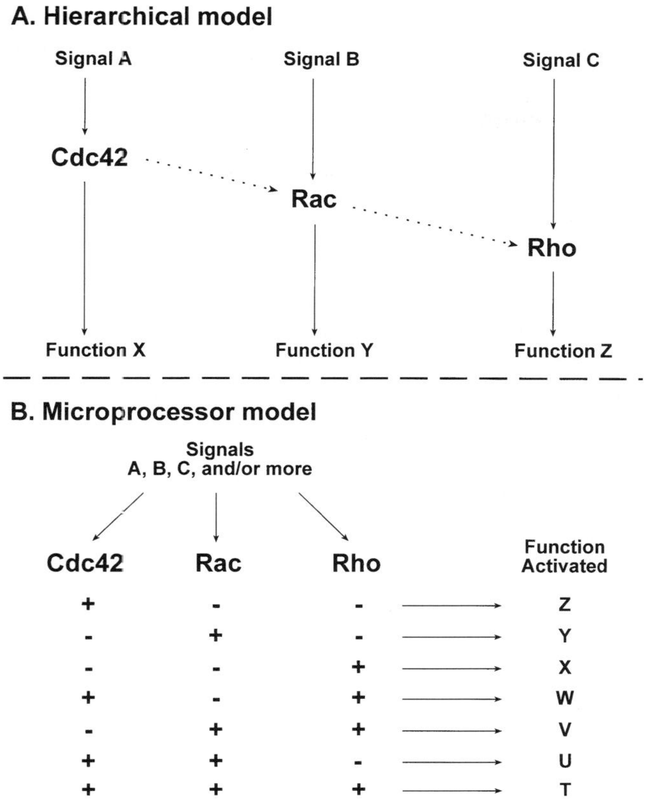

Fig. 3. Models of the activation of Rho GTPases The model is described in the text.

observations. Based on our own studies, our group has expanded this hierarchical paradigm into an alternative model of Rho GTPase action (Fig. 2B). In this scenario, cellular functions are stimulated by Rho GTPases in the context of interaction and/or competition of multiple family members, somewhat reminiscent of pathways in a digital microprocessor. For example, a cellular effect resulting from the inactivation of Rho does not simply represent only a 'turning off' of the Rho pathway, but also may represent an increase in the now unopposed activities of Rac and/or Cdc42 [28]. Additionally, functions modulated by Rho GTPases can be visualized as occurring in response to patterns of active/inactive GTPases seen in conjunction with any number of bound/unbound GEFs, GAPs, and effectors. The particular patterns of GTPases and ancillary factors that lead to a specific cellular function are likely dependent on cell type as well as the time and subcellular location of activation signal.

Clostridial toxins have proven to be extremely valuable probes in understanding the structure and function of a number of Rho family GTPases. We believe that even more information will be obtained as the enzymatic activities of the toxins are detailed and the investigation of Rho-signalling progresses along a biochemically-based track. In the case of the C. *difficile* toxins, further work will also provide valuable insight into the pathogenesis of C. *difficile* colitis and even potentially provide the basis for novel approaches to treatment and prevention.

Acknowledgments

I wish to particularly thank Drs. Chang Hahn, Jonathan Moorman and Joel Moss for invaluable discussions and collaborations related to the topics and concepts presented in this manuscript. The hypothetical model of Rho activation outlined in this paper is the joint concept of myself, Dr. Moorman and Dr. Hahn. Research in my laboratory is supported by grants from the N1H (R01-54572) and CTR (CTRC612).

References

1. Aktories K, Just I: Monoglucosylation of low-molecular-mass GTP-binding Rho proteins by Clostridial cytotoxins. Trends Cell Biol 5: 441–443, 1995
2. Aktories K: Rho proteins: Targets for bacterial toxins. Trends Microbiol 5: 282–288, 1997
3. von Eichel-Streiber C, Boquet P, Sauerborn M, Thelestam M: Large clostridial cytotoxins: A family of glycosyltransferases modifying small GTP-binding proteins. Trends Microbiol 4: 375–382, 1996
4. Macara IG, Lounsgury KM, Richards SA, McKiernan C, Bar-Sagi D: The Ras superfamily of GTPases. FASEB J 10: 625–630, 1996
5. Khosravi-Far R, Der CJ: The Ras signal transduction pathway. Cancer Metast Rev 13: 67–89, 1994
6. Symons M: Rho family GTPases: The cytoskeleton and beyond. Trends Biochem Sci 21: 178–181, 1996
7. Ridley AJ: Rho: Theme and variations. Curr Biol 6: 1256–1264, 1996
8. Narumiya S: The small GTPase Rho: Cellular functions and signal transduction. J Biochem 120: 215–228, 1996
9. Lamarche N, Hall A: GAPs for rho-related GTPases. Trends Genet 10: 436–440, 1994
10. Olson MR: Guanine nucleotide exchange factors for the Rho GTPases. J Mol Med 74: 563–571, 1996
11. Cerione RA, Zheng Y: The Dbl family of oncogenes. Curr Opin Cell Biol 8: 216–222, 1996
12. Exton JH: Cell signalling through guanine-nucleotide-binding regulatory proteins and phospholipases. Eur J Biochem 243: 10–20, 1997
13. Denhardt DT: Signal-transducing protein phosphorylation cascades mediated by Ras/Rho proteins in the mammalian cell: The potential for multiplex signaling. Biochem J 318: 797–747, 1996
14. Lim L, Manser E, Leung T, Hall C: Regulation of phosphorylation pathways by p21 GTPases. Eur J Biochem 242: 171–185, 1996
15. Schweins T, Wittinghofer A: Structures, interactions, and relationships. Curr Biol 4: 547–550, 1994
16. Takai Y, Sasaki T, Tanaka K, Nakanishi H: Rho as a regulator of the cytoskeleton. Trends Biochem Sci 20: 227–231, 1995
17. Hall A: Small GTP-binding proteins and the regulation of the actin cytoskeleton. Annu Rev Cell Biol 10: 31–54, 1994

18. Bobak D, Moorman J, Guanzon A, Gilmer L, Hahn C: Inactivation of the small GTPase Rho disrupts cellular attachment and induces adhesion-dependent and adhesionindependent apoptosis. Oncogene 15: 2179–2189, 1997
19. Moorman JP, Bobak DA, Hahn CS: Inactivation of the small GTP-binding protein Rho induces multinucleate cell formation and apoptosis in murine T lymphoma EL4. J Immunol 156: 4146–4153, 1996
20. Moorman JP, Bobak DA: *Clostridium difficile* infections. In: R.L. Guerrant, D.J. Krogstad, J.H. Maguire, D.H. Walker, P.F. Weller (eds). Tropical Infectious Diseases: Principles, Pathogens, and Practice. New York, 1998, in press
21. Moncrief JS, Lyerly DM, Wilkins TD: Molecular biology of the *Clostridium difficile* toxins. Molecular Biology and Pathogenesis of the Clostridia: Academic Press, 1998, in press
22. Pothoulakis C: Pathogenesis of *Clostridium difficile*-associated diarrhea. Eur J Gastroen Hepat 8: 1041–1047, 1996
22a. Ciesla WP, Bobak DA: *Clostidium difficile* toxins A and B are cation-dependent UDP-glucose hydrolases with differing catalytic activities. J Biol Chem 272: 16021–16026, 1998
23. Chant J, Stowers L: GTPase cascades choreographing cellular behavior: Movement, morphogenesis, and more. Cell 81: 14, 1995
24. Nobes CD, Hall A: Rho, Rac, and Cdc42 GTPases regulate the assembly of multimolecular focal complexes associated with actin stress fibers, lamellipodia, and filopodia. Cell 81: 53–62, 1995
25. Zigmond SH: Signal transduction and actin filament organization. Curr Opin Cell Biol 8: 66–73, 1996
26. Vojtek AB, Cooper JA: Rho family members: activators of MAP kinase cascades. Cell 82: 527–529, 1995
27. Kozma R, Samer S, Ahmed S, Lim L: Rho family GTPases and neuronal growth cone remodelling: Relationship between increased complexity induced by Cdc42Hs, Rac1, and acetylcholine and collapse induced by RhoA and lysophosphatidic acid. Mol Cell Biol 17: 1201–1211, 1997
28. Moorman JP, Luu D, Wickman J, Bobak DA, Hahn SC: A balance of signalling by Rho family small GTPases RhoA, Rac1, and Cdc42 coordinates cytoskeletal morphology but not cell survival. Oncogene, 1999, in press

Molecular and Cellular Biochemistry **193**: 43–51, 1999.

Role of brefeldin A-dependent ADP-ribosylation in the control of intracellular membrane transport

Maria Giuseppina Silletta, Antonino Colanzi, Roberto Weigert, Maria Di Girolamo, Ivana Santone, Giusy Fiucci, Alexander Mironov, Maria Antonietta De Matteis, Alberto Luini and Daniela Corda

Istituto di Ricerche Farmacologiche 'Mario Negri', Consorzio Mario Negri Sud, Department of Cell Biology and Oncology, Via Nazionale, Santa Maria Imbaro (Chieti), Italy

Abstract

The fungal toxin brefeldin A (BFA) dissociates coat proteins from Golgi membranes, causes the rapid disassembly of the Golgi complex and potently stimulates the ADP-ribosylation of two cytosolic proteins of 38 and 50 kDa. These proteins have been identified as the glycolytic enzyme glyceraldehyde-3-phosphate dehydrogenase (GAPDH) and a novel guanine nucleotide binding protein (BARS-50), respectively. The role of ADP-ribosylation in mediating the effects of BFA on the structure and function of the Golgi complex was analyzed by several approaches including the use of selective pharmacological blockers of the reaction and the use of ADP-ribosylated cytosol and/or enriched preparations of the BFA-induced ADP-ribosylation substrates, GAPDH and BARS-50.

A series of blockers of the BFA-dependent ADP-ribosylation reaction identified in our laboratory inhibited the effects of BFA on Golgi morphology and, with similar potency, the ADP-ribosylation of BARS-50 and GAPDH. In permeabilized RBL cells, the BFA-dependent disassembly of the Golgi complex required NAD^+ and cytosol. Cytosol that had been previously ADP-ribosylated (namely, it contained ADP-ribosylated GAPDH and BARS-50), was instead sufficient to sustain the Golgi disassembly induced by BFA.

Taken together, these results indicate that an ADP-ribosylation reaction is part of the mechanism of action of BFA and it might intervene in the control of the structure and function of the Golgi complex. (Mol Cell Biochem **193**: 43–51, 1999)

Key words: brefeldin A, ADP-ribosylation, Golgi complex, G proteins, membrane transport, inhibitors of ADP-ribosylation reaction

Introduction

The function of the Golgi apparatus is to participate in the biosynthesis of many proteins and lipids, then rapidly transport these molecules in the *cis-trans* direction and distribute them to a variety of final destinations. The molecular mechanisms involved in this process have not been completely clarified. Brefeldin A (BFA), a fungal toxin that causes the rapid disassembly of the Golgi complex, has been shown to be a very useful tool to understand the molecular mechanisms modulating Golgi dynamics [1]. BFA, in addition to causing the detachment of coat proteins from Golgi membranes [2], potently stimulates the endogenous ADP-ribosylation of two cytosolic proteins of 38 kDa (glyceraldehyde-3-phosphate dehydrogenase, GAPDH), and 50 kDa [3, 4]. The 50 kDa BFA-dependent ADP-ribosylation substrate (BARS-50) has been proposed to be a novel G protein involved in membrane transport [5, 6]. Interestingly, the concentrations of BFA able to activate the ADP-ribosylation reaction are similar to those inhibiting ARF binding to Golgi membranes [3]. Moreover, the ability of several BFA analogues to inhibit ARF binding to Golgi membranes parallels the ability (and potency) of the same analogues to stimulate the ADP-ribosylation of GAPDH and BARS-50 [4, 5, 7]. These observations strongly suggest that ADP-ribosylation plays a role in the mechanism of action of BFA, and thus that this reaction might be

Address for offprints: D. Corda, Consorzio Mario Negri Sud, Via Nazionale, 66030 Santa Maria Imbaro, (Chieti), Italy involved in the control of Golgi dynamics. We have undertaken a series of experiments to directly examine the role of ADP-ribosylation in the maintenance of the Golgi structure. These include the use of blockers of the BFA-dependent ADP-ribosylation reaction, the inhibition of this reaction by depleting cells of NAD^+ (the ADP-ribose donor) and the use of enriched preparations of the BFA-substrate BARS-50 and/or cytosol, both in the native and ADP-ribosylated forms, in experiments aimed at reconstituting *in vitro* the BFA effect [7, 8]. The data obtained by these different approaches indicate that NAD^+ and ADP-ribosylation of BARS-50 participate in the mechanisms controlling the structure/function of the Golgi complex.

Materials and methods

Cell culture

Rat basophilic leukemia (RBL)-2H3 cells were grown in MEM supplemented with 16% Fetal Calf serum (FCS) and 1 mM L-glutamine; CHO cells were cultured in DME supplemented with 10% FCS; FRTL5 cells were cultured in F-12 Coon's modified medium, a hormone mixture and 5% Calf serum (CS), as previously described [3, 5, 8].

Antibodies and other reagents

NAD^+, NADH, BFA, were obtained from Sigma (St. Louis, MO, USA). Tran^{35}S-label (a mixture of ^{35}S-labeled methionine and cysteine) was purchased from ICN (Irvine, CA, USA). Tissue culture materials including plasticware, chamber slides (Nunc, Roskilde, Denmark) and medium (GIBCO) were purchased from Mascia Brunelli S.p.A. (Milano, Italy). ATP was purchased from Boehringer Mannheim (Mannheim, Germany). The temperature sensitive strain (ts045) of vesicular stomatitis virus (VSV) was kindly provided by Dr. K. Simons (EMBL, Heidelberg, Germany). Ilimaquinone was a gift from Dr. V. Malhotra (University of California, San Diego, CA, USA). The P5D4 mAb against the G protein of VSV was obtained from Sigma (St. Louis, MO, USA). Rabbit anti-α-mannosidase II (Man II) antibody was kindly provided by Dr. K. Moremen (University of Georgia, Athens, GA, USA). All other chemicals were obtained from commercial sources at the highest available purity.

Cell permeabilization

RBL and CHO cells were permeabilized by 1 U/ml of streptolysin O (SLO, Biomerieux, Marcy l'Etoile, France) as previously descibed [8]. The permeabilization buffer consisted of 25 mM HEPES-KOH, pH 6.95, 125 mM K^+ acetate, 2.5 mM Mg^{2+} acetate, 10 mM glucose, 1 mM DTT, 1 mM EGTA and 0.5 μM taxol. The extent of cell permeabilization was checked by Trypan blue (and propidium iodide) staining and by the leakage of the cytosolic enzyme lactic dehydrogenase. 95% of cells were stained with Trypan blue or propidium iodide and more than 75% of the lactic dehydrogenase activity was recovered in the supernatant of the permeabilized cell monolayer. Rat brain cytosol was prepared according to Malhotra *et al.* [9].

BFA-dependent ADP-ribosylation

Postnuclear supernatants were prepared from confluent RBL [3] or Fisher rat thyroid (FRTL5) cells [5], rapidly frozen and used in the ADP-ribosylation assay as previously described [3, 5]. ADP-ribosylation of permeabilized RBL cells and cytosol (prepared from rat brain as in ref. [3]) was performed as described [8]. The maximal ADP-ribosylation of BARS-50 was usually obtained using 200 μM NAD^+ and 100 μM BFA for 60 min at 37°C [8]. The radioactivity bound to BARS-50 and GAPDH was evaluated by quantitative electronic autoradiography using an Instant Imager (Packard).

VSV infection and assay of VSV-G protein transport

RBL cells were grown to confluence in 12-well plates and infected at the nonpermissive temperature of 40°C with the ts045 temperature mutant [10] of VSV in MEM containing 10% FCS and 5 μg/ml actinomycin D. One hour later the medium was replaced with MEM and the incubation continued for an additional 2 h. To label proteins, cells were kept for 10 min in DME lacking methionine and cysteine, pulse-labeled with ^{35}S-methionine (50 μCi/ml) for 10 min, then shifted to 33°C in MEM containing 2 mM methionine with or without BFA-dependent ADP-ribosylation inhibitors for various times. In some experiments ADP-ribosylation inhibitors were given in pre-treatment during the starving and/or the pulse period. In this case a slight (dicumarol) and a significant (ilimaquinone) inhibition of viral protein synthesis were observed. To evaluate the effect of ADP-ribosylation inhibitors on the BFA-induced retrograde transport of resident Golgi enzymes into the ER, the cells were chased at the non-permissive temperature of 40°C for 15, 60 or 90 min in the presence of BFA alone or in combination with ADP-ribosylation inhibitors. At the end of the various chase periods, the cells were lysed by direct addition of a buffer containing 0.5% SDS, 1% β-mercaptoethanol and anti-proteases (1 mM PMSF, 0.5 mM o-phenantrolin, 10 μg/ml leupeptin, 1 μM pepstatin, 1 mM DTT) and heated for 10 min at 90°C. Sodium acetate (100 mM) and $CaCl_2$ (10 mM) were added to the samples and the pH was adjusted to 5.5. Samples were fractionated in three aliquots and incubated overnight at 37°C with neuraminidase (5 mU/sample) and/or endoglycosidase H (20 U/samples). Concentrated (5 ×) Laemmli sample buffer was added, and samples were analyzed by SDS-PAGE and autoradiography.

Fluorescence microscopy

For immunostaining and fluorescence microscopy (performed as described in ref. [8]), RBL or CHO cells fixed in 4% paraformaldehyde in phosphate buffered saline (PBS), were stained with FITC-conjugated Helix pomatia lectin (100 µg/ml in PBS containing 0.2% BSA) or incubated with primary anti-Man II antibody for 1 h at room temperature, washed thoroughly with PBS and incubated with specific FITC-, TRITC- or Cy3-conjugated secondary antibody, then examined using a Zeiss Axiophot microscope equipped with a Plan-Neofluar 40 × objective [8].

Partial purification of BARS-50

A 45-fold enriched preparation of BARS-50 was obtained as previously described [8]. Briefly, rat brain cytosol [9] was precipitated with 35% saturated $(NH_4)_2SO_4$. The precipitate was dissolved in 25 mM HEPES pH 8.0 containing 5% glycerol, 0.5 M $(NH_4)_2SO_4$ and 1 mM DTT (buffer A) and applied to a phenyl Sepharose HP column (Pharmacia) equilibrated with buffer A. Proteins were eluted with a linear gradient of buffer A minus $(NH_4)_2SO_4$. The fractions containing BARS-50 were identified by the BFA-dependent ADP-ribosylation assay [3]. These fractions (which did not contain GAPDH) were concentrated and dialyzed against buffer B (25 mM HEPES, pH 7.2, 50 mM K and 1 mM Mg acetate) overnight, to reach a final protein concentration of 2-3 mg/ml.

Results

Substrates and features of the BFA-dependent ADP-ribosylation reaction

BFA induces the ADP-ribosylation of two specific substrates of 38 and 50 kDa, identified as the multifunctional protein GAPDH and BARS-50, respectively (Fig. 1 A, and refs [3–6]). BARS-50 is a novel G protein as indicated by the fact that it binds GTP and its ADP-ribosylation is inhibited by the $\beta\gamma$ subunit of heterotrimeric G proteins [5]. However, BARS-50 does not seem to belong to the Gs or Gi protein families, as suggested by the observation that this protein is

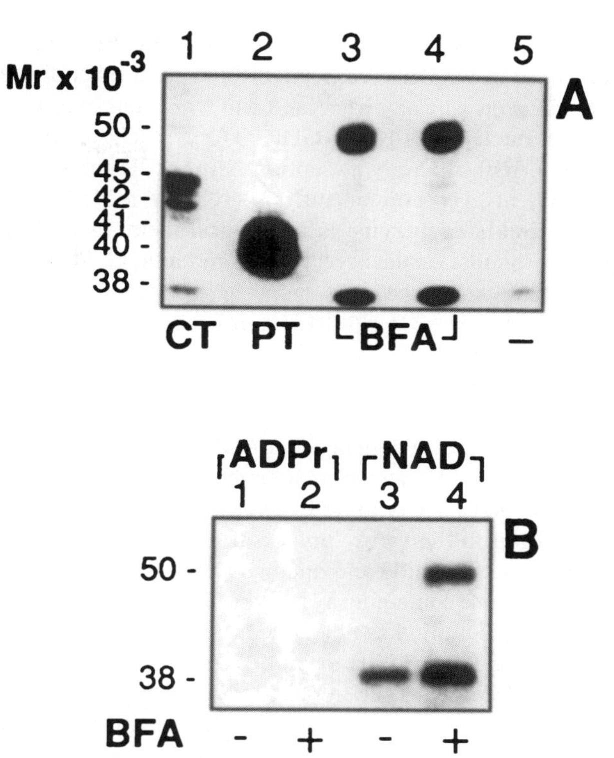

Fig. 1. The BFA-dependent ADP-ribosylation of GAPDH and BARS-50 requires an ADP-ribosyltransferase. *Panel A:* Membrane proteins from FRTL-5 cells were ADP-ribosylated in the presence of cholera (CT) (lane 1) or pertussis toxin (PT) (lane 2), and post nuclear supernatant from FRTL-5 cells ADP-ribosylated in the absence (lane 5) or presence (lanes 3 and 4) of BFA, and analyzed in an 8% polyacrylamide gel with 4M urea [5]. *Panel B:* post nuclear supernatant from FRTL-5 cells ADP-ribosylated in the absence (lanes 1 and 3) or presence (lanes 2 and 4) of BFA, using either free ADP ribose (lanes 1 and 2) or NAD⁺ (lanes 3 and 4). The Mr (kDa) are indicated. Similar results were observed in at least five different experiments.

not a substrate of the ADP-ribosylation induced by the transferases of cholera or pertussis toxins (Fig. 1A). In two-dimensional electrophoresis BARS-50 can be separated into several related protein spots, all substrates for BFA-dependent ADP-ribosylation [5, 6], suggesting that either BARS-50 undergoes a variety of different post-translational modifications or, alternatively, represents a group of very similar proteins. The enzyme involved in the reaction is a transferase, since the BFA-dependent ADP-ribosylation of either GAPDH or BARS-50 cannot be sustained by free ADP-ribose, requiring instead the presence of NAD^+ (Fig. 1B, and refs [3, 5]).

The BFA-dependent reaction has recently been characterized by using a quantitative assay which employs as substrate pure GAPDH [7, 11]. BFA activates the mono-ADP-ribosylation of GAPDH (and BARS-50, EC_{50} of 17 ± 3 µg/ml) by increasing the V_{max} (510 ± 150 pmol/h/mg) but not the Km (154 ± 13 µM) for NAD^+ [7]. The BFA-stimulated

ADP-ribosylation of GAPDH is inhibited by Zn^{2+} and Cu^{2+} (IC_{50}s ~75 and 500 μM, respectively), whereas other divalent cations such as Ca^{2+}, Mg^{2+} and Mn^{2+} are inactive, as are several nucleotides (GTP, GDP, GTPγS, GDPβS, ATPγS, cAMP), with the only exception being ATP, that was inhibitory at high concentrations (IC_{50} ~1 mM) [7]. The BFA-stimulated enzyme is an integral membrane protein present in most tissues (see Table 1 and refs. [3, 7]).

Characterization of specific inhibitors of the BFA-dependent ADP-ribosylation reaction

Inhibitors of the BFA-dependent ADP-ribosylation reaction were identified by screening a large number of compounds known to block bacterial ADP-ribosyltransferases [7, 11, 12]. In addition to the weak 'universal' ADP-ribosylation inhibitors nicotinamide and analogs [13–15], two classes of molecules, one containing a coumarin and the other a quinone ring, were identified as potent and specific inhibitors of the BFA-dependent ADP-ribosylation *in vitro*. In particular, dicumarol (known as an inhibitor at the NADH binding site of the oxido-reductase that catalyzes the reduction of vitamin K; ref. [16]) and ilimaquinone, a metabolite isolated from a marine sponge that causes a reversible breakdown of the Golgi complex [17], were used in both *in vitro* and *in vivo* experiments because of their relatively high potency and low toxicity (Table 2 and refs [7, 8, 18]). Dicumarol and ilimaquinone inhibited the BFA-induced ADP-ribosylation reaction with IC_{50}s of ~180 and 30 μM, respectively (Table 2), by a non-competitive mechanism with respect to both BFA and NAD^+ (Fig. 2 and ref. [7]). The inhibitory effect of dicumarol on the ADP-ribosylation of BARS-50 and GAPDH is shown in Fig. 3. Figure 4 shows the effects of the inhibitors as antagonists of the action of BFA on the structure of the Golgi complex *in vivo*. A concentration of dicumarol (200 μM) with near-maximal effects in ADP-ribosylation assays *in vitro* markedly inhibited the ability of BFA to induce the redistribution of the Golgi complex as revealed with Helix pomatia lectin. In contrast, warfarin, a structural analog of dicumarol devoid of ADP-ribosylation inhibitory activity, was also inactive as a BFA antagonist *in vivo* (Fig. 4). Ilimaquinone, like dicumarol, inhibited the BFA-induced Golgi disassembly, in addition to causing, in line with previous reports [17], the breakdown of the Golgi complex into initially large fragments (at 15 min) which became progressively smaller and distributed in the cytosol (45 min). BFA when administered concomitantly with, or 15 min after, ilimaquinone, lost its ability to induce the typical diffuse ER-like redistribution of Golgi markers and the ilimaquinone

Table 1. Distribution of the BFA-dependent mono-ADP-ribosylating enzyme in different tissues and subcellular fractions.

Tissue	[ADP-ribosylated GAPDH] (pmoles/mg)
Brain	152 ± 10
Liver	85 ± 6
Heart	55 ± 6
Spleen	35 ± 3
Lung	35 ± 4
Kidney	3 ± 2
Golgi	4.6 ± 0.1
Smooth E. R.	4.6 ± 1.4
Rough E.R.	1.1 ± 0.3
Mitochondria	3.3 ± 0.1

The ADP-ribosylation assay was carried out using either pure GAPDH (10 μg) and 10 μg of membranes from the different tissues (as enzyme source) or 50 μg of rat brain cytosol (as GAPDH source) and 8 μg of subcellular fractions enriched in Golgi, smooth ER, rough ER and mitochondria (as enzyme source) (see ref. [7] for details). Samples were then analyzed by SDS-PAGE and the labeling of GAPDH was quantified by electronic autoradiography (instant Imager, Packard). Data are means ± S.D. of at least three determinations.

Table 2. Potency of quinones and coumarins on the BFA-dependent ADP-ribosylation of GAPDH *in vitro* and on the BFA-induced Golgi redistribution *in vivo*.

Compound	*In vitro* IC_{50} (μM)	*In vivo* IC_{50} (μM)
QUINONES		
1. Ilimaquinone	30	40
2. Benzoquinone	30	*
3. Naphthoquinone	100	*
4. Menadione	200	*
5. Coenzyme Q0	100	*
6. Duroquinone	300	*
7. AA861	300	*
8. Dimethylinaphthoquinone	500	*
9. Coenzyme Q2	>1 mM	>1 mM
10. Coenzyme Q4	>1 mM	>1 mM
11. Coenzyme Q6	>1 mM	>1 mM
12. Coenzyme Q9	>1 mM	>1 mM
13. Coenzyme Q10	>1 mM	>1 mM
14. Vitamin K1	>1 mM	>1 mM
15. Vitamin K2	>1 mM	>1 mM
COUMARINS AND RELATED COMPOUNDS		
16. Coumermycin A1	60	70
17. Dicumarol	180	150
18. Novobiocin	300	300
19. Warfarin	>1 mM	>1 mM
20. Coumarin	>1 mM	>1 mM
21. Coumarin 35	>1 mM	>1 mM
22. Coumarin 138	>1 mM	>1 mM
23. Coumarin 152	>1 mM	>1 mM
24. 4-Hydroxycoumarine	>1 mM	>1 mM
25. Esculetin	>1 mM	>1 mM

This table is a modified version of Table 1 in ref. [7]. The IC_{50} of each molecule was determined as described in the legend to Fig. 5. S.D. never exceeded 10% for the *in vitro* data and 15% for the *in vivo* experiments. *Not determined; See ref. [7] for details.

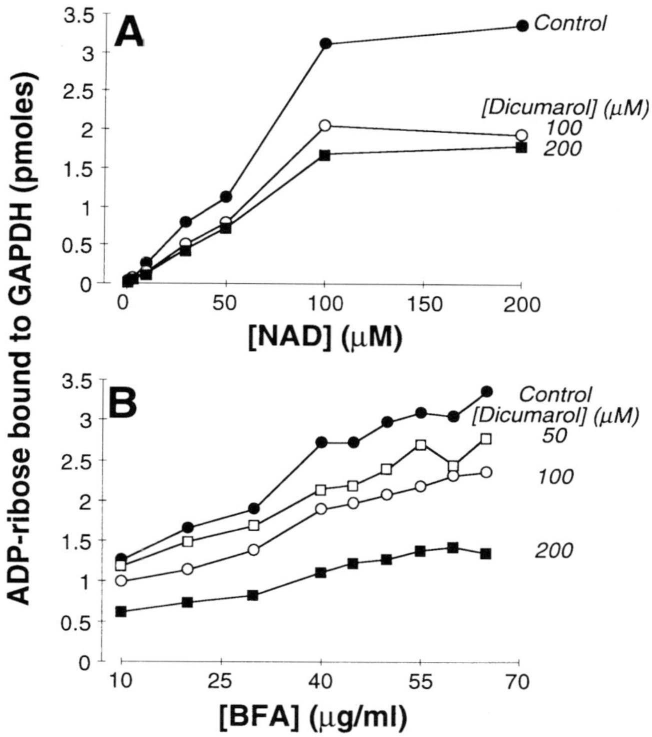

Fig. 2. NAD and BFA dependence of the inhibitory effect of dicumarol on the BFA-dependent ADP-ribosylation of GAPDH. GAPDH (10 μg) was incubated with 10 μg of rat brain membranes and [^{32}P]-NAD$^+$ for 1 h at 37°C (see Materials and methods) using a range of [^{32}P]-NAD$^+$ (A) concentrations [7] and BFA (B). Dicumarol was used at 50 μM (open squares), 100 μM (open dots) or 200 μM (full squares). Full dots show control samples. Data are means of four determinations.

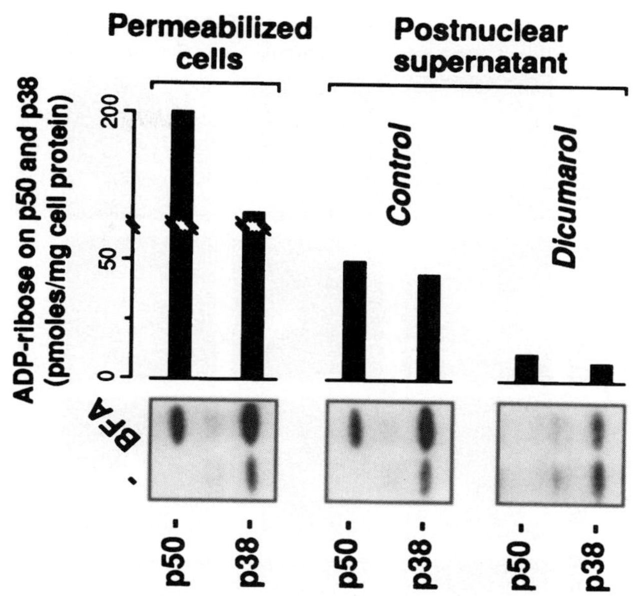

Fig. 3. ADP-ribosylation of GAPDH (p38) and BARS-50 (p50) in permeabilized cells. RBL cells permeabilized with 3U/ml SLO and stimulated by BFA (50 μg/ml) in the presence of [^{32}P]-NAD$^+$ for 60 min at 37°C. For comparison, an RBL-derived post nuclear supernatant was incubated for 60 min in the same buffer and the same concentration of BFA. The effect of 200 μM dicumarol on BFA-dependent ADP-ribosylation in postnuclear supernatant is also shown. At the end of the incubation cell proteins were separated by SDS-PAGE. The radioactivity bound to BARS-50 and GAPDH in a BFA-dependent manner was evaluated by electronic autoradiography and by subtracting the basal signal from that induced by BFA. The data shown in the graph are means of three determinations; S.D.s never exceeded 10% of the means. Similar results were obtained in four experiments.

phenotype with larger Golgi fragments remained dominant (Fig. 5 j-1, refs [17, 19]).

Since the effect of these inhibitors is to prevent an active process (Golgi redistribution), it was important to verify that their action is selectively mediated by a block of ADP-ribosylation, rather than by non-specific effects. The inhibitory effects of ilimaquinone and dicumarol were not due to general cell toxicity as indicated by experiments showing that cell viability and ATP levels were not affected by these drugs [18]. Further evidence that dicumarol does not act non-specifically as an inhibitor of Golgi disassembly came from another line of observations showing that BFA (at high concentrations or applied for a long time) completely overcomes the inhibition by dicumarol (Fig. 4f). This indicates that the drug does not paralyze the machinery responsible for BFA-induced Golgi redistribution and suggests that it acts by antagonizing selectively the action of BFA. Finally, a number of inhibitors of ADP-ribosylation acting with widely differing potencies *in vitro* [7] were tested also as antagonists of BFA *in vivo,* and the profiles of inhibitory potency of these drugs in the two assays were compared. The rationale of this experiment is that if there is a strong similarity between the two potency profiles, one can infer that the two effects of the inhibitors (*in vivo* antagonism of BFA and *in vitro* block of ADP-ribosylation) are very likely to be causally related. Fig. 5 shows that the two profiles were indeed strikingly similar, thus suggesting a strong link between the two actions of the drugs [7].

Finally, the ability of the ADP-ribosylation inhibitors to prevent the Golgi redistribution induced by BFA was studied by biochemical means. Normally, the disassembly of the Golgi complex results in the redistribution of Golgi enzymes into the ER. This effect can be evaluated by assessing the acquisition of endoglycosidase H resistance (due to re-modeling of sugar moieties operated by the Golgi enzyme Man II) by the ts045 VSV-G protein at the restrictive temperature (40°C), at which the protein is retained inside the ER and can therefore be modified by Man II only if the enzyme reaches the ER [20]. Under these experimental conditions the BFA-induced acquisition of endoglycosidase H resistance of the VSV-G protein was partially inhibited by

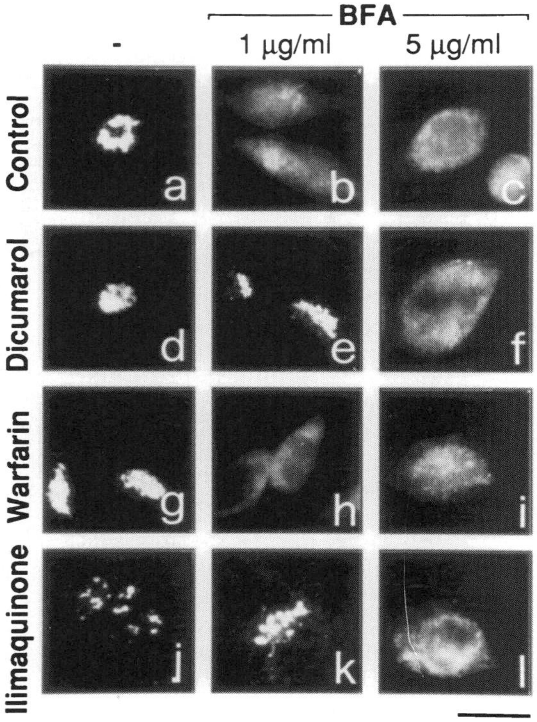

Fig. 4. ADP-ribosylation inhibitors prevent the BFA-induced redistribution of the Golgi complex. Intact RBL cells were pretreated with control buffer (a-c), or 200 µM dicumarol (d-f), 400 µM warfarin (g-i), or 40 µM ilimaquinone (j-l) for 15 min before treatment with 1 µg/ml (b, e, h, k) or 5 µg/ml BFA (c, f, i, l) for 10 min. The cells were fixed, permeabilized with saponin and labeled with FITC-conjugated Helix pomatia lectin. Dicumarol (e) and ilimaquinone (k) were able to prevent the effects of BFA (1 µg/ml) on the Golgi complex with a potency identical to that with which they inhibit BFA-dependent ADP-ribosylation *in vitro.* Warfarin, a dicumarol related compound inactive as an inhibitor of ADP-ribosylation [7], could not prevent the BFA effects *in vivo* (h). High (5 µg/ml) concentrations of BFA were able to overcome the inhibitory effects of ADP-ribosylation inhibitors on Golgi redistribution (see f and l). Similar results were obtained following the distribution of the anti-Man II antibody [8]. Experiments were repeated 3 times in duplicate with similar results. Bar, 5 µm.

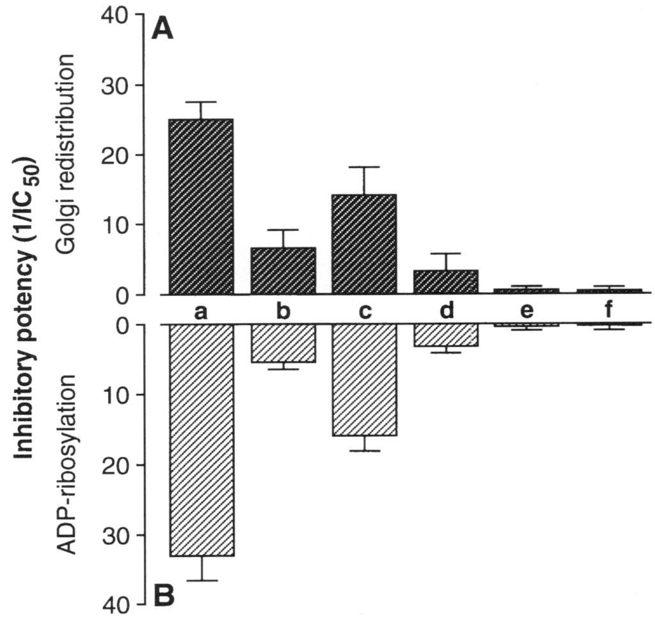

Fig. 5. Potency profiles of selected molecules as inhibitors of BFA-induced ADP-ribosylation (lower half) and Golgi disassembly (upper half). Rat brain cytosol (50 µg/sample) was incubated with rat brain membranes (10 µg/sample) for 1 h at 37°C in the presence of $[^{32}P]$-NAD^+, 30 µg/ml BFA and increasing concentrations of the drugs [7]. Ilimaquinone (a), dicumarol (b), coumermycin A, (c), novobiocin (d), nicotinamide (e), warfarin (f). The IC_{50}, expressed in µM units, was the drug concentration inhibiting 50% of the BFA-dependent signal. The IC_{50} in the Golgi redistribution assay was evaluated in RBL cells treated with 1 µg/ml BFA and is defined as the concentration of inhibitor that prevented the BFA-induced redistribution of the Man II or Helix pomatia staining in 50% of the BFA-sensitive cells (90% of cells respond to BFA under control conditions). The percentage of cells exhibiting BFA-induced redistribution was evaluated by an expert observer in a blind fashion by examining 50 randomly chosen cells per sample. The data presented are representative of three separate experiments performed in duplicate.

dicumarol (Fig. 6), and ilimaquinone (not shown), indicating that Man II redistribution into the ER had been prevented. The lack of a stronger effect of dicumarol was probably due to the fact that in order to induce a clearly measurable endoglycosidase H resistance by the VSV-G protein a rather high concentration of BFA was used in these experiments (2 µg/ml), a concentration that partially overcomes the inhibitory effect of dicumarol [8, 18].

The effects of BFA on the Golgi complex require NAD^+, ADP-ribosylated cytosol or BARS-50

In another approach to investigate the role of ADP-ribosylation in the action of BFA, the effects of NAD^+ on Golgi structure were examined in permeabilized cells that maintained the morphology of the Golgi complex, but lost the low molecular weight soluble molecules, including NAD^+ [8, 21]. Cytosolic factors were provided by added rat brain cytosol that had been NAD^+-depleted by extensive dialysis. In control intact RBL cells, BFA caused the expected redistribution of the Golgi markers (as seen by Helix Pomatia lectin); in permeabilized cells, however, the toxin was completely inactive [8]. Strikingly, the toxic effect of BFA could be completely reconstituted by adding NAD^+ to the permeabilization medium, together with dialyzed cytosol [8]. Thus, NAD^+ is necessary for the action of BFA on the Golgi complex, in line with the involvement of the ADP-ribosylation reaction in the effects of BFA.

Next, it was important to demonstrate that ADP-ribosylation can occur also in permeabilized cells. RBL cells, treated exactly as for morphological experiments, were exposed to

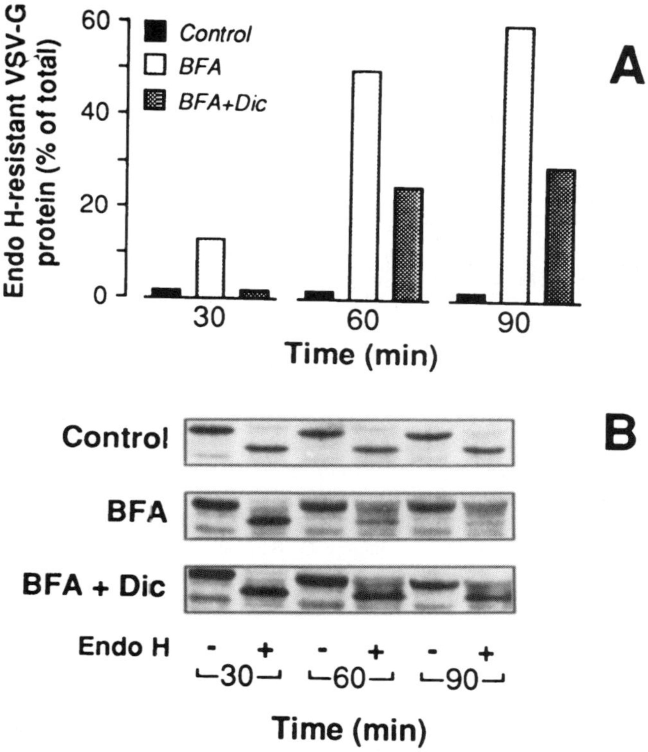

Fig. 6. Dicumarol prevents the BFA-induced retrograde transport of Golgi enzymes into the ER. RBL cells were infected with ts045 VSV at the non-permissive temperature of 40°C; they were pulse-labeled and kept at 40°C in the absence (control) or presence of 2 μg/ml BFA (BFA) alone or in combination with 200 μM dicumarol (BFA + Dic). At the end of the incubation the cells were lysed, treated with endoglycosidase H (Endo H) and analyzed by SDS-PAGE and autoradiography *(Panel B)*. The radioactivity bound to the VSV-G protein was quantified with an Instant Imager (Packard) *(Panel A)*. The Endo H resistant form of the protein is expressed as percent of the total (Endo H resistant and sensitive forms). The data presented in the graph *(Panel A)* are means of three determinations; S.D.s. never exceeded 10% of the means. Similar results were obtained in four experiments.

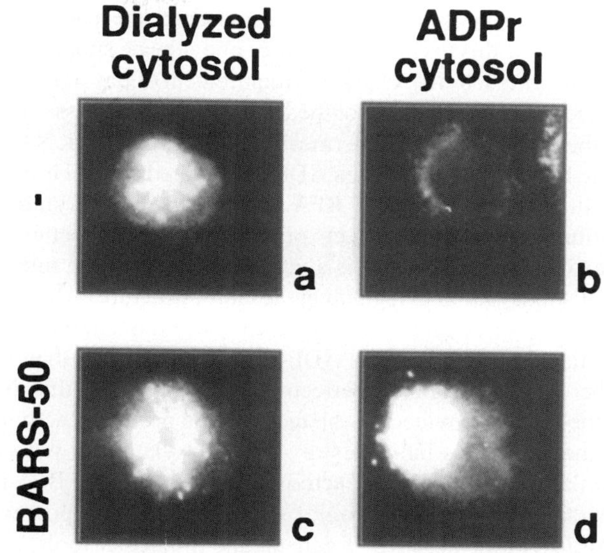

Fig. 7. Pre-ADP-ribosylated cytosol sustains the BFA-induced redistribution of the Golgi complex in permeabilized cells. RBL cells permeabilized with 3 U/ml SLO, were incubated for 20 min at 37°C in a medium containing BFA (10 μg/ml) and NAD^+ (150 μM) in the presence of control (a, c) or pre-ADP-ribosylated (b, d) cytosol (1 mg/ml) [8]. An enriched BARS-50 preparation (see Materials and methods) was added to samples c and d. The cells were fixed and labeled with FITC-conjugated Helix pomatia lectin. Similar results were obtained in 3 independent experiments.

SLO and BFA in the presence of radioactive NAD^+ (30 μM), and the labeling of GAPDH and BARS-50 was evaluated after 1 h of incubation at 37°C. Figure 3 shows that the two proteins are efficiently ADP-ribosylated under these conditions. This observation, together with the above finding that in permeabilized cells BFA requires NAD^+ to disassemble the Golgi structure, is clearly in line with the involvement of NAD^+ and ADP-ribosylation in the toxin-induced disassembly of the organelle [8].

To prove more directly that ADP-ribosylation is involved in the action of BFA, permeabilized RBL cells were treated with BFA and either control or ADP-ribosylated cytosol (Fig. 7). The ADP-ribosylated cytosol substituted for the requirement of NAD^+ and sustained the BFA-induced redistribution of the Golgi complex (Fig. 7b). However, when ADP-ribosylated cytosol was added together with an enriched preparation of native BARS-50 (see Materials and methods), the BFA effect was completely prevented, while GAPDH was inactive, suggesting that the element of the ADP-ribosylated cytosol sustaining the effect of BFA is the modified BARS-50 (Fig. 7 and ref. [8]).

Finally, similar experiments in permeabilized cells were carried out to further verify the specificity of action of the inhibitors and begin to establish a link between their effects and the selective ADP-ribosylation of BARS-50 and GAPDH. Dicumarol inhibited the Golgi redistribution induced by BFA and NAD^+ in the presence of control cytosols; however, when dicumarol was tested in the presence of ADP-ribosylated cytosol, its inhibitory effect was markedly prevented [8]. These data suggest that dicumarol requires unmodified BARS-50 or GAPDH or both (GAPDH, however, was only partially ADP-ribosylated) to exert its effects and that the previous ADP-ribosylation of these proteins renders them unable to mediate the drug's action.

Discussion

The body of evidence presented in this and previous papers [3–8, 11, 12, 18] indicates that the endogenous ADP-ribosylation induced by BFA is a component of the mechanism of action of this toxin. Several lines of evidence support this conclusion [6–8, 12]. One is the requirement of physiological concentrations of NAD^+ for the BFA-dependent disassembly of the Golgi complex in permeabilized cells. Second, NAD^+ is no longer necessary when ADP-ribosylated cytosol is used to support the effect of BFA. Third, ADP-ribosylation inhibitors prevent the effects of BFA on the architecture of the Golgi complex. This is a valid indication of the role of ADP-ribosylation in regulating the Golgi structure only if the actions of these inhibitors *in vivo* can be demonstrated to be mediated by the block of ADP-ribosylation rather than by other unrelated (or toxic) effects. The best evidence that this is the case is provided by the fact that the profile of activity of these drugs as inhibitors of ADP-ribosylation *in vitro* is similar to their profile of activity as antagonists of BFA *in vivo* (Fig. 5). The criterion of similarity between potency profiles, in the absence of genetic and molecular tools, has been classically used by pharmacologists to establish identity or similarity between binding sites studied *in vitro* and drug receptor sites involved in cellular functions *in vivo* [13]. The similarity between the two profiles in Fig. 5 is indeed impressive and represents therefore strong, albeit correlative, evidence that the *in vivo* activity of the inhibitors as BFA antagonists is linked to their property to block ADP-ribosylation. Fourth, dicumarol largely lost its inhibitory effect when a cytosol containing pre-ADP-ribosylated BARS-50 and GAPDH was used to support the BFA-induced Golgi redistribution in permeabilized cells [8], suggesting that the inhibitor exerts its effects through one or both of these proteins [8].

Together, these results indicate that the inhibitors of ADP-ribosylation prevent the Golgi-disassembling effects of BFA through their effect on the BFA-dependent ADP-ribosylation reaction. The Golgi apparatus, in spite of its complexity, is a highly dynamic organelle, as evidenced most dramatically by the rapid and reversible effects of BFA. Hopefully, future work on the physiological significance of the NAD^+-dependent regulation and the molecular role(s) of the ADP-ribosylation protein substrates will offer new inroads into the problem of the dynamic control of the structure/function of the Golgi complex.

Acknowledgments

We thank Drs. G. Lenaz and M. Cavazzoni for helpful comments and discussion, Dr. K.K Moremen for donating the anti-Man II polyclonal antibody, Dr. K. Simons for providing the ts045 VSV, Dr. V. Malhotra for the gift of ilimaquinone, Mr. G. Di Tullio and Ms. A. Fusella, C. Valente and C. Cericola for expert technical help, Ms. R. Bertazzi for preparation of the figures. This research was supported in part by grants from the Italian National Research Council (Convenzione C.N.R.-Consorzio Mario Negri Sud and Progetto Finalizzato A.C.R.O. contract No96.00755.PF39) and the Italian Association for Cancer Research (AIRC, Milano, Italy). R.W is the recipient of a fellowship from the Centro di Formazione e studi per il Mezzogiorno (FORMEZ).

References

1. Lippincott-Schwartz J, Yuan LC, Bonifacino JS, Klausner RD: Rapid redistribution of Golgi proteins into the ER in cells treated with Brefeldin A: Evidence for membrane cycling from Golgi to ER. Cell 56: 801–813, 1989
2. Klausner RD, Donaldson JG, Lippincott-Schwartz J: Brefeldin A: Insights into the control of membrane traffic and organelle structure. J Cell Biol 116: 1071–1080, 1992
3. De Matteis MA, Di Girolamo M, Colanzi A, Pallas M, Di Tullio G, McDonald W, Moss J, Santini G, Bannykh S, Corda D, Luini A: Stimulation of endogenous ADP-ribosylation by brefeldin A. Proc Natl Acad Sci USA 91: 1114–1118, 1994
4. Colanzi A, Di Girolamo M, Santini G, Sciulli G, Santarone S, Pallas M, Di Tullio G, Bannykh S, Corda D, De Matteis MA, Luini A: Brefeldin A, an inhibitor of vesicular traffic, stimulates the ADP-ribosylation of two cytosolic proteins. In: D. Corda, H. Hamm and A. Luini (eds). GTPase-Controlled Molecular Machines. Ares-Serono Symposia Publications, Rome, 1994, pp. 197–217
5. Di Girolamo M, Silletta MG, De Matteis MA, Braca A, Colanzi A, Pawlak D, Rasenick MM, Luini A, Corda D: Evidence that the 50-kDa substrate of brefeldin A-dependent ADP-ribosylation binds GTP and is modulated by the G-protein $\beta\gamma$ subunit complex. Proc Natl Acad Sci USA 92: 7065–7069, 1995
6. Silletta MG, Di Girolamo M, Fiucci G, Weigert R, Mironov A, De Matteis MA, Luini A, Corda D: Possible role of BARS-50, a substrate of brefeldin A-dependent mono-ADP-ribosylation, in intracellular transport. Adv Exp Med Biol 419: 321–330, 1997
7. Weigert R, Colanzi A, Mironov A, Buccione R, Cericola C, Sciulli MG, Santini G, Flati S, Fusella A, Donaldson J, Di Girolamo M, Corda D, De Matteis MA, Luini A: Characterization and development of chemical inhibitors of the brefeldin A-induced mono-ADP-ribosylation. J Biol Chem 272: 14200–14207, 1997
8. Mironov A, Colanzi A, Silletta MG, Fiucci G, Flati S, Fusella A, Polishchuk R, Mironov A Jr, Di Tullio G, Weigert R, Malhotra V, Corda D, De Matteis MA, Luini A: Role of NAD^+ and ADP-Ribosylation in the maintenance of the Golgi structure. J Cell Biol 139: 1109–1118, 1997
9. Malhotra V, Serafini T, Orci L, Shepherd JC, Rothman JE: Purification of a novel class of coated vesicles mediating biosynthetic protein transport through the Golgi stack. Cell 58: 329–336, 1989
10. Bergmann JE: Using temperature-sensitive mutants of VSV to study membrane protein biogenesis. Meth Cell Biol 32: 85–110, 1989
11. Weigert R, Colanzi A, Limina C, Cericola C, Di Tullio G, Mironov A, Santini G, Sciulli G, Corda D, De Matteis MA, Luini A: Characterization of the endogenous mono-ADP-ribosylation stimulated by brefeldin A. Adv Exp Med Biol 419: 337–342, 1997
12. Colanzi A, Mironov A, Weigert R, Limina C, Flati S, Cericola C, Di Tullio G, Di Girolamo M, Corda D, De Matteis MA, Luini A: Brefeldin A-induced ADP-ribosylation in the structure and function of the Golgi complex. Adv Exp Med Biol 419: 331–335, 1997
13. Moss J, Vaughan M: ADP-ribosylation of guanyl nucleotide-binding proteins by bacterial toxins. Adv Enzymol 61: 303–379, 1988

14. Sestili P, Spadoni G, Balsamini C, Scovassi I, Cattabeni F, Duranti E, Cantoni O, Higgins D, Thomson C: Structural requirements for inhibitors of poly(ADP-ribose) polymerase. J Cancer Res Clin Oncol 116: 615–622, 1990
15. Okazaki IJ, Moss J: Common structure of the catalytic sites of mammalian and bacterial toxin ADP-ribosyltransferases. Mol Cell Biochem 138: 177–181, 1994
16. Ma Q, Cui K, Xiao F, Lu AYH, Yang CS: Identification of a glycine-rich sequence as an NAD(P)H-binding site and tyrosine 128 as a dicumarol-binding site in rat liver NAD(P)H: Quinone oxidoreductase by site-directed mutagenesis. J Biol Chem 267: 22298–22304, 1992
17. Takizawa PA, Yucel JK, Veit B, Faulkner DJ, Deerinck T, Soto G, Ellisman M, Malhotra V: Complete vesiculation of Golgi membranes and inhibition of protein transport by a novel sea sponge metabolite, Ilimaquinone. Cell 73: 1079–1090, 1993
18. Mironov AA, Polishchuk RS, Colanzi A, Santone I, Mironov AA Jr, Fusella A, Lupetti P, Silletta MG, Weigert R, Dallai R, Malhotra V, Corda D, De Matteis MA, Luini A: Agents selectively fragmenting the Golgi tubular structure. J Cell Biol (submitted).
19. Veit B, Yucel JK, Malhotra V: Microtubule independent vesiculation of Golgi membranes and the reassembly of vesicles into Golgi stacks. J Cell Biol 122: 1197–1206, 1993
20. Doms RW, Russ G, Yewdell JW: Brefeldin A redistributes resident and itinerant Golgi proteins to the endoplasmic reticulum. J Cell Biol 109: 61–72, 1989
21. Bhakdi S, Weller I, Walev E, Martin D, Palmer JM: A guide to the use of poreforming toxins for controlled permeabilization of cell membranes. Med Microbiol Immunol 182: 167–175, 1989

Molecular and Cellular Biochemistry **193**: 53–60, 1999.

A dual approach in the study of poly (ADP-ribose) polymerase: *In vitro* random mutagenesis and generation of deficient mice

Carlotta Trucco,[1] Véronique Rolli,[1] F. Javier Oliver,[1] Eric Flatter,[1] Murielle Masson,[1] Françoise Dantzer,[1] Claude Niedergang,[1] Bernard Dutrillaux,[2] Josiane Ménissier-de Murcia[1] and Gilbert de Murcia[1]

[1]*Ecole Supérieure de Biotechnologie de Strasbourg, UPR 9003 du Centre National de la Recherche Scientifique "Cancérogénèse et Mutagénèse Moléculaire et Structurale", Illkirch-Graffenstaden;* [2]*CENFAR, Département de Radiobiologie, Fontenay-aux-Roses, France*

Abstract

A dual approach to the study of poly (ADP-ribose)polymerase (PARP) in terms of its structure and function has been developed in our laboratory. Random mutagenesis of the DNA binding domain and catalytic domain of the human PARP, has allowed us to identify residues that are crucial for its enzymatic activity.

In parallel PARP knock-out mice were generated by inactivation of both alleles by gene targeting. We showed that: (i) they are exquisitely sensitive to γ-irradiation, (ii) they died rapidly from acute radiation toxicity to the small intestine, (iii) they displayed a high genomic instability to γ-irradiation and MNU injection and, (iv) bone marrow cells rapidly underwent apoptosis following MNU treatment, demonstrating that PARP is a survival factor playing an essential and positive role during DNA damage recovery and survival. (Mol Cell Biochem **193**: 53–60, 1999)

Key words: DNA binding protein, NAD metabolism, cellular response to DNA damage, γ-rays, alkylating agents, genomic instability, apoptosis

Introduction

During evolution, eukaryotic cells have developed a complex DNA damage surveillance network to protect the genetic information. Studies carried out in the recent years have demonstrated the existence of different repair pathways. One of the enzymes involved in the cellular response to DNA damage is poly(ADP-ribose)polymerase (PARP, EC 2.4.2.30) which specifically detects and signals DNA strand-breaks produced by genotoxic agents.

To understand the function of PARP at the molecular level, different complementary approaches have been carried out in our laboratory. First, a large scale screening procedure in recombinant *Escherichia coli* has been developed to identify and to characterize random loss-of-function mutations in human poly(ADP-ribose)polymerase [1, 2]. Second, we have developed animal and cellular models deficient in PARP by inactivating both alleles by homologous recombination [3] in mice. Altogether, these new powerful tools should facilitate the elucidation of its role in the maintenance of genomic integrity and stability in cells under conditions of genotoxic stress.

Results

Random mutagenesis of the human poly(ADP-ribose) polymerase DNA-binding domain (DBD)

PARP is a multifunctional, highly conserved enzyme; its enzymatic activity is stimulated more than 500-fold upon binding to DNA strand-breaks. The protein (113 kDa) has a modular structure comprising three main distinct regions (see

Address for offprints: G. de Murcia, Ecole Supérieure de Biotechnologie de Strasbourg, UPR 9003 du Centre National de la Recherche Scientifique, "Cancérogénèse et Mutagénèse Moléculaire et Structurale", Boulevard Sébastien Brant, F-67400 Illkirch-Graffenstaden, France

Fig. 1A): a N-terminal DNA-binding domain (DBD) (modules A, B and C) bearing the two zinc fingers (FI, FII) acting as a molecular nick sensor [4, 5], a central automodification domain containing auto-ADP-ribosylation sites implicated presumably in the regulation of PARP-DNA interactions (module D) and a C-terminal catalytic domain (modules E, F) involved in the nick-binding dependent poly (ADP-ribose) synthesis [6].

To understand the molecular mechanism underlying PARP enzymatic activity stimulation by DNA strand-breaks, we performed mutagenesis of the PARP DBD by PCR to generate a random library of PARP* mutants [1]. These mutants were characterized both for their inability to be catalytically stimulated by DNA strand-breaks and for their DNA binding capacity. The mutated cDNA fragments (encoding for the amino acids, aa 1–333) were cloned back in a prokaryotic expression vector instead of the wild-type (wt) sequence and the recombinant PARP, specifically mutated in the DBD region was subsequently overproduced in *E. coli*. Screening for PARP activity on replica filter by activity-blot revealed colonies with no enzymatic activity; these colonies were subsequently amplified and the residual polymerizing activity was quantified in cleared lysates prepared from PARP* mutants. Of 1600 colonies screened, four mutants were isolated having a single base substitution that retained less than 1.5% of the wt enzymatic activity (Table 1A and Fig. 1B). Two mutations were located in module A between the two zinc fingers (L77P and K97R) while two other mutations (K249E and G313E) lied in module C whose function is unknown (Fig. 1B). PARP* mutants were tested for their DNA-binding capacity using a South-Western blot assay [7]. Surprisingly,

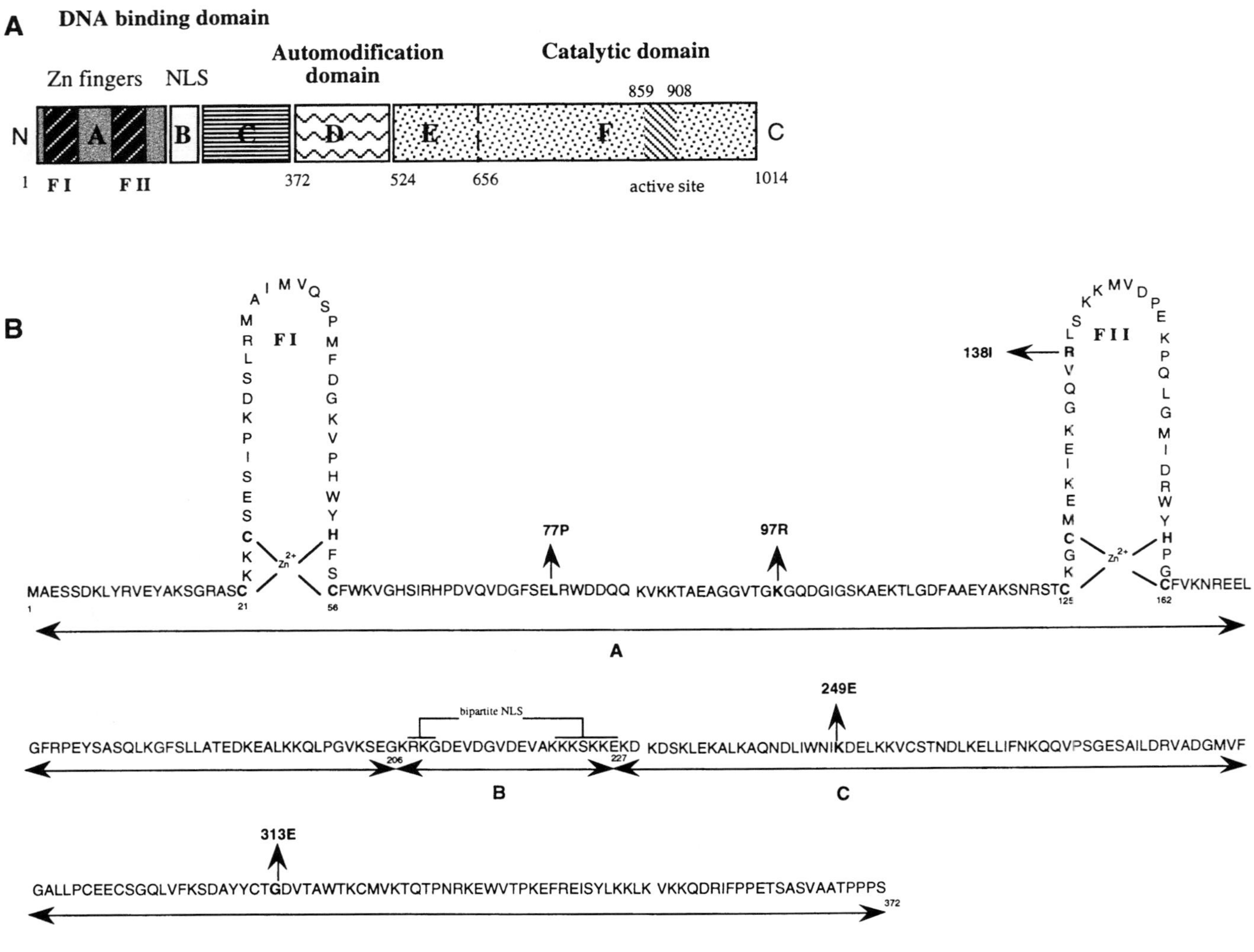

Fig. 1. (A) modular organization of human PARP. (B) Location of mutations of the DNA binding domain (DBD), in the amino-terminal sequence of human PARP. (FI and FII: zinc fingers I and II; Cys and His involved in zinc coordination are indicated in bold as are the mutations R1381, L77R, K97R, K249E and G313E).

Table 1. Characterization of PARP mutants obtained by random mutagenesis.

A. Mutations of the DNA-binding domain

protein	enzymatic activity(%)	mutated module
wt PARP	100	
L77P	< 0.5	A
K97R	< 0.5	A
K249E	< 0.5	C
G313E	1.5	C

B. Selected mutations of the catalytic domain

protein	enzymatic activity(%)	polymer length	relative branching
wt PARP	100	normal	1
L713F*	900	normal	1
R847C	75	normal	0.5
E923G	20	normal	0.33
C972R	16	normal	0.33
Y986S	11	short	1
Y986H	14	normal	15
E988K	1.25	mono-ADP-ribose	–
L999P	30	normal	0.5

* Alves Miranda *et al.* (1995)

all PARP* mutants obtained by PCR mutagenesis retain a full capacity to bind to nicked DNA compared to the wt enzyme.

We and others have previously identified critical residues that impair the stimulation of poly ADP-ribosylation by DNA strand breaks. These residues, unsurprisingly, were the cysteine and histidine ligands of the two zinc ions [7, 8], and arginine-138 which is located in a putative α-helix of the second zinc finger [9]. Our data suggest that the identified loss-of-function mutations may either induce a strong change in the tertiary structure of the enzyme (i.e. L77P) or play an important role in self-association and/or in the hetero-dimerization with other proteins. The latter hypothesis was tested using the Far Western technique, and no difference was observed in homo- or hetero-dimerization experiments (data not shown) indicating that these mutations do not affect possible protein-protein interactions involving the DBD [10]. However these results indicate that the DNA-dependent activation of PARP requires not only a capacity to bind to DNA breaks, but also a number of crucial residues to maintain a correct conformation of the N-terminal region necessary to transfer an 'activation signal' to the catalytic domain.

Random mutagenesis of the human poly(ADP-ribose)polymerase catalytic domain (module F)

The mutational analysis of the PARP catalytic domain was carried out under the same experimental conditions as for the PARP DBD. Briefly, a random library of PARP* mutants, specifically mutated in their catalytic domain (aa 686–1014), was generated by PCR and screened for loss-of-function by activity-blot [11]. Twenty-six mutants diversely affected in their enzymatic activity, by a single amino acid substitution, were isolated and subsequently analysed.

The distribution of the mutations on the amino acid sequence was clearly non-random, falling into two distinct regions: (i) the 'signature' sequence of PARP (residues 859–908, Fig. 1A) [12], thus confirming the importance of this highly conserved region in the recognition of the substrate and/or in catalysis, and (ii) the region around E988, whose role in poly(ADP-ribose) elongation has been demonstrated previously [13]. PARP* mutants, partially purified from clear lysates, were characterized by (i) quantitative analysis comparing their enzymatic activity to that of the wt PARP, (ii) qualitative analysis of the reaction products on poly-acrylamide gel and, (iii) measurement of the degree of polymer branching using 2D-TLC [14]. PARP* mutants were therefore classified in four categories: (i) those which catalyse the synthesis of ADP-ribose polymers, similar to that formed by the wt PARP, both in terms of polymer average length and branching frequency (this was the case for the majority of the mutants, i.e. F897S, Y989H); (ii) those which were unable to catalyse the synthesis of long polymers (i.e. Y986S); (iii) those which were affected in the branching frequency (i.e. Y986H) and, (iv) E988K, which was severely affected in the elongation reaction and, in fact, converted into a mono ADP-ribosylating enzyme (Table 1B).

Crystals and co-crystals of the chicken PARP catalytic domain have been obtained [15] and the structure has been solved recently at 2.4 Å resolution, in the presence and in the absence of the nicotinamide analogue PD 128763 [16]. The catalytic domain is composed of a N-ter region formed of a bundle of 6 α-helices (residues 662–784) (Fig. 2), and a C-terminal region (residues 785–1010) made principally of two perpendicular β-sheets containing the NAD^+ binding cavity, as suggested by the presence of the competitive inhibitor PD 128763. Interestingly, the PARP 'signature' containing the block of 50 amino acids (aa 859–908) folded in a β-α-loop-β-α motif (shown in orange in Fig. 2) turned out to be structurally superimposable to most of the active site of mono ADP-ribosylating toxins like diphtheria toxin, pertussis toxin, exotoxin A from *P. aeruginosa,* cholera toxin, and heat-labile enterotoxin. Inactivating mutations obtained by random mutagenesis were therefore localized with respect to the crystal structure of the PARP catalytic domain, thus allowing us to determine essential residues and regions.

As shown in Fig. 2, most of the mutations are located in the sheet domain of the structure. Nine out of 26 (35%) of inactivating substitutions map in the crevice delineated by the motif β-α-loop-β-α, thus confirming the crucial role of this super-secondary structure in NAD^+ binding and/or catalysis

Fig 2

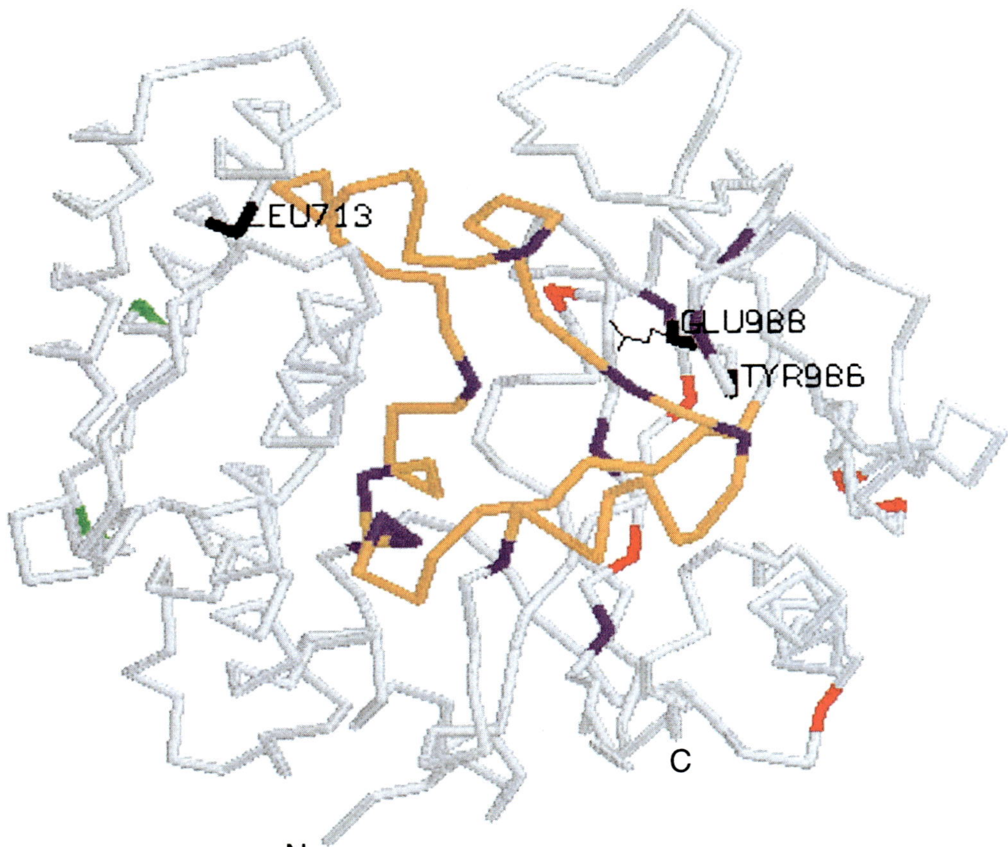

Fig. 2. Distribution of the mutations inactivating the poly-ADP-ribosylating reaction on the three-dimensional structure of the chicken PARP catalytic domain. View of the backbone (residues 662–1010). The α-helical subdomain (left) and the C-terminal β-sheet subdomain (right) may be distinguished. The active site is represented in orange (residues 859–908, motif β-α-loop-β-α). Residues affected by random mutagenesis are represented in different colours: green (C699 and R704) and red (S808, R847, E923, L964, C972 and Y1001) residues are those localized at the surface of the protein; those in purple have no particular characteristic.

[16, 17]. Another group of inactivating substitutions is found around the active-site glutamate E988, further demonstrating the importance of this region for catalysis since both mutants (E988 and Y986) lose their enzymatic activity (k_{cat} is reduced) but retained their binding affinity for NAD^+ (K_m unaffected) [13], (Rolli *et al.* unpublished results). E988 is clearly implicated in the elongation step, while Y986 , which is not conserved in toxins, is involved in the elongation and/or the branching of polymer as indicated by the dramatic increase (15 fold) in branching frequency.

Interestingly, it was possible to classify the mutations into two groups at the surface of the catalytic domain (Fig. 2). The first group includes C699R and R704W located in the α-helical domain lining the active site pocket. The presence in the same region of a gain-of-function mutation, L713F [18], located 7.1Å from the active site loop, suggests that a movement of this loop can be induced by substitutions in this domain, thus decreasing (G699R, R704W) or increasing (L713F) the access of the catalytic cleft to the substrate. Alternatively, it is tempting to speculate that the α-helical domain, absent in ADP-ribosylating toxins, could relay the activation signal issued on binding to damage DNA by intra-molecular contacts, giving rise to a more favourable NAD^+ binding mode. The second group of mutations is located in the β-sheet region (S808P, R847C, E923C, L964P, C972R, Y1001C, Fig. 2) and could constitute an interaction interface with another PARP molecule, or with another acceptor protein.

Knock-out mice and derived cells lines are exquisitely sensitive to the mono functional methylating agent MNU and to γ-rays

To establish a cellular and an animal model deficient in PARP activity, we have disrupted the PARP gene by homologous recombination [3]. Mutant mice were totally devoid of PARP as judged by western- and activity-blot analysis using fibroblasts isolated from 13.5 day old embryos (MEF) (Fig. 3).

Mice lacking PARP activity are viable and fertile as reported previously [19]. However they displayed slightly alterated phenotypes: (i) their average litter size (4.5 ± 2.4 pups) was smaller than those of PARP +/+ mice (7.4 ± 1.9 pups), (ii) KO mice have a body weight reduced by about 23% compared to their +/+ and +/– litter mates, six weeks after birth (Fig. 4). Despite the decrease in body size, the outward appearance of the PARP mice showed normal proportions.

The unique property of PARP, to bind to DNA strand-breaks generated by monofunctional methylating agents or ionizing radiation, led us to test the sensitivity of PARP null mice to these two genotoxic agents. Injection of a single dose of 75 mg/kg (bw) of N-Methyl-N-Nitrosourea (MNU) resulted in 100% mortality in PARP –/– and only 43% in PARP +/+ mice during 6 weeks of observation, indicating that mice lacking PARP are acutely sensitive to the alkylating agent (data not shown). PARP +/+ and PARP –/– mice were irradiated with two doses of γ-irradiation at various ages and were monitored for mortality. Following a radiation dose of 4 Gy at 6 days of age, about half of PARP null mice died within 4 weeks compared to the 14% of mortality of wt mice (Fig. 5). Following a dose of 8 Gy, at 6–8 weeks of age, 62% of PARP null mice died 4 days post-irradiation and thus all mutant mice were dead by 9 days post-irradiation. In contrast, only 50% of wt mice were dead by 15 days post-irradiation (Fig. 5). To better understand the cause of the precocious lethality of the PARP –/– mice, we irradiated wt and mutant mice with 8 Gy and an autopsy was performed 3 days post-irradiation. In all mutant mice the small intestine was distended by intraluminal fluid accumulation and histological sections through the duodenum indicated that the villi of irradiated PARP –/– mice were considerably shortened. Interestingly, *Atm*-deficient mice [20], as well as SCID mice [21] are as sensitive to ionizing radiation as PARP null mice and exhibit the same phenotype following γ-irradiation. The selective acute toxicity of specific tissues of these mice to γ-irradiation underline the cardinal function of DNA screening programmes in rapidly proliferating tissues.

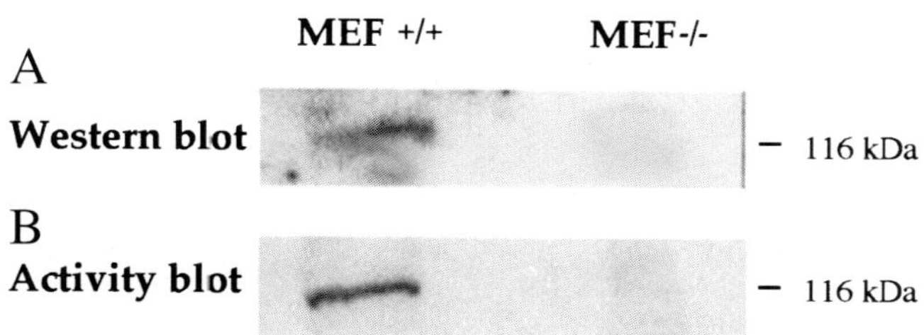

Fig. 3. Western blot analysis and poly ADP-ribosylation activity on mouse embryonic fibroblasts. Fibroblasts were isolated from 13.5 day old PARP–/– and PARP+/+ embryos. Crude extract of 10^6 cells were separated on SDS-acrylamide gel. Proteins were probed either with a polyclonal antibody raised against the human PARP (A) or with [α^{32}P]NAD+ in an activity assay (B).

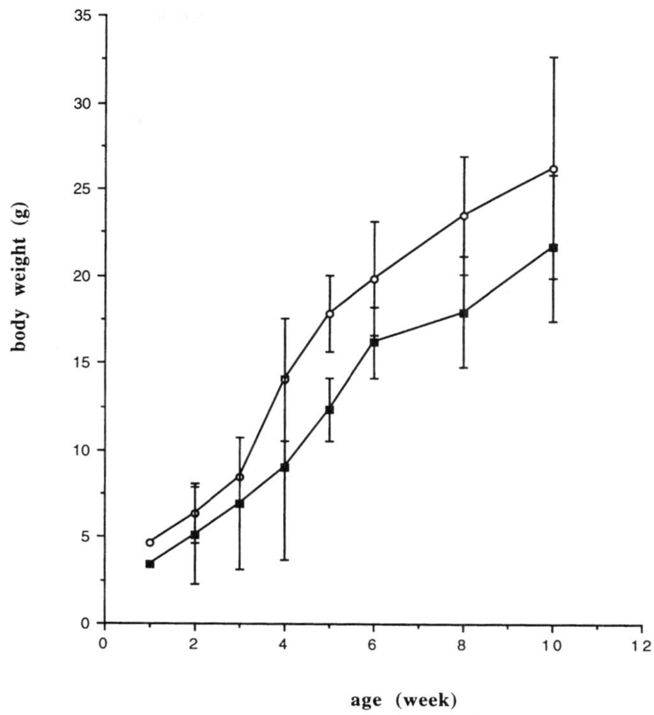

Fig. 4. Body weight of PARP +/+ and PARP –/– mice during ten weeks of observation (+/+, open circles, –/– filled squares).

Previous studies demonstrated that inhibition of PARP activity in cultured cells increases genomic instability following DNA damage [22, 23]. An *in vivo* study of sister chromatid exchanges (SCES) and chromosome breakage was therefore undertaken in PARP +/+ and –/– mice. As expected, in mutant mice an increase in reciprocal exchange of segments of sister-chromatids (Fig. 6A) and the formation of chromatid and chromosome breaks (Fig. 6B) was observed after treatment with MNU and γ-irradiation respectively. These data confirm that the absence of PARP led to a considerable increase of recombination events shown by an enhanced frequency of SCEs and by abnormal rejoining of breakages after treatment with DNA damaging agents.

To determine the specific role of PARP during the execution phase of programmed cell death [24], bone marrow cells were isolated from both PARP + / + and PARP –/– mice and challenged with 2 mM MNU. Four hours following treatment, PARP –/– cells showed the typical DNA fragmentation of programmed cell death, while wt cells did not (Fig. 7). Similar results were obtained using splenocytes exposed to the same dose of MNU [3]. These results do not contradict previous published work [25] as different apoptogenic stimuli

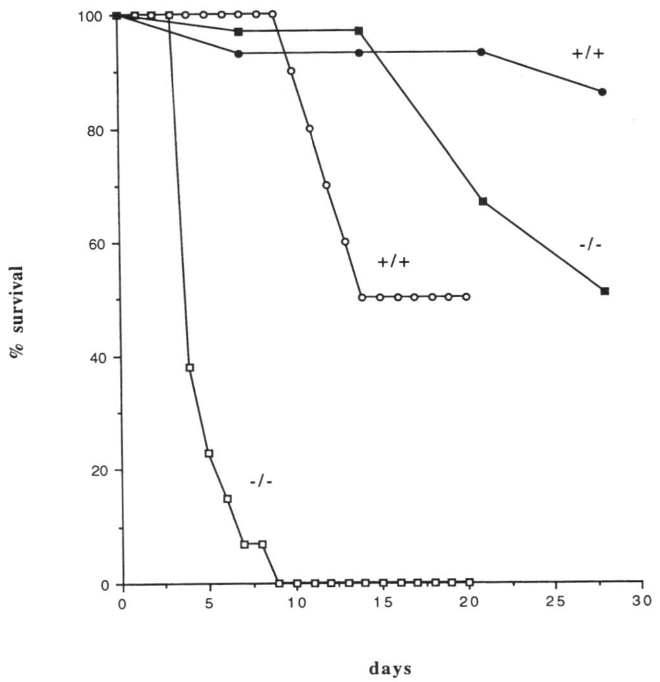

Fig. 5. Survival of PARP +/+ (circles) and PARP –/– (squares) mice after γ-irradiation with 8 Gy, at 6-8 weeks of age (open symbols) and after γ-irradiation with 4 Gy, at six days of age (filled symbols).

not involving DNA damage, were used. It is well documented that apoptosis might be triggered by different effectors acting on different intracellular targets. We found that only DNA lesions which are repaired throught the base excision pathway (BER) lead to an increased apoptotic demise in PARP –/– cells. This is exemplified in Interleukin-3-dependent bone marrow cells isolated from PARP +/+ and PARP –/– mice: following MNU exposure, the latter died rapidly by apoptosis (Fig. 7) whereas they were equally sensitive to cell death induced by interleukin-3 removal (data not shown). Cleavage of PARP and other nuclear proteins and repair enzymes, during the execution phase of apoptosis is a key feature to ensure the irreversibility of the process. Having PARP out of the context, might somehow favour a more rapid conclusion of the cell death programme.

Conclusions and prospects

The link between the structure and function studies of PARP is essential to better understand the physiological role of this protein. Analysis by *in vitro* random mutagenesis of the DNA binding domain and the catalytic domain has allowed us to identify crucial residues, whose mutation affects the enzymological properties, as well as the reaction products. From the crystal structure we have learned that the striking similarity

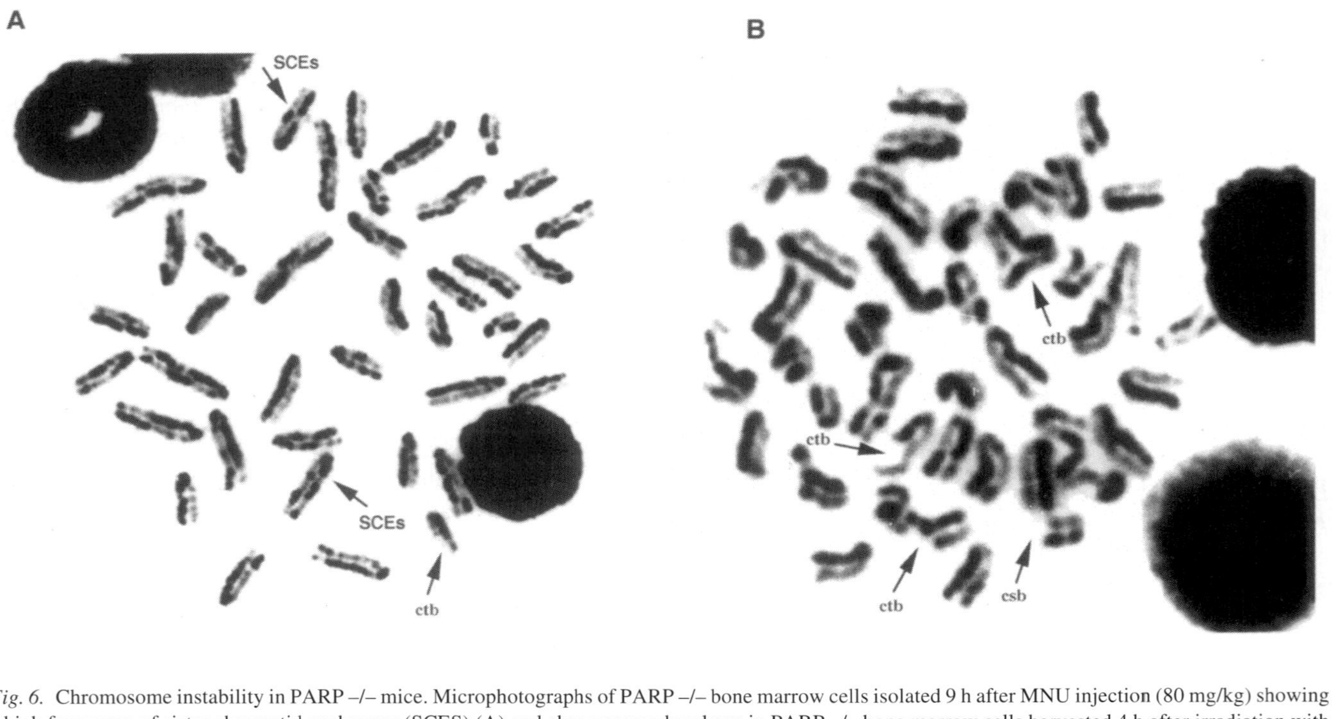

Fig. 6. Chromosome instability in PARP –/– mice. Microphotographs of PARP –/– bone marrow cells isolated 9 h after MNU injection (80 mg/kg) showing a high frequency of sister chromatid exchanges (SCES) (A) and chromosome breakage in PARP –/– bone marrow cells harvested 4 h after irradiation with 2 Gy (B) . (ctb: chromatid breaks, csb: chromosome breaks).

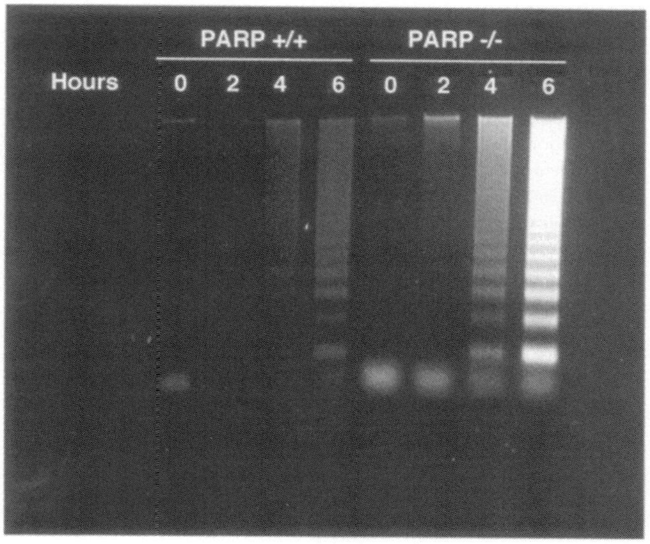

Fig. 7. PARP deficiency activates the programmed cell death following MNU treatment. Time course induction of apoptosis in PARP +/+ and PARP–/– bone marrow cells exposed to 2 mM MNU.

between the NAD+-binding fold of PARP and that of prokaryotic toxins is probably related to a common catalytic mechanism, at least for the first step of the reaction, involving a highly conserved active-site glutamate. Clearly, more work will be necessary to understand the nick-detection mechanism conducted by the N-terminal domain on a damaged template. Again, the combination of the mutational analysis and the determination of the crystal structure of this module will certainly help to unravel this essential cellular function.

The disruption of the PARP gene by homologous recombination, reveals undoubtedly a serious DNA repair deficiency in PARP –/– cells. The role of PARP in the BER pathway is likely to be mediated through its direct or indirect association with the repair proteins: XRCC1, DNA polymerase β, and DNA ligase III [26]. It is tempting to speculate that in the absence of PARP these proteins are not efficiently recruited at DNA strand-breaks and therefore DNA resynthesis and ligation could be compromised, leading to chromosomal instability, cell cycle arrest and cell death. The increased DNA damage induced apoptosis in cells lacking PARP suggests that PARP inhibitors, based on the crystal structure of the PARP catalytic domain [16], could be useful as chemo- and radio-potentiation agents. In that case, possible side effects of these inhibitors could be evaluated in PARP deficient mice.

Acknowledgements

We are indebted to M. Ricoul, M. Mark, A. Dierich, M. LeMeur, and C. Waltzinger for their help in the PARP KO mice program and to G. Schulz and A. Ruf for fruitful discussions. We wish to thank Prof. P. Chambon for his continuous interest and support. This work was supported by the CNRS Action Concertée 'Radiations ionisantes', by Association pour la Recherche Contre le Cancer (ARC), by the Ligue Nationale Contre le Cancer and by Electricité de France. C.T and F.J.O. were supported by a EC fellowship (Human Capital Mobility) and an ARC fellowship respectively.

References

1. Trucco C, Flatter E, Fribourg S, de Murcia C, and Ménissier-de Murcia J: Mutations in the amino-terminal domain of the human poly(ADP-ribose) polymerase that affect its catalytic activity but not its DNA binding capacity. FEBS Lett 399: 313–316, 1996
2. Rolli V, O'Farrell M, Ménissier-de Murcia J, de Murcia C: Random mutagenesis of the poly(ADP-ribose) polymerase catalytic domain reveals amino acids involved in polymer branching. Biochemistry 36: 12147–12154, 1997
3. Ménissier-de Murcia J, Niedergang C, Trucco C, Ricoul M, Dutrillaux B, Mark M, Olivier FJ, Masson M, Dierich A, LeMeur M, Walztinger C, Chambon P, de Murcia C: Requirement of poly(ADPribose) polymerase in recovery from DNA damage in mice and in cells. Proc Natl Acad Sci USA 94: 7303–7307, 1997
4. Ménissier de Murcia J, Molinete M, Gradwohl G, Simonin F, de Murcia G: Zinc-binding domain of poly(ADP-ribose)polymerase participates in the recognition of single strand breaks on DNA. J Mol Biol 210: 229–233, 1989
5. Schreiber V, Molinete M, Boeuf H, de Murcia G, Ménissier-de Murcia J: The human poly(ADP-ribose) polymerase nuclear localization signal is a bipartite element functionally separate from DNA binding and catalytic activity. EMBO J 11: 3263–2369, 1992
6. Simonin F, Ménissier-de Murcia J, Poch O, Muller S, Gradwohl G, Molinete M, Penning C, Keith G, de Murcia G: Expression and site-directed mutagenesis of the catalytic domain of human poly(ADP-ribose)polymerase in *Escherichia coli*. Lysine 893 is critical for activity. J Biol Chem 265: 19249–19256, 1990
7. Gradwohl G, Ménissier-de Murcia J, Molinete M, Simonin F, Koken M, Hoeijmakers JH, de Murcia G: The second zinc-finger domain of poly(ADP-ribose) polymerase determines specificity for singlestranded breaks in DNA. Proc Natl Acad Sci USA 87: 2990–2994, 1990
8. Ikejima M, Noguchi S, Yamashita R, Ogura T, Sugimura T, Gill DM, Miwa M: The zinc fingers of human poly(ADP-ribose) polymerase are differentially required for the recognition of DNA breaks and nicks and the consequent enzyme activation. Other structures recognize intact DNA. J Biol Chem 265: 21907–21913, 1990
9. Molinete M, Vermeulen W, Burkle A, Ménissier-de Murcia J, Kupper JH, Hoeijmakers JH, de Murcia G: Overproduction of the poly(ADP-ribose) polymerase DNA-binding domain blocks alkylationinduced DNA repair synthesis in mammalian cells. EMBO J 12: 2109–2117, 1993
10. Griesenbeck J, Oei SL, Mayer-Kuckuk P, Ziegler M, Buchlow G, Schweiger M: Protein-protein interaction of the human poly(ADP-ribose)polymerase depends on the functional state of the enzyme. Biochemistry 36: 7297–7304, 1997
11. Simonin F, Briand JP, Muller S, de Murcia G: Detection of poly(ADP-ribose) polymerase in crude extracts by activity-blot. Anal Biochem 195: 226–231, 1991
12. de Murcia C, Ménissier-de Murcia J: Poly(ADP-ribose) polymerase: A molecular nick-sensor. Trends Biochem Sci 19: 172–176, 1994

13. Marsischky CT, Wilson BA, Collier RJ: Role of glutamic acid 988 of human poly-ADP-ribose polymerase in polymer formation. Evidence for active site similarities to the ADP-ribosylating toxins. J Biol Chem 270: 3247–3254,1995
14. Keith G, Desgres J, de Murcia G: Use of two-dimensional thin-layer chromatography for the components study of poly(adenosine diphosphate ribose). Anal Biochem 191: 309–313, 1990
15. Jung S, Miranda EA, Ménissier-de Murcia J, Niedergang C, Delarue M, Schulz CE, de Murcia G: Crystallization and X-ray crystallographic analysis of recombinant chicken poly(ADP-ribose) polymerase catalytic domain produced in Sf9 insect cells. J Mol Biol 244: 114–116, 1994
16. Ruf A, Ménissier-de Murcia J, de Murcia G, Schulz GE: Structure of the catalytic fragment of poly(ADP-ribose)polymerase from chicken. Proc Natl Acad Sci USA 93: 7481–7485, 1996
17. Kim H, Jacobson MK, Rolli V, Ménissier-de Murcia J, Reinbolt J, Simonin F, Ruf A, Schulz G, de Murcia G: Photoaffinity labelling of human poly(ADP-ribose) polymerase catalytic domain. Biochem J 322: 469–475, 1997
18. Miranda EA, Dantzer F, O'Farrell M, de Murcia G, Ménissier-de Murcia J: Characterization of a gain-of-function mutant of poly(ADP-ribose) polymerase. Biochem Biophys Res Commun 212: 317–325, 1995
19. Wang ZQ, Auer B, Stingl L, Berghammer H, Haidacher D, Schweiger M, Wagner EW: Mice lacking ADPRT and poly(ADP-ribosyl)ation develop normally but are susceptible to skin disease. Genes Dev 9: 509–520, 1995
20. Barlow C, Hirotsune S, Paylor R, Liyanage M, Eckhaus M, Collins F, Shilo Y, Crawley JN, Ried T, Tagle D, Wynshaw-Boris A: *Atm*-deficient mice: A paradigm of ataxia telangectasia. Cell 86: 159–171, 1996
21. Biedermann KA, Sun J, Giaccia AJ, Tosto LM, Brown JM: *scid* mutation in mice confers hypersensivity to ionizing radiation and a deficiency in DNA double-strand break repair. Proc Natl Acad Sci 88: 1394–1397, 1991
22. Schreiber V, Hunting D, Trucco C, Gowans B, Crunwald D, de Murcia G, Ménissier-de Murcia J: A dominant-negative mutant of human poly(ADP-ribose) polymerase affects cell recovery, apoptosis, and sister chromatid exchange following DNA damage. Proc Natl Acad Sci USA 92: 4753–4757, 1995
23. Oikawa A, Tohda H, Kanai M, Miwa M, Sugimura T: Inhibitors of poly(adenosine diphosphate ribose) polymerase induce sister chromatid exchanges. Biochem Biophys Res Comm 97: 1311–1316, 1980
24. Tewari M, Quan LT, O'Rourke K, Desnoyers S, Zeng Z, Beidler DR, Poirier CC, Salvesen CS, Dixit VM: Yama/CPP32β, a mammalian homolog of CED-3, is a CrmA-inhibitable protease that cleaves the death substrate poly(ADP-ribose) polymerase. Cell 81: 801–809, 1995
25. Leist M, Single B, Künstle G, Volbracht C, Hentze H, Nicotera P: Apoptosis in the absence of poly(ADP-ribose) polymerase. Biochem Biophys Res Commun 233: 518–522, 1997
26. Masson M, Niedergang C, Schreiber V, Ménissier-de Murcia J, de Murcia G: XRCC1 is specifically associated with poly(ADP-ribose) polymerase and negatively regulates its activity following DNA damage. Mol Cell Biol 18: 3563–3571, 1998

Molecular and Cellular Biochemistry **193**: 61–68, 1999.

The RT6 (Art2) family of ADP-ribosyltransferases in rat and mouse

Rita Bortell,[1] Toshihiro Kanaitsuka,[1] Linda A. Stevens,[2] Joel Moss,[2] John P. Mordes,[1] Aldo A. Rossini[1] and Dale L. Greiner[1]

[1]*Department of Medicine, University of Massachusetts Medical Center, Worcester, MA;* [2]*Pulmonary-Critical Care Medicine Branch, National Heart, Lung, and Blood Institute, NIH, Bethesda, MD, USA*

Abstract

Recent evidence suggests that a new member of the mono-ADP-ribosyltransferase/NAD glycohydrolase family, RT6, may be important in immune regulation. RT6 is expressed in two allelic forms and is present on post-thymic T cells in the rat. RT6-expressing T cells in the rat may have a regulatory role, a conclusion based on their ability to prevent autoimmune diabetes in the BB rat model of insulin-dependent diabetes mellitus. This observation led to investigation of RT6 at a molecular and biochemical level resulting in the determination that RT6 protein exists as both glycosylated and non-glycosylated glycosylphosphatidylinositol (GPI)-linked cell surface molecules. RT6, like many GPI-linked proteins, can mediate cell signal transduction events associated with T cell activation, and is also present in a soluble form in the circulation. The discovery that RT6 is an NAD glycohydrolase and auto-ADP-ribosyltransferase led to the ongoing investigations into the role that enzymatic activity may have in the immunoregulatory function of rat RT6$^+$ T cells. A homologue of rat RT6, termed Rt6, has been identified in the mouse. Rt6 is predominately an ADP-ribosyltransferase enzyme as determined using simple guanidino compounds (e.g. arginine) as ribose acceptors. Abnormalities in mouse Rt6 mRNA are associated with the expression of autoimmunity. In the present manuscript, we review recent data on RT6/Rt6, and discuss the potential mechanisms by which RT6-expressing cells, and perhaps RT6 protein itself, may mediate immune regulation. (Mol Cell Biochem **193**: 61–68, 1999)

Key words: ART, mono-ADP-ribosyltransferase, RT6, immune regulation, autoimmunity

Abbreviations: ART – NAD$^+$:arginine ADP-ribosyltransferase; DP-BB rat – diabetes-prone BB rat; DR-BB rat – diabetes-resistant BB rat; GPI – glycosylphosphatidylinositol; IDDM – insulin-dependent diabetes mellitus; IEL – intraepithelial lymphocyte; NOD – non-obese diabetic; PI-PLC – phosphatidylinositol-specific phospholipase C; TcR – T cell receptor

Introduction

Several proteins with NAD:arginine ADP-ribosyltransferase (ART) activity are expressed in T cells, affect T cell function, and modulate immune system activity. Rat T cells that express the ART designated RT6 are determinants of the expression of autoimmune diabetes [1, 2]. In the mouse, levels of mRNA encoding the ART designated Rt6 also correlate with autoimmunity in at least two strains [3, 4]. This review focuses on the 'RT6' family of ARTs in rat and mouse (Table 1). We use the 'RT6' nomenclature found in most current literature, but, as described below, this nomenclature is to be replaced by 'Art' designations in the near future.

Rat RT6 (ART2^a and ART2^b)

What came eventually to be termed the 'RT6' alloantigenic system in the rat was identified in several laboratories in the 1970's and given the designations PtA, RT-Ly-2, A.R.T.-2, and AgF. An international nomenclature committee in 1983

Address for offprints: D.L. Greiner, 373, Plantation Street, Biotech 2, Suite 218, Worcester, MA 01605, USA

Table 1. Properties of Rat RT6 and mouse Rt6 proteins

	RT6.1 ($ART2^a$)	RT6.2 ($ART2^b$)	Rt6-1 ($Art2^a$)	Rt6-2 ($Art2^b$)
Molecular weight				
Non-glycosylated	26–28 kDa	26–28 kDa	33 kDa	33 kDa
Glycosylated	30–38 kDa	–	35–42 kDa	35–42 kDa
Glycosylation sites	One	None	Two	Two
ART activity (exogenous proteins)	No	No	Yes	Yes
Auto-ART activity	Conflicting data	Yes	Yes	Not known
NAD^+ glycohydrolase activity	Yes	Yes	Low	Low
Signal transduction activity	Yes	Yes	Not known	Not known
GPI-linked	Yes	Yes	Yes	Yes
Lymphoid restricted	Yes	Yes	Yes	Yes

designated the alloantigens identified in these various systems as RT6 [5].

The RT6 alloantigen was first detected as a surface protein on T cells and was used extensively as a marker of T cell development and function in the rat [6]. The discovery of its allelic nature led to its use as a marker of donor-vs. host-origin cells in bone marrow and lymphocyte adoptive transfer systems [7–9]. The expression of RT6 antigen in the peripheral lymphoid compartment was also found to be restricted to mature T cells [10–12]. This remarkable ontogenetic characteristic led to extensive use of RT6 as a marker in studies of rat T cell development and in the fractionation of functional T cell subsets with well defined properties [9–16]. The recent discovery of its enzymatic activity [17] has further enhanced its interest for immunologists. Some of the most important characteristics of RT6 proteins are summarized in Table 1.

Nomenclature and genetics of rat RT6

Two alleles of the rat RT6 gene have been reported [5]. $RT6^a$ encodes the RT6.1 protein and $RT6^b$ encodes the RT6.2 protein. Recognition of the ART activity of RT6 proteins and of the sequence homology of RT6 with other ARTs established RT6 as a member of the ART family [18]. As a result, a new nomenclature has been proposed and recently published [19]. For the rat, $ART2^a$ and $ART2^b$ will replace $RT6^a$ and $RT6^b$, respectively. To facilitate reference to the published literature in this transition period, we refer to the protein products of the $RT6^a$ ($ART2^a$) and the $RT6^b$ ($ART2^b$) alleles as RT6.1 and RT6.2, respectively.

Each allele of RT6 segregates as an autosomal dominant with variable penetrance. The majority of lymphoid cells in the offspring of a cross between homozygous $RT6^a$ and $RT6^b$ parents co-express both RT6.1 and RT6.2 on their surface [20], but a subset of $RT6.2^+$ $RT6.1^-$ cells can be detected [21]. The gene, located on chromosome 1, is linked closely to the genes encoding albinism and a variant hemoglobin [22]. cDNAs for both $RT6^a$ and $RT6^b$ have been sequenced [23, 24], and recombinant proteins of both alleles have been generated using prokaryotic, insect, and mammalian expression systems [25–29]. The cDNA sequence for $RT6^a$ differs from that for $RT6^b$ at 18 nucleotide positions [24]. The RT6 gene comprises eight exons, the seventh of which encodes the entire native protein [30]. Two independent TATA box-containing promoters are associated with exons one and two, respectively. The predominant transcript found in rat spleen cells originates at the transcription site associated with exon two.

Characteristics of cell-associated and soluble RT6 proteins

Predicted amino acid sequences of both $RT6^a$ and $RT6^b$ contain domains characteristic of the attachment sites of glycosylphosphatidylinositol (GPI) cell surface anchors. Studies using phosphatidylinositol-specific phospholipase C (PI-PLC) have established that RT6.1 and RT6.2 are each, in fact, attached to the cell membrane by GPI-linkages [31, 32].

Both isoforms of RT6 are exported to the cell surface as non-glycosylated proteins, each with an apparent MW of ~26–28 kDa. The RT6.1 alloantigen is also detected as a variably glycosylated 30–38 kDa protein [32, 33], consistent with the observation of a single potential glycosylation site in $RT6^a$ but not $RT6^b$ cDNA [23, 24]. No functional activities that distinguish the glycosylated from the non-glycosylated forms of RT6.1 have yet been described.

As is characteristic of many GPI-linked molecules, both RT6.1 and RT6.2 proteins are found in serum in soluble form [34]. In $RT6^a$ rats, the predominant circulating species is the glycosylated form, and in $RT6^b$ rats, the non-glycosylated form. Levels of soluble RT6 protein vary substantially among rat strains; the amount of circulating protein in YOS rats, for example, appears to be at least one order of magnitude greater than that in the DR-BB/Wor rat (unpublished observations). Whether soluble RT6 protein possesses enzymatic activity is not yet known. Surprisingly, soluble RT6 is present in athymic *rnu/rnu* rats despite their severe deficiency of peripheral RT6⁺ T cells (unpublished observations). These observations suggest that peripheral RT6⁺ T cells may not be the sole or even the predominant source of soluble RT6 in the rat.

Cell distribution and ontogeny of rat RT6
RT6 protein is expressed on peripheral T cells and intestinal cells including intraepithelial lymphocytes (IELs) [10, 12, 14, 35, 36]. Thymocytes and bone marrow cells do not express RT6. In peripheral lymphoid tissues, essentially all $RT6^+$ T cells co-express the T cell receptor (TcR) complex [35, 36]. RT6 is expressed on ~80% of $CD8^+$, ~50% of $CD4^+$, and ~35% of $CD45^+$ T cells [10, 37]. Peripheral $RT6^+$ T cells proliferate in response to alloantigens and mitogens and participate in delayed hypersensitivity and graft-vs.-host responses [14, 38, 39]. A putative immunoregulatory subset of $RT6^+$ $CD4^+$ $CD45^{low}$ T cells has been found to express IL-4 mRNA [40]. Cytotoxic T cell activity in the rat appears to be restricted to the $RT6^-$ cell subset [1, 15]. The majority of intestinal IELs in normal rats are $RT6^+$ and co-express CD8 and the αβTcR/CD3 complex [36, 41]. Only a very small subset of $RT6^+$ IELs in normal rats are TcR^-. $RT6^+$ IELs are also present in athymic *rnu/rnu* rats that lack peripheral $RT6^+$ T cells, but the majority of these IELs are TcR^- [36, 41]. Interestingly, $RT6^+$ IELs express ~10 times more cell surface RT6 protein per cell than do peripheral $RT6^+$ T cells [36, 41].

$RT6^+$ T cells are present in the gut at birth but do not appear in peripheral lymphoid tissues until 3–7 days of age [10, 12, 36]. Maximal levels of peripheral $RT6^+$ T cells are reached at ~8–12 weeks of age and are maintained at that level throughout adult life. The development of peripheral $RT6^+$ T cells is thymus-dependent, whereas at least one subset of $RT6^+$ IELs is thymus-independent [10, 11, 36]. Thymus-dependent peripheral T cells are released into the circulation with a $Thyl^+$ $RT6^-$ phenotype [42–45]. Post-thymic differentiation occurs within 3–7 days and is characterized by up-regulation of RT6 and simultaneous down-regulation of Thyl. Essentially all peripheral $RT6^+$ T cells are Thy^-.

Mouse Rt6 (Art2a and Art2b)

The mouse homologue of rat RT6, designated 'Rt6' was not discovered until 1990, some 20 years after the discovery of the rat antigen [46]. The delay was due in large measure to lack of reagents capable of identifying cell surface Rt6. Lack of reagents suitable for flow cytometry and immunohistochemistry continues to impede the analysis of murine $Rt6^+$ cells. Rt5 cDNA was cloned using rat cDNA probes [3, 46]. Availability of the cDNA led to rapid genetic and biochemical characterization of mouse Rt6 protein which, like rat RT6, is an enzyme belonging to the ART family [18]. Table 1 summarizes important characteristics of the murine Rt6 proteins.

Nomenclature and genetics of mouse Rt6
Two genes encoding Rt6 in the mouse have been reported [3, 47]. Rt6 locus-1 encodes the Rt6-1 protein, and Rt6 locus-2 encodes the Rt6-2 protein. Recognition of the ART activity of Rt6 proteins and of the sequence similarities of Rt6 and other ARTs has established Rt6 as another member of the ART family. For the mouse, $Art2^a$ and $Art2^b$ designations will replace Rt6 locus-1 and Rt6 locus-2, respectively [19]. We shall refer to the protein products of the Rt6 locus-1 ($Art2^a$) and the Rt6 locus-2 ($Art2^b$) genes as Rt6-1 and Rt6-2, respectively.

The two mouse Rt6 genes appear to be closely linked on chromosome 7 and may have arisen from a gene duplication event [3]. The mouse Rt6 genes are syntenic with rat RT6 and linked to albinism and variant hemoglobin genes. Rt6 co-segregates with allotypes of the *H1* minor histocompatibility system [48]. Multiple polymorphic forms of the mouse Rt6 genes have been revealed by restriction enzyme analysis [3]. The sequence of BALB/c Rt6 locus-1 cDNA is 87% identical to the 129/Sv mouse Rt6 locus-2 cDNA and 79.7% identical to rat RT6 cDNAs [46]. BALB/c Rt6 locus-1 cDNA contains a Sac1 restriction site that is not present in Rt6-2 cDNA [4]. This site permits quantitative comparison of the expression of Rt6 locus-1 and Rt6 locus-2 mRNAs in individual mice and among different strains of mice. Both Rt6 locus-1 and Rt6 locus-2 are expressed in most mouse strains but can be differentially expressed in different strains. BALB/c mice express predominantly Rt64 mRNA, C57BL/6 mice predominantly Rt6-2 mRNA [3, 4, 47], and NZW mice predominantly Rt6-1 [4]. Recently, the cDNA for the Rt6-1 and Rt6-2 proteins of C57BL/6 and BALB/c mice were cloned, sequenced, and expressed in recombinant systems [49]. The C57BL/6 Rt6-1 cDNA contains a stop codon that results in a truncated protein without enzymatic activity, providing a molecular explanation for the differential expression of Rt6-1 and Rt6-2 in this strain.

Characteristics of mouse Rt6 proteins
Predicted amino acid sequences of both Rt6-1 and Rt6-2 contain two potential glycosylation sites [46, 47]. Glycosylation of recombinant forms of both proteins has been inferred from their appearance on immunoblots [28, 29, 47, 50], but the presence of carbohydrate moieties remains to be demonstrated directly. The Rt6-1 and Rt6-2 cDNAs predict proteins of ~33 kDa in the absence of glycosylation. Rt6 proteins identified by immunoblot analysis of cell extracts and recombinant Rt6 proteins synthesized in insect

and mammalian cells have apparent MWs of ~30–40 k [29]. PI-PLC-treatment appears to release both Rt6-1 and Rt6-2 proteins from mouse lymphoid tissues, consistent with their attachment to the cell surface by GPI anchors [47].

Cell and tissue distribution, and ontogeny of mouse Rt6
Immunoblot analyses have documented the expression of GPI-anchored cell surface Rt6 protein in mouse T cells [47]. Unfortunately, lack of suitable reagents for flow cytometry or immunohistochemical analysis continues to hinder direct study of such cells. Studies of the tissue distribution and ontogeny of Rt6 have been forced to rely on analyses of message. Rt6 mRNA is present at low levels in the thymus, at intermediate levels in peripheral lymphoid tissues, and at higher levels in the intestine [3, 4]. Cell depletion studies indicate that Rt6 mRNA expression is restricted to T cells in peripheral lymphoid tissues, but the ontogeny and functional capabilities of mouse T cells that express Rt6 mRNA are not known. Expression of mouse Rt6 mRNA appears to be developmentally regulated [4].

RT6 and autoimmunity

More than a decade passed between the discovery of the RT6 alloantigenic system and the recognition that RT6$^+$ T cells have immunoregulatory functions. Those functions were discovered in studies of insulin-dependent diabetes mellitus (IDDM) in the rat and led to the conclusion that a subset of peripheral RT6$^+$ T cells has the ability to suppress autoreactive effector cells.

Role of RT6$^+$ regulatory cells in the BB rat model of IDDM
The regulatory function of RT6$^+$ T cells in autoimmunity was first demonstrated in studies of the diabetes-prone (DP)-BB rat [1, 2]. DP rats develop spontaneous autoimmune IDDM and thyroiditis. They are also T lymphocytopenic. It was first observed that they could be protected from spontaneous IDDM by transfusions of whole blood [8, 51, 52]. It was later recognized that the lymphopenia of DP-BB rats was in large measure due to a severe deficiency of peripheral RT6$^+$ T cells [53] and that prevention of IDDM by transfusion was mediated by a population of RT6$^+$ CD4$^+$ T cells [54]. A single transfusion of as few as 50×10^6 lymphocytes can prevent IDDM provided that RT6$^+$ T cells became engrafted early in life in the pre-diabetic DP recipient [55]. Lymphopenia and deficiency of RT6$^+$ T cells in the DP-BB rat is not due to a defect in the RT6 gene *per se,* but rather to homozygosity for *lyp*, a gene of unknown function on chromosome 4 [56].

The role of the RT6$^+$ T cell subset in the pathogenesis of rat autoimmunity was confirmed in analyses of the coisogenic diabetes-resistant (DR)-BB rat. Derived from DP-BB forebears, DR-BB rats have been bred for disease resistance [57]. They are not only free of spontaneous autoimmune disease, but they are also non-lymphopenic and have normal numbers of circulating RT6$^+$ T cells [1]. Treatment of DR-BB rats with a cytotoxic monoclonal antibody (mAb) against RT6.1$^+$ T cells was found to lead to the development of autoimmune diabetes and thyroiditis within two to four weeks [1, 58]. This result was interpreted to suggest that the DR-BB rat shared the genetic predisposition of the DP animal to autoimmunity, but was kept free of disease by its population of RT6$^+$ 'suppressor T cells.'

The hypothesis that an RT6$^+$ T cell subset acts as suppressor cells received additional support from adoptive transfer studies. Injection of T lymphocytes from diabetic RT6-depleted DR-BB rats into histocompatible athymic *rnu/rnu* recipients rapidly induced diabetes that was preventable by co-administration of RT6$^+$ T cells [59]. The phenotype of the protective cell subset has been identified as RT6$^+$ CD4$^+$ CD45^{low} (unpublished observations). This phenotype is similar to that of a regulatory cell population in a PVG rat model of IDDM induced by thymectomy and irradiation [40].

Recently, it has been documented that both the autoreactive and regulatory T cells in BB rats originate in the thymus [60]. Transfer of DR-BB thymocytes to histocompatible *rnu/rnu* recipients resulted in the generation of both RT6$^-$ and RT6$^+$ T cell populations. These adoptive recipients did not develop IDDM. If, however, recipients were treated with an anti-RT6.1 mAb to deplete the regulatory RT6.1$^+$ T cell subset emerging from the thymocyte inoculum, the recipients eventually developed both IDDM and thyroiditis [60]. The result was again consistent with the hypothesis that the expression of autoimmunity in rats is a function of RT6$^-$ effector cells acting in the absence of RT6$^+$ regulatory cells.

The involvement of RT6$^+$ T cells in rat models of autoimmunity is not restricted to the BB and PVG rat strains. RT6$^+$ T cells also appear to play a role in the expression of mercury-induced autoimmune glomerulonephritis in the BN rat [61]. Following $HgCl_2$ injections that induce disease, the relative proportion of peripheral RT6$^+$ T cells falls and autoantibodies and kidney pathology ensue.

Possible role of Rt6 in mouse models of autoimmunity
The importance of RT6$^+$ T cells in rat autoimmunity is well documented. Rt6 may play a corresponding role in mouse autoimmunity, but the available data are more limited. In the nonobese spontaneously diabetic (NOD) mouse, levels of Rt6 mRNA correlate inversely with the development of autoimmune insulitis and hyperglycemia [3]. The finding of decreased levels of Rt6 mRNA in the NOD mouse is especially interesting because this animal, rather than being lymphopenic like the DP rat, has increased numbers of T cells in the circulation [62].

Two defects in Rt6 have also been observed in the NZW and the (NZW × NZB)F_1 models of systemic lupus erythematosus [4]. The NZW mouse lacks the Rt6 locus-2 gene, and (NZW × NZB)F_1 mice display reduced levels of Rt6 mRNA in spleen and intestine. Reduced levels of Rt6 mRNA are not observed in the parental NZB strain, suggesting that the F_1 has inherited a dominant factor from the NZW strain that results in reduced expression of Rt6. Interestingly, although both NZB and (NZW × NZB)F_1 mice develop lupus, the disease is much more rapid in onset and more severe in the F_1 mouse in which Rt6 mRNA expression is reduced [4].

RT6 and Rt6: Recent discoveries and possible mechanisms of immunomodulation

How $RT6^+$ T cells regulate immune function is not known but is the subject of growing interest. Recent discoveries relating RT6 and Rt6 to signal transduction and to enzymatic activity suggest that several possible mechanisms could be involved.

Cell signal transduction capabilities of rat RT6

Many GPI-linked proteins have cell signal transduction capabilities [63–65]. Our laboratory has recently found that cell surface RT6 can mediate cell signal transduction events [66]. Antibody-mediated cross-linking, in the presence of phorbol ester, enhanced expression of the α-subunit of the interleukin-2 (IL-2) receptor and increased the proliferation of rat T-cells cultured in the presence of recombinant IL-2 (rIL-2) and/or rIL-4. Co-immunoprecipitation demonstrated the association of RT6.1 and RT6.2 proteins with five tyrosine-phosphorylated proteins including $p60^{fyn}$ and $p56^{lck}$, members of the *src* tyrosine kinase family. Signal transduction through RT6 appeared to increase the phosphorylation of the five co-immunoprecipitated proteins and also to increase the amount of $p60^{fyn}$ and $p56^{lck}$ that could be co-immunoprecipitated. These observations suggest that RT6-mediated signaling events could modulate immune function by priming T-cells to respond to exogenous cytokines. Whether mouse Rt6 proteins possess similar capabilities is not yet known.

Rat RT6 proteins have NAD⁺ glycohydrolase and auto-ADP-ribosyltransferase activities

RT6 proteins were discovered to possess enzymatic activity in 1994 after the observation of cDNA sequence similarities between rat RT6 and the catalytic site consensus sequence of bacterial toxin ADP-ribosyltransferase [17]. Recombinant RT6.2 protein synthesized in mammalian cells demonstrated NAD^+ glycohydrolase activity, cleaving NAD^+ and releasing free ADP-ribose and nicotinamide [17]. This observation was rapidly confirmed using RT6.2 protein produced in insect cells [28]. Additional studies revealed that RT6.2 protein also possesses auto-ADP-ribosyltransferase activity [25, 28]. Rat RT6.1 proteins synthesized in a transfected rat thymic lymphoma cell line and in a bacterial expression system have also been shown to possess NAD^+ glycohydrolase activity, but there are conflicting reports regarding their auto-ADP-ribosyltransferase activity [25, 27].

The K*m* for the NAD^+ glycohydrolase activity of recombinant rat RT6.2 synthesized in insect cells was reported to be ~25 μM, but neither rat RT6.1 nor RT6.2 catalyze the ADP-ribosylation of exogenous proteins [25–29, 67]. Based on their inability to ADP-ribosylate exogenous protein, rat RT6.1 and RT6.2 have been classified as ARTs with primarily NAD^+ glycohydrolase activity. Interestingly, substitution of a single amino acid, glutamine for glutamic acid at position 207, produces an RT6.1 protein that has reduced NAD^+ glycohydrolase activity and can ADP-ribosylate agmatine and arginine [26, 67].

Recombinant mouse Rt6 proteins are ADP-ribosyltransferases

Like rat RT6 proteins, deduced amino acid sequences of mouse Rt6 proteins share similarities with the family of eukaryotic ARTs [18, 68, 69]. Recombinant mouse Rt6-1 protein and an Rt6-2 fusion protein have been shown to be typical ARTs with the ability to ADP-ribosylate exogenous acceptor molecules like agmatine [28, 29, 50]. The Rt6-2 fusion protein also has NAD^+ glycohydrolase activity, but it is substantially less than its ART activity [50]. The amino acid at position 207 in mouse Rt6, as in rat RT6, appears to be an important determinant of the enzyme's catalytic activity. Changing glutamate-207 to glutamine converts mouse Rt6 from an ADP-ribosyltransferase to an NAD^+ glycohydrolase [67].

RT6, Rt6 and mechanisms of immune system modulation

Cell biological studies of RT6 together with newer biochemical and molecular analyses suggest at least four potential mechanisms by which cells expressing RT6 or Rt6 could modulate immune reactivity.

First, RT6 may mark a population of T cells that regulate immune system activity indirectly, via mechanisms that do not depend on the expression or absence of RT6 *per se*. The mechanism could involve $RT6^+$ T cell-mediated deviation of the cytokine profile of effector $RT6^-$ T cells, resulting in a Th2 (protective) rather than a Th1 (destructive) type cellular response [70]. This possibility is supported by the observation that $RT6^+$ $CD4^+$ $CD45^{low}$ cells express IL-4 mRNA and prevent autoimmunity in two different models of IDDM [1, 40].

Second, the cell signaling activities of RT6 could modulate the immune system. RT6-mediated modification of the signals received by the cell through the TcR or other receptors could influence the type and quantity of cytokine receptors expressed. Alternatively, RT6-mediated signals could induce the cell to undergo differentiation. The observation that cross-linking cell surface RT6 induces the expression of IL-2 and/or IL-4 receptors supports these possibilities [66].

Third, the NAD^+ glycohydrolase activity of RT6 may influence the function of activated T cells. In support of this hypothesis, it has been observed that addition of the enzyme's substrate, NAD^+, to stimulated rat $RT6^+$ T lymphocyte cultures inhibits cell proliferation [66].

Finally, the ADP-ribosyltransferase activities of RT6 and Rt6 may be important mechanisms. Rat RT6 undergoes auto-ADP-ribosylation that could modulate the signal transduction properties or other activities of the protein itself [25, 27, 28]. Mouse Rt6 can ADP-ribosylate exogenous proteins [28, 29, 50, 67]. It has been observed that ADP-ribosylation of cell surface proteins inhibits the proliferation and function of a murine cytotoxic $CD8^+$ T cell clone [71]. One protein known to undergo ADP-ribosylation is LFA-1, a molecule important in adhesion and costimulation [72]. ADP-ribosylated LFA-1 is a less adhesive ligand than the native protein. It has also been observed that a GPI-anchored ADP-ribosyltransferase is released by activation from cytotoxic T cells [73].

Conclusion

The discovery that a subset of $RT6^+$ T cells has the ability to prevent autoimmune diabetes and thyroiditis in the rat generated exciting expectations that similar cells would be discovered in humans. Those expectations were initially disappointed when it was found that the human gene homologous to RT6 is not expressed.

Continuing study of the biochemistry and molecular and cell biology of rat RT6 and mouse Rt6 has, however, placed these proteins in the larger family of ARTs, many of which are expressed on human cells. Understanding the signals transduced by cell surface RT6 and the biochemistry of its enzymatic activity may yet provide information on the regulation of immune processes relevant to human disease. For that reason, intensive study of the ART2 family of proteins continues.

Acknowledgement

This work was supported in part by grants DK41235, DK25306, DK36024 from the National Institutes of Health, an institutional Diabetes and Endocrinology Research Center Grant, and by Research Grants from the Juvenile Diabetes Foundation International. The contents of this publication are solely the responsibility of the authors and do not necessarily represent the official views of the National Institutes of Health.

References

1. Mordes J, Bortell R, Doukas J, Rigby M, Whalen BJ, Zipris D, Greiner DL, Rossini AA: The BB/Wor rat and the balance hypothesis of autoimmunity. Diabetes/Metab Rev 2: 103–109, 1996
2. Mordes JP, Greiner DL, Rossini AA: Animal models of autoimmune diabetes mellitus. In: D LeRoith, SI Taylor, JM Olefsky (eds). Diabetes Mellitus. A Fundamental and Clinical Text. Lippincott-Raven, Philadelphia, 1996, pp 349–360
3. Prochazka M, Gaskins HR, Leiter EH, Koch-Nolte F, Haag F, Thiele H-G: Chromosomal localization, DNA polymorphism, and expression of *Rt-6,* the mouse homologue of rat T-lymphocyte differentiation marker *RT6.* Immunogenetics 33: 152–156, 1991
4. Koch-Nolte F, Klein J, Hollmann C, Kühl M, Haag F, Gaskins HR, Leiter E, Thiele H-G: Defects in the structure and expression of the genes for the T cell marker *Rt6* in NZW and (NZB × NZW)F_1 mice. Int Immunol 7: 883–890, 1995
5. Lubaroff DM, Butcher G, DeWitt C, Gill TJ III, Günther E, Howard J, Wonigeit K: Fourth international workshop on alloantigenic systems in the rat. Transplant Proc 15: 1683, 1983
6. Greiner DL, Mordes JP, Angelillo M, Handler ES, Mojcik CF, Nakamura N, Rossini AA: Role of regulatory $RT6^+$ T-cells in the pathogenesis of diabetes mellitus in BB/Wor rats. In: E. Shaffir, A.E. Renold (eds). Frontiers in Diabetes Research: Lessons from Animal Diabetes II. John Libbey, London, 1988, pp 58–67
7. Greiner DL, Goldschneider I, Lubaroff DM: Identification of thymocyte progenitors in hemopoietic tissues of the rat. I. A quantitative assay system for thymocyte regeneration. Thymus 6: 181–199, 1984
8. Rossini AA, Mordes JP, Greiner DL, Nakano K, Appel MC, Handler ES: Spleen cell transfusion in the BB/W rat: Prevention of diabetes, NMC restriction, and long term persistence of transfused cells. J Clin Invest 77: 1399–1401, 1986
9. Goldschneider I, Kornschlies KL, Greiner DL: Studies of thymocytopoiesis in rats and mice. I. Kinetics of appearance of thymocytes using a direct intrathymic adoptive transfer assay for thymocyte precursors. J Exp Med 163: 1–17, 1986
10. Mojcik CF, Greiner DL, Medlock ES, Kornschlies KL, Goldschneider I: Characterization of RT6 bearing rat lymphocytes. I. Ontogeny of the $RT6^+$ subset. Cell Immunol 114: 336–346, 1988
11. Mojcik CF, Greiner DL, Goldschneider I: Characterization of RT6-bearing lymphocytes. II. Developmental relationships of $RT6^-$ and $RT6^+$ T cells. Develop Immunol 1: 191–201, 1991
12. Thiele H-G, Koch F, Kashan A: Postnatal distribution profiles of $Thy\text{-}1^+$ and $RT6^+$ cells in peripheral lymph nodes of DA rats. Transplant Proc 19: 3157–3160, 1987
13. Mojcik CF, Greiner DL, Medlock ES, Goldschneider I: Development of T cell subsets in the rat expressing the RT-6 alloantigen. Fed Proc 44: (abstr) 1301, 1985
14. Greiner DL, Reynolds CW, Lubaroff DM: Maturation of functional T-lymphocyte subpopulations in the rat. Thymus 4: 77–90, 1982
15. Chen-Woan M, McGregor DD, Harris WV, Greiner DL: Monoclonal antibody analysis of Listeria monocytogenes-induced cytotoxic lymphocytes. Immunology 57: 505–513, 1986
16. Angelillo M, Greiner DL, Crisá L, Kitagawa Y, Heyderman JI, Mordes JP, Rossini AA: $RT6^+$ T cell developmental defects in BB/Wor rats. In: E. Shaffir (ed). Frontiers in Diabetes Research II. Lessons from Animal Diabetes III. Smith-Gordon, Edinburgh, 1990, pp 114–119

17. Takada T, Iida K, Moss J: Expression of NAD glycohydrolase activity by rat mammary adenocarcinoma cells transformed with rat T cell alloantigen RT6.2. J Biol Chem 269: 9420–9423, 1994
18. Koch-Nolte F, Haag F, Kastelein R, Bazan F: Uncovered: the family relationship of a T-cell-membrane protein and bacterial toxins. Immunol Today 17: 402–405, 1996
19. Haag F, Koch-Nolte F: The vertebrate gene family of mono(ADP-ribosyl)transferases: Proposal for a unified nomenclature. Adv Exp Med Biol 419: 459–462, 1997
20. Angelillo M, Greiner DL, Mordes JP, Handler ES, Nakamura N, McKeever U, Rossini AA: Absence of RT6⁺ T cells in diabetes-prone BioBreeding/Worcester rats is due to genetic and cell developmental defects. J Immunol 141: 4146–4151, 1988
21. Thiele H-G, Haag F, Nolte F: Asymmetric expression of RT6.1 and RT6.2 alloantigens in ($RT6^a \times RT6^b)F_1$ rats is due to a pretranslational mechanism. Transplant Proc 25: 2786–2788, 1993
22. Greiner DL, Barton RW, Goldschneider I, Lubaroff DM: Genetic linkage and cell distribution analysis of T cell alloantigens in the rat. J Immunogenet 9: 43–50, 1982
23. Koch F, Haag F, Kashan A, Thiele H-G: Primary structure of rat RT6.2, a nonglycosylated phosphatidylinositol-linked surface marker of postthymic T cells. Proc Natl Acad Sci USA 87: 964–967, 1990
24. Haag F, Koch F, Thiele H-G: Nucleotide and deduced amino acid sequence of the rat T-cell alloantigen RT6.1. Nucleic Acids Res 18: 1047, 1990
25. Haag F, Andresen V, Karsten S, Koch-Nolte F, Thiele H-G: Both allelic forms of the rat T cell differentiation marker RT6 display nicotinamide adenine dinucleotide (NAD)-glycohydrolase activity, yet only RT6.2 is capable of automodification upon incubation with NAD. Eur J Immunol 25: 2355–2361, 1995
26. Maehama T, Hoshino S, Katada T: Increase in ADP-ribosyltransferase activity of rat T lymphocyte alloantigen RT6.1 by a single amino acid mutation. FEBS Lett 388: 189–191, 1996
27. Maehama T, Nishina H, Hoshino S, Kanaho Y, Katada T: NAD^+-dependent ADP-ribosylation of T lymphocyte alloantigen RT6.1 reversibly proceeding in intact rat lymphocytes. J Biol Chem 270: 22747–22751, 1995
28. Rigby MR, Bortell R, Stevens LA, Moss J, Kanaitsuka T, Shigeta H, Mordes JP, Greiner DL, Rossini AA: Rat RT6.2 and mouse Rt6 locus 1 are NAD^+:arginine ADP-ribosyltransferases with auto-ADP-ribosylation activity. J Immunol 156: 4259–4265, 1996
29. Moss J, Stevens LA, Cavanagh E, Okazaki IJ, Bortell R, Kanaitsuka T, Mordes JP, Greiner DL, Rossini AA: Characterization of mouse Rt6.1 NAD:arginine ADP-ribosyltransferase. J Biol Chem 272: 4342–4346, 1997
30. Haag FA, Kuhlenbäumer G, Koch-Nolte F, Wingender E, Thiele H-G: Structure of the gene encoding the rat T cell ecto-ADP-ribosyltransferase RT6. J Immunol 157: 2022–2030, 1996
31. Koch F, Thiele H-G, Low MG: Release of the rat T cell alloantigen RT-6.2 from cell membranes by phosphatidylinositol-specific phospholipase C. J Exp Med 164: 1338–1343, 1986
32. Crisá L, Sarkar P, Waite DJ, Haag F, Koch-Nolte F, Rajan TV, Mordes JP, Handler ES, Thiele H-G, Rossini AA, Greiner DL: An $RT6^a$ gene is transcribed and translated in lymphopenic diabetes-prone BB rats. Diabetes 42: 688–695, 1993
33. Crisá L, Greiner DL, Mordes JP, MacDonald RG, Handler ES, Czech W, Rossini AA: Biochemical studies of RT6 alloantigens in BB/Wor and normal rats: Evidence for intact but unexpressed $RT6^a$ structural gene in diabetes prone BB rats. Diabetes 39: 1279–1288, 1990
34. Waite DJ, Handler ES, Mordes JP, Rossini AA, Greiner DL: The RT6 rat lymphocyte alloantigen circulates in soluble form. Cell Immunol 152: 82–95, 1993
35. Schwinzer R, Hedrich HJ, Wonigeit K: T cell differentiation in athymic nude rats (rnu/rnu): Demonstration of a distorted T cell subset structure by flow cytometry analysis. Eur J Immunol 19: 1841–1847, 1989
36. Waite DJ, Appel MC, Handler ES, Mordes JP, Rossini AA, Greiner DL: Ontogeny and immunohistochemical localization of thymus dependent and thymus independent RT6⁺ cells in the rat. Am J Pathol 148: 2043–2056, 1996
37. Woda BA, Padden C, McFadden ML: T helper (T_H) cells in the BioBreeding Worcester (BB/Wor) rat are subdivided into distinct subpopulations by the OX22 and RT6 antibodies. Diabetes 37 (Suppl. I): (abstr) 205A, 1988
38. Hunt HD, Lubaroff DM: Interaction of T lymphocyte subpopulations bearing the RT-6 alloantigen. Transplant Proc 17: 1861–1863, 1985
39. Ernst DN, Lubaroff DM: Membrane antigen phenotype of sensitized T lymphocytes mediating tuberculin-delayed hypersensitivity in rats. Cell Immunol 88: 436–452, 1984
40. Fowell D, Mason D: Evidence that the T cell repertoire of normal rats contains cells with the potential to cause diabetes. Characterization of the CD4⁺ T cell subset that inhibits this autoimmune potential. J Exp Med 177: 627–636, 1993
41. Fangmann J, Schwinzer R, Wonigeit K: Unusual phenotype of intestinal intraepithelial lymphocytes in the rat: Predominance of T cell receptor a/b⁺/CD2⁻ cells and high expression of the RT6 alloantigen. Eur J Immunol 21: 753–760, 1991
42. Hosseinzadeh R, Goldschneider I: Recent thymic emigrants in the rat express a unique antigenic phenotype and undergo post-thymic maturation in peripheral lymphoid tissues. J Immunol 150: 1670–1679, 1993
43. Zadeh HH, Greiner DL, Wu DY, Tausche F, Goldschneider I: Abnormalities in the export and fate of recent thymic emigrants in diabetes-prone BB/W rats. Autoimmunity 24: 35–46, 1996
44. Groen H, Klatter FA, Brons NHC, Mesander G, Nieuwenhuis P, Kampinga J: Abnormal thymocyte subset distribution and differential reduction of CD4⁺ and CD8⁺ T cell subsets during peripheral maturation in diabetes-prone BioBreeding rats. J Immunol 156: 1269–1275, 1996
45. Groen H, Klatter FA, Brons NHC, Wubbena AS, Nieuwenhuis P, Kampinga J: High-frequency, but reduced absolute numbers of recent thymic migrants among peripheral blood T lymphocytes in diabetes-prone BB rats. Cell Immunol 163: 113–119, 1995
46. Koch F, Haag F, Thiele H-G: Nucleotide and deduced amino acid sequence for the mouse homologue of the rat T-cell differentiation marker RT6. Nucleic Acids Res 18: 3636, 1990
47. Hollmann C, Haag F, Schlott M, Damaske A, Bertuleit H, Matthes M, Kühl M, Thiele H-G, Koch-Nolte F: Molecular characterization of mouse T-cell ecto-ADP-ribosyltransferase Rt6: Cloning of a second functional gene and identification of the Rt6 gene products. Mol Immunol 33: 807–817, 1996
48. Koch-Nolte F, Hollmann C, Kühl M, Haag F, Prochazka M, Leiter EH, Thiele H-G: Molecular polymorphism in the Rt6 genes of laboratory mice correlates with the allotypes of the H1 minor histocompatibility system. Immunogenetics 41: 152–155, 1995
49. Kanaitsuka T, Bortell F, Stevens LA, Moss J, Sardinha D, Rajan TV, Zipris D, Mordes JP, Greiner DL, Rossini AA: Expression In BALB/c and C57BL/6 mice of Rt6-1 and Rt6-2 ADP-ribosyltransferases that differ in enzymatic activity: C57BL/6 Rt6-1 is a natural transferase knockout. J Immunol 1997, (in press)
50. Koch-Nolte F, Petersen D, Balasubramanian S, Haag F, Kahlke D, Willer T, Kastelein R, Bazan F, Thiele H-G: Mouse T cell membrane proteins Rt6-1 and Rt6-2 are arginine protein mono(ADP-ribosyl)transferases and share secondary structure motifs with ADP-ribosylating bacterial toxins. J Biol Chem 271: 7686–7693, 1996
51. Rossini AA, Mordes JP, Pelletier AM, Like AA: Transfusions of whole blood prevent spontaneous diabetes in the BB/W rat. Science 219: 975–977, 1983
52. Rossini AA, Faustman D, Woda BA, Like AA, Szymanski I, Mordes JP: Lymphocyte transfusions prevent diabetes in the Bio-Breeding/Worcester rat. J Clin Invest 74: 39–46, 1984

53. Greiner DL, Handler ES, Nakano K, Mordes JP, Rossini AA: Absence of the RT-6 T cell subset in diabetes-prone BB/W rats. J Immunol 136: 148–151, 1986
54. Mordes JP, Gallina DL, Handler ES, Greiner DL, Nakamura N, Pelletier A, Rossini AA: Transfusions enriched for W3/25+ helper/inducer T lymphocytes prevent spontaneous diabetes in the BB/W rat. Diabetologia 30: 22–26, 1987
55. Burstein D, Mordes JP, Greiner DL, Stein D, Nakamura N, Handler ES, Rossini AA: Prevention of diabetes in the BB/Wor rat by a single transfusion of spleen cells: Parameters that affect the degree of protection. Diabetes 38: 24–30, 1989
56. Jacob HJ, Pettersson A, Wilson D, Mao Y, Lernmark Å, Lander ES: Genetic dissection of autoimmune type I diabetes in the BB rat. Nature Genet 2: 56–60, 1992
57. Butler L, Guberski DL, Like AA: Genetics of diabetes production in the Worcester colony of the BB rat. In: E. Shaffir, A.E. Renold (eds). Frontiers in diabetes research: Lessons from animal diabetes II. John Libbey, London, 1988, pp 74–78
58. Greiner DL, Mordes JP, Handler ES, Angelillo M, Nakamura N, Rossini AA: Depletion of RT6.1+ T lymphocytes induces diabetes in resistant BioBreeding/Worcester (BB/W) rats. J Exp Med 166: 461–475, 1987
59. Whalen BJ, Greiner DL, Mordes JP, Rossini AA: Adoptive transfer of autoimmune diabetes mellitus to athymic rats: Synergy of CD4+ and CD8+ T cells and prevention by RT6+ T cells. J Autoimmun 7: 819–831, 1994
60. Whalen BJ, Rossini AA, Mordes JP, Greiner DL: DR-BB rat thymus contains thymocyte populations predisposed to autoreactivity. Diabetes 44: 963–967, 1995
61. Kosuda LL, Wayne A, Nahounou M, Greiner DL, Bigazzi PE: Reduction of the RT6.2+ subset of T lymphocytes in brown Norway rats with mercury-induced renal autoimmunity. Cell Immunol 135: 154–167, 1991
62. Leiter EH, Serreze DV, Prochazka M: The genetics and epidemiology of diabetes in NOD mice. Immunol Today 11: 147–149, 1990
63. Kroczek RA, Gunter KC, Seligmann B, Shevach EM: Induction of T cell activation by monoclonal anti-Thy-1 antibodies. J Immunol 136: 4379–4384, 1986
64. Hahn AB, Soloski MJ: Anti-Qa-2-induced T cell activation. The parameters of activation, the definition of mitogenic and nonmitogenic antibodies, and the differential effects on CD4+ vs. CD8+ T cells. J Immunol 143: 407–413, 1989
65. Rock KL, Reiser H, Bamezai A, McGrew J, Benacerraf B: The LY-6 locus: A multigene family encoding phosphatidylinositol-anchored membrane proteins concerned with T-cell activation. Immunol. Rev 111 (review): 195–224, 1989
66. Rigby M, Bortell R, Greiner DL, Czech MP, Klarlund JK, Mordes JP, Rossini AA: The rat T-cell surface protein RT6 is associated with src family tyrosine kinases and generates an activation signal. Diabetes 45: 1419–1426, 1996
67. Hara N, Tsuchiya M, Shimoyama M: Glutamic acid 207 in rodent T-cell RT6 antigens is essential for arginine-specific ADP-ribosylation. J Biol Chem 271: 29552–29555, 1996
68. Okazaki IJ, Moss J: Mono-ADP-ribosylation: A reversible post-translational modification of proteins. Adv Pharmacol 35: 247–280, 1996
69. Takada T, Iida K, Moss J: Conservation of a common motif in enzymes catalyzing ADP-ribose transfer. Identification of domains in mammalian transferases. J Biol Chem 270: 541–544, 1995
70. Racke MK, Bonomo A, Scott DE, Cannella B, Levine A, Raine CS, Shevach EM, Röcken M: Cytokine-induced immune deviation as a therapy for inflammatory autoimmune disease. J Exp Med 180: 1961–1966, 1994
71. Wang J, Nemoto E, Kots AY, Kaslow RK, Dennert G: Regulation of cytotoxic T cells by ecto-nicotinamide adenine dinucleotide (NAD) correlates with cell surface GPI-anchored/arginine ADP-ribosyltransferase. J Immunol 153: 4048–4058, 1994
72. Nemoto E, Yu YJ, Dennert G: Cell surface ADP-ribosyltransferase regulates lymphocyte function-associated molecule-1 (LFA-1) function in T cells. J Immunol 157: 3341–3349, 1996
73. Nemoto E, Stohlman S, Dennert G: Release of a glycosylphosphatidylinositol-anchored ADP-ribosyltransferase from cytotoxic T cells upon activation. J Immunol 156: 85–92, 1996

Molecular and Cellular Biochemistry **193**: 69–74, 1999.

Mapping the role of NAD metabolism in prevention and treatment of carcinogenesis

Elaine L. Jacobson,[1,2,3,4] W. Melissa Shieh[1] and Arnold C. Huang[2]

[1]*Department of Clinical Sciences;* [2]*Multidisciplinary Program in Nutritional Sciences;* [3]*Advanced Science and Technology Commercialization Center;* [4]*Lucille P. Markey Cancer Center, University of Kentucky, Lexington, KY, USA*

Abstract

Studies presented here show that cellular NAD, which we hypothesize to be the relevant biomarker of niacin status, is significantly lower in humans than in the commonly studied animal models of carcinogenesis. We show that nicotinamide and the resulting cellular NAD concentration modulate expression of the tumor suppressor protein, p53, in human breast, skin, and lung cells. Studies to determine the optimal NAD concentrations for responding to DNA damage in breast epithelial cells reveal that DNA damage appears to stimulate NAD biosynthesis and that recovery from DNA damage occurs several hours earlier in the presence of higher NAD or in cells undergoing active NAD biosynthesis. Finally, analyses of normal human skin tissue from individuals diagnosed with actinic keratoses or squamous cell carcinomas show that NAD content of the skin is inversely correlated with the malignant phenotype. Since NAD is important in modulating ADP-ribose polymer metabolism, cyclic ADP-ribose synthesis, and stress response proteins, such as p53, following DNA damage, understanding how NAD metabolism is regulated in the human has important implications in developing both prevention and treatment strategies in carcinogenesis. (Mol Cell Biochem **193**: 69–74, 1999)

Key words: skin, lung and breast cancer, p53 expression, cyclic ADP-ribose, poly(ADP-ribose) metabolism, niacin status

Introduction

Considerable evidence now indicates that the NAD content of cells influences cellular responses to genomic damage by multiple mechanisms. NAD is directly consumed for the synthesis of ADP-ribose polymers and cyclic ADP-ribose. The metabolism of ADP-ribose polymers appears to be involved with responses that can lead to normal cellular recovery, apoptosis, or necrosis [1, 2]. Cyclic ADP-ribose is a potent calcium releasing agent that also may mediate signalling pathways leading to apoptosis or necrosis [3, 4]. Finally, recent studies have shown that the NAD content of cells modulates the expression of stress proteins that play important roles in responses to genomic damage [5], including the tumor suppressor protein, p53 [6]. Consequently, NAD metabolism is a target for both prevention and treatment of cancer. The scheme shown in Fig. 1 outlines relationships in NAD metabolism that remain poorly defined. While it is known that dietary nicotinamide and nicotinic acid serve as precursors of NAD in many human tissues, much less is known about the conversion of tryptophan to NAD. The liver and perhaps the kidney are capable of the latter pathway. Since 60 mg of tryptophan consumed in protein is often assumed to be converted to 1 mg of niacin, tryptophan accounts for nearly one half of calculated niacin consumption in Western diets [7]. However, tryptophan does not appear to be a source of tissue NAD in humans under conditions of restricted niacin intake over a period of a few weeks [8]. Thus, it is likely that niacin intake may be significantly less than reported [7].

Measuring the relationship of dietary intake of NAD precursors and the circulating levels of various precursors that supply the tissues has been technically difficult. Recently, our laboratory has developed methods that allow measurement of nicotinic acid and nicotinamide in fasted serum samples [9]. Preliminary studies that control niacin in animal models suggest that a given concentration of precursor can produce vastly different effects on NAD in the various tissues [10–13]. The optimal intracellular NAD content for eliciting protective biochemical responses following DNA damage

Address for offprints: E.L. Jacobson, A323A ASTeCC Building, University of Kentucky, Lexington, KY 40506-0286, USA

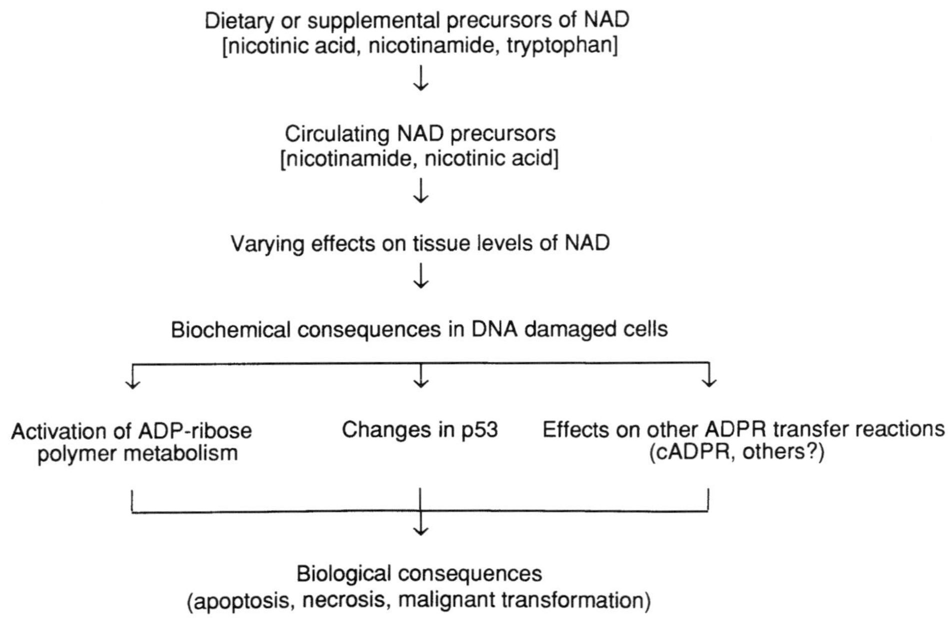

Fig. 1. Mapping the role of NAD metabolism in cancer prevention and treatment.

has not been defined. Given the relationship between NAD content and biochemical pathways involved in carcinogenesis, we are attempting to determine the interrelationships shown in Fig. 1 to understand how niacin metabolism and tissue NAD impact the prevention and treatment of cancer.

Results and discussion

Species and tissue differences in requirements for NAD precursors

The first step in characterizing the relationships of NAD metabolism shown in Fig. 1 for humans is the establishment of reliable methods for assessing niacin status. Biochemical assessment of niacin status has traditionally involved measurement of urinary oxidation products of nicotinamide. However, discovery of reactions that use NAD as a substrate in ADP-ribose transfer reactions and the relationship between NAD content and expression of stress response proteins such as p53, have led us to hypothesize that intracellular NAD may be a more relevant measure of niacin status. A biochemical measure of niacin status where the ratio of NAD to NADP in blood cells reflects dietary intake has been developed from the metabolic ward studies of Fu *et al.* and our observations on the distribution of NAD in whole blood [8, 14, 15]. We refer to this ratio as niacin number.

Since studies of carcinogenesis frequently employ rodents as animal models, we have compared niacin status as measured by blood cell pyridine nucleotides in mice, rats, and humans as a function of niacin intake. The data of Table 1 are compiled from our laboratory and others as indicated. The values are shown as niacin number and as NAD per ml of blood, because many earlier studies have only measured the latter. We have compiled the data in both formats for means of comparison. The NAD content of human blood in an individual consuming the RDA for niacin is less than 50% of that in rats and less than 25% of that in mice consuming proportional dietary niacin. These data indicate that the niacin status of humans is considerably lower than that of rodents. Additionally, analyses from a large Western population within The Malmö Diet and Cancer Study reveal a large variability in the human population. The data in Table 1 represent the values for 95% of a population where n is equal to 1300. A range of NAD content of 3 fold is observed. Approximately 15–20% of individuals in this population have significant niacin deficiency [16]. Because pharmacological doses of niacin are frequently administered as therapy for hypercholesterolemia, we have been able to measure the effect of introducing niacin supplementation on human blood cell NAD. As shown in Table 1, a mean niacin number of 175 in this population increased nearly 4 fold to 663 after 2 months of niacin therapy. It is interesting that this therapy elevates human niacin status to that of mice. However, in contrast to

Table 1. Comparison of blood cell niacin status in humans and rodents

Species	nmol NAD/ml blood	Niacin number (NAD/NADP× 100)
Human, controlled intake at RDA[a,b,h]	41	175
Human, range in large Western population[a,h]		84–236
Human, niacin depletion[b]		62
Human, niacin supplementation[c,h]	184	667
Rat	71–75[d], 119–122[e], 83[f]	321–357[d]
Rat, niacin depletion	27[d], 71[e], 43[f]	124[d]
Mouse[g,h]	179	702
Mouse, niacin supplementation[g,h]	157	683
Mouse, niacin depletion[d]	unchanged	unchanged

[a]Jacobson, E.L., unpublished observations; [b]Fu, *et al.*, J Nutr 119: 1949–1955, 1989 (ref. [8]); [c]Jacobson, E.L., Robins, H.I. and The University of Wisconsin Lipid Clinic, unpublished observations; [d]James Kirkland, unpublished observations; [e]Zhang, JZ., *et al.*, J Nutr 123: 1349–1355, 1993 (ref. [11]); [f]Shibata, K. and Murata, K, Nutr Int 2: 177–181, 1986 (ref. [13]); [g]Jacobson, E.L., *et al.*, Proc Am Assoc Cancer Res 37: 279, 1996 (ref. [10]); [h]NAD and NADP were determined from whole blood as previously described by Jacobson, E.L. and Jacobson, M.K., Meth Enzymol 280: pp 221–230, Academic Press, New York, 1997 (ref. [14]).

humans, mice supplemented with pharmacological doses of niacin maintain the same blood cell NAD content. When all 3 species are placed on diets deficient in niacin, blood cell NAD drops to about 30% of control values in humans and to about 50% of control values in rats. Interestingly, mice are resistant to niacin deficiency [17]. These species differences are likely related to the efficiency with which each converts tryptophan to niacin. Despite the mechanism, the data in Table 1 clearly show that humans maintain a lower niacin status than rodents when each are given tryptophan and preformed niacin at the recommended daily intake for humans. Therefore, humans appear to be at increased risk for niacin deficiency relative to rats or mice. These differences are critically important to interpretation of data derived from model systems in studying the role of niacin in cancer prevention and treatment. Pursuing information regarding dietary intake, circulating precursors, and the resulting NAD in blood cells and other tissues will be essential to optimizing requirements for niacin in humans.

Tissues at risk for NAD depletion include breast, lung and skin

While blood cell NAD can be used as a clinical measure of niacin status, we also have been interested in how different tissues respond to niacin intake. We predict that tissues that undergo cell division and/or are exposed to oxidative stress will be at increased risk for niacin deficiency. Consequently, our initial studies are focusing on human tissues with significant cell turnover and/or exposure to oxidative stress, such as breast, lung, and skin. The data in Table 2 show that human fibroblasts or epithelial cells from each of these tissues become severely NAD depleted if grown for 4 to 5 population doublings in the absence of nicotinamide. This phenomenon is completely reversible, but requires 6–24 h in excess nicotinamide. Further, nicotinamide depletion has no effect on cell growth rates until NAD content drops to less than 10% of normal. Human breast epithelium undergoes significant cell growth in phase with the menstruation cycle. Skin also undergoes continuous replacement with new cells, and the lung epithelium is believed to turn over, although it is not clear whether this is in response to oxidative stress or is a normal periodic replacement. Thus, growth of these tissues in limiting niacin would be expected to result in depleted NAD pools as is observed in cultured cells from these tissues (Table 2). Studies employing niacin deficiency in growing rats confirm this prediction [11–13].

Modulation of p53 expression by NAD in breast, lung and skin cells

NAD depletion in cultured cells derived from all of the tissues mentioned above is accompanied by decreased expression of the tumor suppressor protein, p53. These data were obtained by Western blot analyses of cell extracts separated by polyacrylamide gel electrophoresis (PAGE), as indicated in the legend to Table 2, and are presented as a ratio of p53 to actin to correct for any quantitative errors in loading of the protein on gels or during transfer to nitrocellulose for blotting. NAD depletion varied between the different cell lines due to varying periods of growth in niacin deficient medium and different growth rates and also due to the fact that in breast cells, nicotinamide was not completely removed from the medium because of a requirement for bovine pituitary extract, which appears to contain nicotinamide. In all cases, p53 expression relative to actin was decreased 36 to 70%, due solely to restriction of nicotinamide in the medium. This represents the first report of nicotinamide modulation of p53 expression in

Table 2. NAD content and p53 expression in nicotinamide depleted human cells

Cell type	Population doublings in limiting nicotinamide	NAD pmol/10^6 cells	p53/actin
IMR-90 human diploid lung fibroblasts	0	2164	1.73
	4	476	0.52
CF3 human diploid skin fibroblasts	0	1023	3.33
	4	384	1.27
Human primary mammary epithelial cells	0	3241	0.82
	5	513	0.53

The normal human lung diploid fibroblast strain IMR-90 (ATCC CCL 186) and the normal human foreskin diploid fibroblast strain CF-3 (gift from Dr. R. Dell'Orco) were cultured in Dulbecco's modified Eagle's medium (Gibco) in the presence of 10% fetal bovine serum (Hyclone) at 37°C in a humidified atmosphere of 5% CO_2. These cell lines were subcultured once per week, employing a 1:4 split, and fresh medium was applied every 4 days. For depletion of nicotinamide, commercially prepared medium free of nicotinamide was used in combination with exhaustively dialyzed fetal calf serum. Human mammary epithelial cells (Clonetics) were grown in a commercially prepared medium (Clonetics) or an optimized formulation of MCDB 170 basal medium containing 14 mM sodium bicarbonate and serum-free supplements (SFS) [20]. For nicotinamide depletion experiments, MCDB 170 lacking nicotinamide was custom prepared in this laboratory from stock solutions according to the procedures of McKeehan [21]. Stock cell cultures were maintained in modified MCDB 170 (Clonetics) with SFS and bovine pituitary extract (BPE), which was purchased from Clonetics. Isoproterenol (Sigma) was added at 1.0 μM in some experiments to maintain HMEC in a continuously dividing state [20]. Dishes of cells to be analyzed were washed with phosphate buffered saline, and NAD, cell number, and protein were determined. NAD was extracted with 0.5 M $HClO_4$ and neutralized with 1.0 M KOH/0.33 M potassium phosphate, pH 7.3 and quantified as described previously [14]. Protein was quantified from the $HClO_4$ precipitate by the method of Bradford [19]. Cell number was determined by a particle counter. For Western blot analyses of p53 and actin, cells were detached from the culture dishes using a 0.25% trypsin-EDTA solution (Sigma) for 1 min. The trypsin was then inactivated by adding 3 ml of culture medium containing 10% serum and centrifuged at 500 × g for 5 min at 4°C. After the supernatant was removed, the cell pellet was washed twice with PBS and re-suspended in a lysis buffer, containing 150 mM NaCl, 0.5% sodium deoxycholate, 0.1% SDS, 0.1% Triton x-100, 0.5% Nonidet P-40, 0.1% EDTA, 1 mM phenylmethylsulfonyl fluoride, and 0.01 mg/ml each of aprotinin and leupeptin, at a ratio of 50 μl per 10^6 cells. The solution was then sonicated on a Virsonic Cell Disrupter to lyse the cells. Cell lysates were subjected to SDS-12.5% PAGE, followed by transfer of proteins to a nitrocellulose membrane (Schleicher and Schuell). The membrane was then probed with monoclonal antibody Ab-3 (Oncogene Science) against p53 and monoclonal antibody Ab-1 (Oncogene Science) against actin. The secondary antibody was an anti-mouse antibody linked to horseradish peroxidase (Amersham). Final detection was by enhanced chemiluminescence (Amersham) on Hyperfilm-ECL (Amersham). The densities of the resulting bands were determined by an Ambis Model 4000 Optical Imaging System. The ratio of responses (p53/actin) was calculated to control for gel loading differences.

human cells and shows that the phenomenon appears to occur in several different cell types. Such a relationship of nicotinamide, NAD, and p53 expression has been reported previously for cells derived from Chinese hamster ovary cells [6]. In those earlier studies, decreased p53 expression correlated with NAD depletion and also occurred in chemically mutated cells carrying greatly reduced poly(ADP-ribose) polymerase (PARP) activity. Thus, the mechanism by which nicotinamide depletion affects p53 metabolism may involve utilization of NAD by PARP. However, further studies employing the PARP null genotype ongoing in our laboratory should resolve this question. Regardless of the mechanism, the modulation of p53 expression by nicotinamide in these human tissues has important implications, since diminished p53 function is strongly associated with tumorigenesis in breast, lung and skin [18].

NAD utilization and biosynthesis after DNA damage in breast cells

In order to define optimal NAD for responses to DNA damage, we have investigated the effects of various concentrations of nicotinamide in the culture medium of primary human mammary epithelial cells (HMEC). Cells were grown for 5 population doublings in medium containing bovine pituitary extract with no added nicotinamide or in the same medium containing 50 or 500 μM nicotinamide. As can be seen in Fig. 2, this resulted in cell populations with 0.75, 2.4, and 3.1 nmol of NAD/10^6 cells (0 time point), respectively. When cells were grown in the absence of nicotinamide (0.75 nmol NAD/10^6 cells) and were then replenished with 50 μM nicotinamide 24 h prior to extraction, the NAD increased to 2.27 nmol/10^6 cells, demonstrating that the depletion is reversible. The four resulting populations of cells (depleted, replenished, 50 and 500 μM nicotinamide cultures) containing a range of NAD were treated with 50 μM N-methyl-N′-nitro-N-nitrosoguanidine (MNNG) for 30 min and analyzed for NAD over the next 24 h. NAD decreased in response to MNNG in all cases except where the cells were grown in the absence of added nicotinamide. In those cells, the small remaining NAD pool appears to be inaccessible to NAD consuming reactions and could represent that sequestered in the mitochondria. A significant increase in NAD occurred between 6 and 9 h after treatment. This represents an apparent stimulation of NAD biosynthesis by

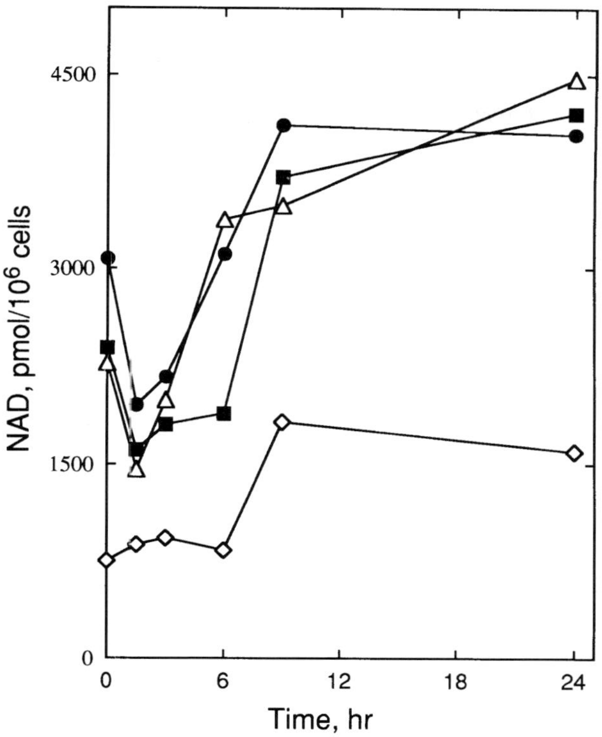

Fig. 2. Effect of various concentrations of nicotinamide in the growth medium on NAD content and recovery following DNA damage. HMEC of passage number 9–10 were seeded into 100 mm dishes (9.3×10^4 cells/dish) in Clonetics medium. Approximately 12 h after plating, the medium was removed and replaced with MCDB 170/SFS (see legend to Table 2) with (50 or 500 μM) or without nicotinamide. The cells were fed, and cell number, protein, and NAD were measured every 48 h. After 3–4 population doublings (10–12 days), medium containing 50 μM MNNG in 0.5% DMSO was added to the cells for 30 min at 37°C. Dishes of cells to be analyzed were washed with phosphate buffered saline, and NAD, cell number, and protein were determined. NAD was extracted with 0.5 M $HClO_4$ and neutralized with 1.0 M KOH/0.33 M potassium phosphate pH 7.3 and quantified as described previously [14]. Protein was quantified from the $HClO_4$ precipitate by the method of Bradford [19]. Cell number was determined by a particle counter. Closed squares represent cells grown in 50 μM nicotinamide; closed circles represent cells grown in 500 μM nicotinamide; open triangles represent cells grown in the absence of nicotinamide until 24 h prior to treatment, at which time 50 μM nicotinamide was added; and open diamonds represent cells grown in the absence of added nicotinamide.

the DNA damage, since no decrease in NAD occurred in response to MNNG. Cells grown in control medium containing 50 μM nicotinamide also demonstrated a burst of NAD biosynthesis during this time, while cells grown in 500 μM nicotinamide or those that had been given nicotinamide after depletion and analyzed 24 h later appeared to initiate NAD biosynthesis earlier, between 3 and 6 h after DNA damage. All cell cultures containing at least 50 μM nicotinamide at the time of DNA damage showed significantly elevated NAD 24 h later, reaching approximately 4.0–4.4 nmol NAD/10^6 cells as compared to 2.4 nmol NAD/10^6 cells prior to DNA damage. The percent elevation was inversely related to the initial NAD content. Thus, in these four conditions of niacin status, cells actively undergoing NAD biosynthesis and/or those with excess nicotinamide recovered from DNA damage more quickly.

Completion of these studies should identify optimal precursor levels for responding to DNA damage and increase our understanding of the regulation of NAD biosynthesis following DNA damage in breast tissue. Using this data in combination with that obtained from *in vivo* relationships of circulating precursors and breast tissue NAD should allow optimization of dietary niacin intake for prevention of breast cancer. Optimizing niacin status also may be very important during chemotherapy of breast cancer since the most common limitation to aggressive therapy is loss of bone marrow stem cells. It is likely that optimizing the NAD pool will contribute to protection of these cells, while not affecting therapy of breast cancer cells.

Inverse correlation of skin NAD with malignant phenotype

Because we have shown nicotinic acid supplementation to decrease incidence of UV induced skin tumors in mice, concomitantly with elevation of skin NAD [10], we have begun to examine how niacin status in human skin affects carcinogenic processes. We have developed methods sensitive enough to measure NAD in human skin shavings. In collaboration with the Arizona Cancer Center, we have measured NAD in skin samples from subjects diagnosed with actinic keratoses (AK) and squamous cell carcinomas (SCC). Skin from these subjects was obtained from normal, sun protected regions, such as the underarm or buttocks, as well as from the AK or SCC areas. AK commonly occurs in fair skinned individuals overexposed to sunlight and are known to be premalignant lesions. The data in Fig. 3 show the results of these analyses. Normal skin from subjects with AK have significantly higher NAD than the normal skin of SCC subjects (3.7 ± 0.6 vs. 1.7 ± 0.8 pmol/μg protein, $p = 0.0006$). These preliminary data correlate a decreased NAD in the skin with the occurrence of the malignant phenotype. This study is now being expanded to a larger population. We have also examined the NAD in AK and SCC tissues and compared them to the normal tissues. Even though AK tissues have a higher average NAD than SCC tissues, the difference is not statistically significant within this sample size. Also, when the NAD of affected tissue is compared to normal tissue of the same individual, no significant difference exists within this population. The association of lower NAD with malignancy in skin supports the hypothesis that niacin maybe an important preventive factor in cancer.

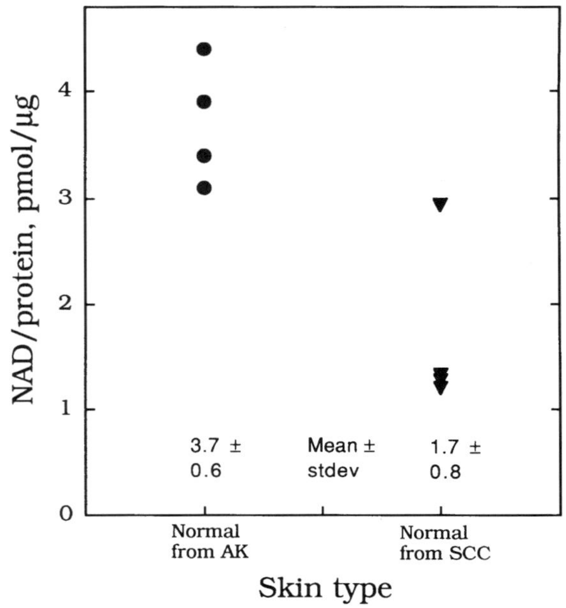

Fig. 3. NAD from human skin shavings. Skin shavings were shipped on dry ice and stored at –80°C. The skin sample was added to a mortar containing 1–2 ml of liquid nitrogen and quickly ground to a powder, which was transferred to a preweighed vial. The vial was weighed again to obtain the sample weight. NAD was extracted using 0.1 ml of ice-cold 1 M NaOH with mechanical agitation. Within 2 min, the alkaline solution was neutralized with 0.027 ml of 2.0 M phosphoric acid, 0.013 ml of 2 mM PES was added to the sample, and the solution was allowed to stand at room temperature in the dark for 5 min to oxidize NADH. The protein was precipitated with 0.047 ml of 2.0 M $HClO_4$ and allowed to stand on ice for 15 min. The insoluble fraction was collected by centrifugation at 14,000 rpm for 5 min. The supernatant (0. 175 ml) was transferred to another vial and neutralized with 55 µl of 2.0 M KOH. Following centrifugation, the supernatant was assayed for NAD [14].

Acknowledgements

This work was supported in part by NIH grants CA 65579 and 43984.

References

1. Jacobson EL, Smith JY, Wielckens K, Hilz H, Jacobson MK: Cellular recovery of dividing and confluent C3H10T1/2 cells from N-methyl-N-nitro-N-nitrosoguanidine in the presence of ADP-ribosylation inhibitors. Carcinogenesis 6: 715–718, 1985
2. Shah GM, Shah RG, Poirier GG: Different cleavage pattern for poly(ADP-ribose) polymerase during necrosis and apoptosis in HL-60 cells. Biochem Biophys Res Commun 229: 838–844, 1996
3. Vu CQ, Coyle DL, Tai H-H, Jacobson EL, Jacobson MK: Intramolecular ADP-ribose transfer reactions and calcium signalling. In: F. Haag, F. Koch-Nolte (eds). ADP-Ribosylation in Animal Tissues: Structure, Function, and Biology of Mono-ADP-Ribosyltransferases and Related Enzymes. Plenum Press, New York, 1997, pp 381–388
4. Vu CQ, Coyle DL, Jacobson EL, Jacobson MK: Intracellular signaling by cyclic ADP-ribose in oxidative cell death. FASEB J 11: A1116, 1997
5. Wright SC, Wei QS, Kinder DH, Larrick JW: Biochemical pathways of apoptosis: nicotinamide adenine dinucleotide-deficient cells are resistant to tumor necrosis factor or ultraviolet light activation of the 24-kD apoptotic protease and DNA fragmentation. J Exp Med 183: 463–471, 1996
6. Whitacre CM, Hashimoto H, Hashimoto S, Tsai M-L, Chatterjee S, Berger SJ, Berger NA: Involvement of NAD-poly(ADP-ribose) metabolism in p53 regulation and its consequences. Cancer Res 55: 3697–3701, 1995
7. Nutrition monitoring in the United States: A progress report from the joint nutrition monitoring evaluation committee. Vol DHHS Publication No. (PHS)86-1255, 1986
8. Fu CS, Swendseid ME, Jacob RA, McKee RW: Biochemical markers for assessment of niacin status in young men: Levels of erythrocyte niacin coenzymes and plasma tryptophan. J Nutr 119: 1949–1955, 1989
9. Jacobson EL, Dame A, Pyrek JS, Jacobson MK: Evaluating the role of niacin in human carcinogenesis. Biochimie 77: 394–398, 1995
10. Jacobson EL, Huang AC, Williams T, Gensler HL: Chemoprevention by niacin in a mouse model of UV-induced skin carcinogenesis. Proc Am Assoc Cancer Res 37: A279, 1996
11. Zhang JZ, Henning SM, Swendseid ME: Poly(ADP-ribose) polymerase activity and DNA strand breaks are affected in tissues of niacin deficient rats. J Nutr 123: 1349–1355, 1993
12. Rawling JM, Jackson TM, Driscoll E, Kirkland JB: Dietary niacin deficiency lowers tissue poly(ADP-ribose) and NAD^+ concentrations in Fischer-344 rats. J Nutr 124: 1597–1603, 1994
13. Shibata K, Murata K: Blood NAD as an index of niacin nutrition. Nutr Inter 2: 177–181, 1986
14. Jacobson EL, Jacobson MK: Tissue NAD as a biochemical measure of niacin status in humans. Meth Enzymol 280: 221–230, 1997
15. Jacobson EL, Jacobson MK: A biomarker for the assessment of niacin nutriture as a potential preventive factor in carcinogenesis. J Int Med 233: 59–62, 1993
16. Jacobson EL: Niacin deficiency and cancer in women. J Am Coll Nutr 12: 412–416, 1993
17. Kirkland JB, Rawling JM: Niacin. In: R. Rucker, J. Suttie, D. McCormick, L. Machlin (eds). Handbook of Vitamins, 3rd ed. Marcel Dekker, New York, 1998 (in press)
18. Donehower LA, Harvey M, Slagle BL, McArthur MJ, Montgomery CA Jr, Butel JS, Bradley A: Mice deficient for p53 are developmentally normal but susceptible to spontaneous tumors. Nature 356: 215–221, 1992
19. Bradford MM: A rapid and sensitive method for the quantitation of microgram, quantities of protein utilizing the principle of protein-dye binding. Anal Biochem 72: 248–254, 1976
20. Hammond SL, Ham RG, Stampfer MR: Serum-free growth of human mammary epithelial cells: Rapid clonal growth in defined medium and extended serial passage with pituitary extract. Proc Natl Acad Sci 81: 5435–5439, 1984
21. McKeehan WL, McKeehan KA, Hammond SL, Ham RG: Improved medium for clonal growth of human diploid fibroblasts at low concentrations of serum protein. *In Vitro* 13: 399–416, 1977

Molecular and Cellular Biochemistry **193:** 75–81, 1999.

Molecular heterogeneity and regulation of poly(ADP-ribose) glycohydrolase

Jean-Christophe Amé,[1] Elaine L. Jacobson[2, 3, 4] and Myron K. Jacobson[1, 3]

[1]*Division of Pharmaceutical Sciences, College of Pharmacy;* [2]*Department of Clinical Sciences;* [3]*Lucille P. Markey Cancer Center;* [4]*Multidisciplinary Ph.D. Program in Nutritional Sciences, University of Kentucky, Lexington, Kentucky, USA*

Abstract

We have recently described the isolation and characterization of bovine cDNA encoding poly(ADP-ribose) glycohydrolase (PARG). We describe here the preparation and characterization of antibodies to PARG. These antibodies have been used to demonstrate the presence of multiple forms of PARG in tissue and cell extracts from bovine, rat, mouse, and insects. Our results indicate that multiple forms of PARG previously reported could result from a single gene. Analysis of PARG in cells in which poly(ADP-ribose) polymerase (PARP) has been genetically inactivated indicates that the cellular content of PARG is regulated independently of PARP. (Mol Cell Biochem **193**: 75–81, 1999)

Key words: poly(ADP-ribose) glycohydrolase, antibody, expression, ADP-ribose polymer metabolism, DNA repair

Introduction

The synthesis and turnover of ADP-ribose polymers is an immediate cellular response to genomic damage (Fig. 1) [1]. This response results from the activation of poly(ADP-ribose) polymerase (PARP) by DNA strand breaks [2, 3]. Once synthesized, polymers are rapidly turned over by the action of poly(ADP-ribose) glycohydrolase (PARG) [4, 5]. An ADP-ribosyl protein lyase has been proposed to catalyze removal of protein-proximal ADP-ribose monomers [6].

Until recently, studies of the structure and function of the enzymes of ADP-ribose polymer metabolism have been mainly limited to PARP [7–11]. PARG has been isolated to apparent homogeneity from bovine thymus [12, 13], guinea pig liver [14], and human placenta [15]. PARG isolated by conventional methods has shown considerable molecular heterogeneity, although the nature of the heterogeneity has not been established. Recently, we reported the isolation and characterization of bovine cDNA encoding PARG [16]. Surprisingly, this cDNA codes for a protein of 111 kDa, nearly twice the size of the isolated protein from bovine thymus. A Northern analysis of poly(A)$^+$ RNA showed a single transcript of nearly 4.3 kb, consistent with the expression of a protein of 111 kDa. Southern hybridization experiments indicated that PARG is encoded by a single copy gene. Expression in *E. coli* confirmed that PARG can be expressed as an active enzyme in either a 111 or a 65 kDa form. However, the question remains whether or not the 65 kDa form of PARG (and possibly other isoforms reported) results from a purification artifact or whether it has a real biological significance.

Studies based on the sequence of bovine PARG raise many new questions concerning the function of PARG in the catabolism of ADP-ribose polymers *in vivo*. The availability of PARG cDNA allows the development new molecular tools to study this enzyme in its cellular context. In this paper, we report results using the first antibody directed against PARG to study features of PARG heterogeneity and possible modes of regulation. We also show that the antibody raised against bovine PARG is able to recognize PARG from other organisms. Thus, this antibody will be valuable in characterizing PARG *in vivo* under defined physiological conditions in many different organisms.

Address for offprints: M.K. Jacobson, College of Pharmacy, University of Kentucky, Lexington, KY 40536-0082, USA

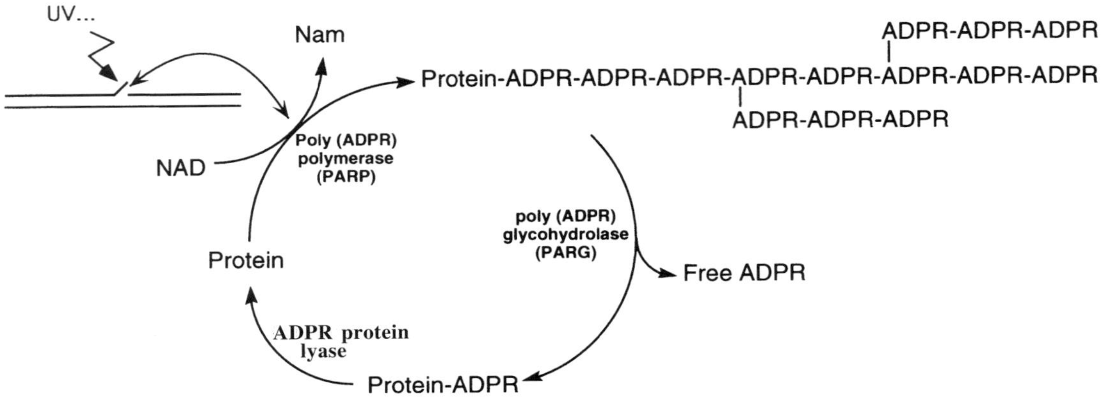

Fig 1. The reversible modification of nuclear proteins by ADP-ribose polymers.

Results

Although PARG was discovered 27 years ago [4], it has been difficult to characterize because of the low abundance of this protein in eukaryotic cells. Until recently, studies of PARG have been made only using conventionally purified enzyme. The availability of the full length cDNA [16] of PARG opened the possibility of new studies by expressing the protein in high levels in heterologous systems.

Cloning and overproduction of the carboxyl-terminal 69 kDa domain of bovine PARG (bPARG) in E. coli

In our previous report [16], we showed that bovine PARG is encoded by a messenger of 4 kb predicting a protein of 111 kDa, almost twice the size of the purified enzyme (65 kDa). We also demonstrated that bPARG can be expressed in *E. coli* as an active enzyme either as a 111 or a 65 kDa protein. This result combined with other evidence implied that the active site of PARG is located in the carboxyl-terminal part of the protein. Figure 2A is a schematic representation of the different clones we have expressed in bacteria. Among them, only the clone designed to express a protein of 69 kDa starting at the amino acid + 380 from the sequence of bovine PARG [16] (bPARG$_{MNDV}$) allowed high level expression as a fusion protein with glutathione-S transferase (GST).

The heterologous expression of bPARG$_{MNDV}$ was conducted as represented in Fig. 2B. The 1.8 kb cDNA encoding the 69 kDa carboxyl-terminal part of bovine PARG was amplified by PCR and cloned in the *Eco*RI site of pGEX-2T vector (Pharmacia) in fusion with GST giving the pGEX-2T-bPARG$_{MNDV}$ plasmid. *E. coli* NM522 cells transformed with the pGEX-2T-bPARG$_{MNDV}$ were induced by addition of IPTG, resulting in expression of a 90 kDa fusion protein. The fusion protein can be conveniently purified using Glutathione-Sepharose (GSH-Sepharose) and the bPARG$_{MNDV}$ can be released by treatment with thrombin while the GST protein remains bound to the beads of GSH-Sepharose. In this manner milligram amounts of protein can be routinely obtained.

Characterization of the purified 65 kDa domain

The purified bPARG$_{MNDV}$ was characterized by activity gel assays [17] by casting polyacrylamide gels with automodified PARP containing [^{32}P]ADP-ribose polymers. Results displayed in Fig. 3 demonstrate that the 65 kDa domain expressed in *E. coli* contained enzymatic activity (Fig. 3, lane 2) migrating with the same apparent molecular weight as the enzyme purified from bovine thymus (Fig. 3, lane 1). Likewise, a construction expressing bPARG$_{MNDV}$ domain in *sf*9 insect cells infected with recombinant baculovirus [18] showed activity (Fig. 3, lane 4) migrating with the same apparent molecular weight.

Generation of antibodies against the 65 kDa domain

Antibodies against bPARG$_{MNDV}$ overexpressed in *E. coli* were raised in rabbits using the procedure described by Vaitukaitis [19]. Specific high affinity antibodies are generated by administration of small doses of immunogens intradermally over a wide anatomic area of the animal. Rabbits were immunized by three injections of 10–50 μg of the Mr 65,000 protein band excised from a preparative SDS-polyacrylamide gel. Titer and affinity of sera harvested weekly were followed by conventional methods. Peak affinity was attained in 8–10 weeks after primary im-

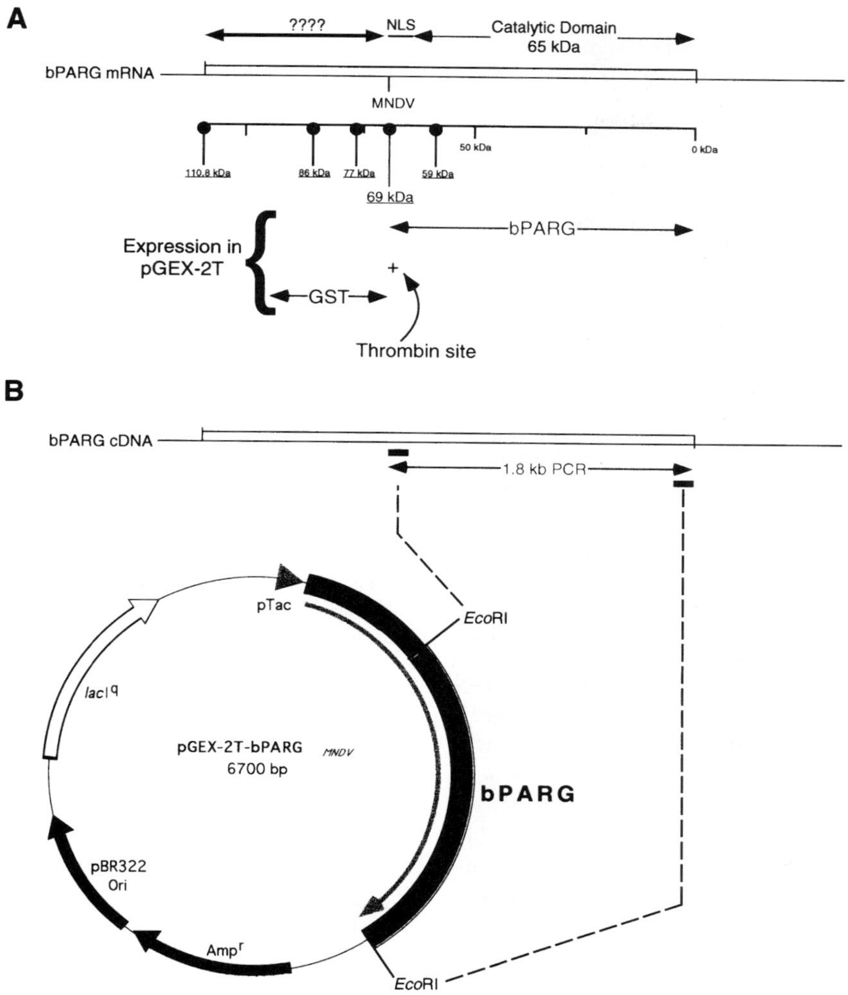

Fig. 2. Schematic representation of the different functional domains of PARG and the cloning procedure for cDNA encoding the carboxyl-terminal 65 kDa domain of the bovine PARG. Part A represents the structure of bovine PARG mRNA containing the open reading frame encoding the 111 kDa PARG protein. The different parts of PARG which have been cloned in expression vectors are represented with the size of the resulting expressed recombinant proteins. The expression of the 65 kDa catalytic domain of PARG (starting at the aa MNDV) in pGEX-2T as a fusion protein with glutathione-S transferase (29 kDa) is detailed. Part B represents the cloning into the *Eco*RI site of pGEX-2T of the 1.8 kb PCR *Eco*RI fragment encoding for the 65 kDa catalytic domain of PARG giving pGEX-2T-bPARG$_{MNDV}$. This construction resulted in the expression of a fused polypeptide consisting of the sequence of GST, amino acids derived from the polylinker and the thrombin site and the 65 kDa domain.

munization. For each animal, a preimmune serum was retained as a control.

Figure 4 shows a Western blot experiment demonstrating the specificity of the resulting PARG anti-serum against the purified bPARG from thymus (Fig. 4, lane 1), *sf*9 protein extract expressing 65 kDa-bPARG$_{MNDV}$ in recombinant baculovirus (Fig. 4, lane 2), recombinant 65 kDa-PARG$_{MNDV}$ purified by treatment with thrombin from GSH-Sepharose (Fig. 4, lane 3), and an *E. coli* crude extract expressing the fusion protein GST-65 kDa-PARG$_{MNDV}$ (Fig. 4, lane 4). The pre-immune serum did not show reactivity against any of these fractions even at a low dilution (1/250) (data not shown).

Conservation of PARG in tissues and organisms

Tissue and cell extracts from different origins were homogenized in a cold hypotonic lysis buffer containing a cocktail of protease inhibitors and sonicated. SDS and β-mercaptoethanol were added to insure inactivation of any remaining active proteases. Thirty μg of protein from each extract was analyzed by Western-blot using the anti-PARG antibody (Fig. 5). In all of the fractions from bovine tissues, PARG was observed as a major band at 65 kDa. However, less intense, discrete proteins of higher molecular weight were also detected. These proteins may correspond to different

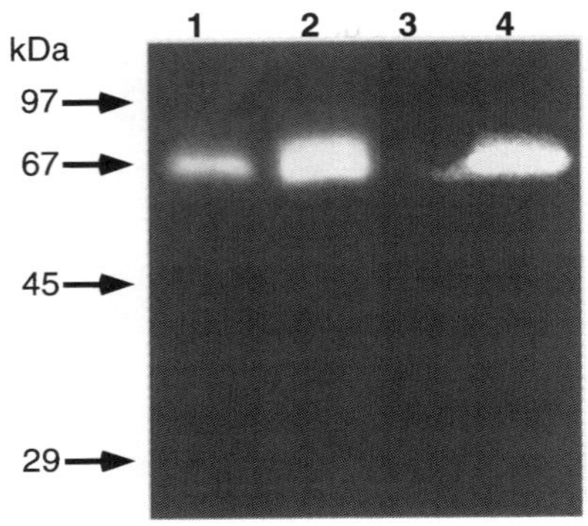

Fig. 3. Analysis of PARG activity of bovine recombinant PARG by zymogram. Activity gel assays for PARG were done by casting polyacrylamide gels with automodified PARP containing [^{32}P]ADP-ribose polymers as described previously [17]. Following electrophoresis in 0.1% SDS-12% polyacrylamide gel, PARG was renatured by incubating the gels at 25°C in 5 volumes of 50 mM sodium phosphate buffer, pH 7.5, 50 mM NaCl, 10% glycerol, 1% Triton X-100, 10 mM β-mercaptoethanol, changing the buffer every 3 h for a total of five changes. After an additional incubation at 37°C for 3 h, gels were incubated in 30% acetic acid, 40% methanol for 30 min and dried. PARG activity was detected following autoradiography as a clear band on a black background. Lane 1, 50 ng of purified bovine thymus PARG; lane 2, recombinant 65 kDa PARG; lane 3, 30 μg of total protein from *sf* 9 insect cell clear lysate; lane 4, recombinant 65 kDa PARG expressed in recombinant baculovirus expression system (30 μg of total protein from a clear lysate of PARG recombinant baculovirus infected *sf* 9 cells).

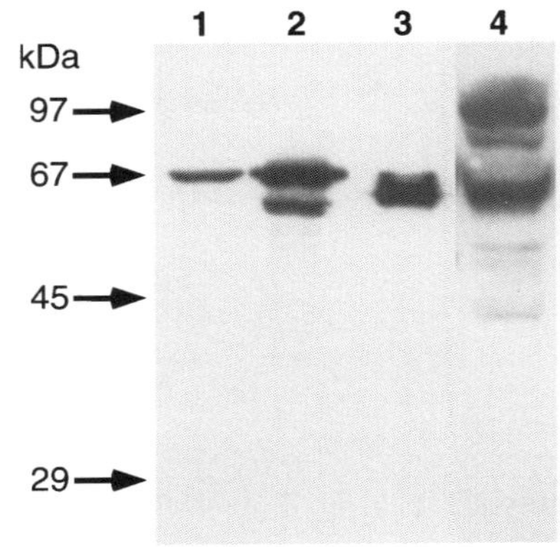

Fig. 4. Characterization of recombinant PARG by Western Blot. Purified PARG (50 ng) from bovine thymus (lane 1), 30 μg of protein of a total extract from PARG recombinant baculovirus infected *sf* 9 cells (lane 2), 150 ng of purified recombinant PARG produced in the bacteria (lane 3) and 30 μg of protein of a crude extract from *E. coli* NM522 transformed with pGEX-2T-bPARG$_{MNDV}$ 2 h after induction by IPTG (lane 4) were separated on a 0.1% SDS-12% polyacrylamide gel, then transferred on nitrocellulose [25], and incubated with a 1/5000 dilution of the rabbit polyclonal antiserum raised against the 65 kDa domain of bPARG. Proteins were revealed by immunofluorescence with the ECL detection kit (Amersham) and autoradiography.

forms of PARG; the band of highest molecular weight (≈ 115 kDa) found in thymus extract likely corresponds to the full length of PARG (111 kDa) as deduced from the cDNA [16]. Multiple species were detected in cell extracts from mouse fibroblasts, rat PC12 cells, and *sf*9 insect cells. This result shows that the sequence of PARG is well conserved phylogenetically. Moreover, the conservation includes multiple molecular forms of the protein.

Regulation of the expression of PARG

In the metabolism of ADP-ribose polymers, the activities of PARP and PARG are closely related. Soon after polymer has been synthesized by PARP following DNA damage, it is extensively degraded by PARG. The net result is that the polymer has a very short half life. The close relationship between the two proteins suggests a possible mode of regulation in which PARG expression depends on the presence of PARP. In order to test if the presence or the absence of PARP influences the expression of PARG, a

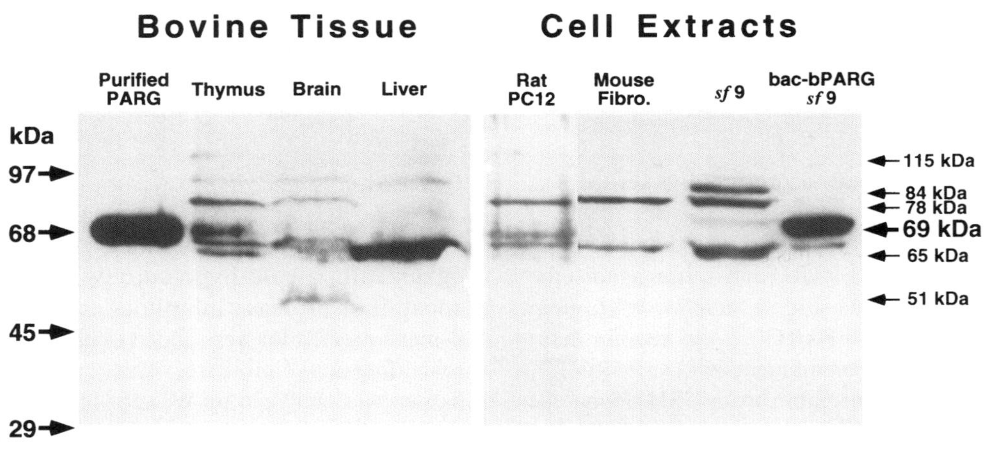

Fig. 5. Characterization of PARG from tissues and cells extracts by immunoblot. Different protein extracts (30 μg) from bovine tissues and cell extracts from rat, mouse and *sf* 9 insect cells were analyzed by Western blot in the same conditions as in Fig. 4. Purified PARG from bovine thymus and recombinant PARG expressed by baculovirus in *sf* 9 cells were loaded as controls.

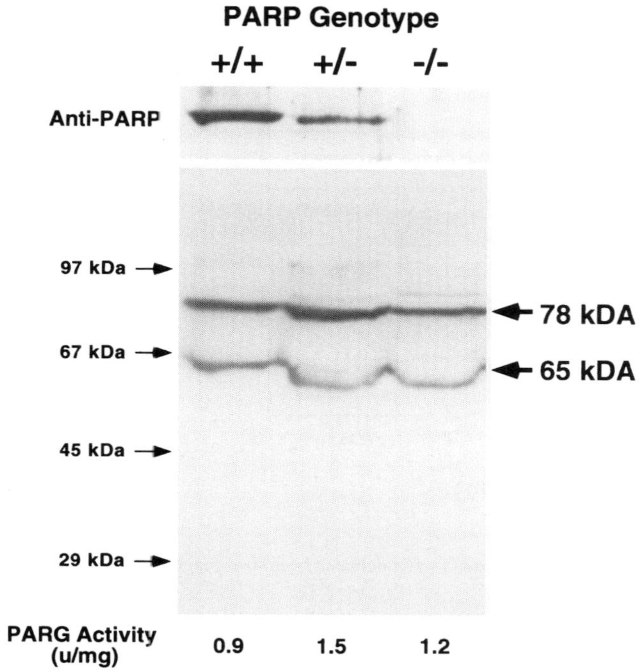

Fig. 6. Characterization of PARG in cell lines where the PARP gene has been disrupted. Cell extract (30 μg) from mouse cell lines with PARP +/+, PARP +/–, and PARP –/– genotypes were analyzed by Western blot in the conditions of Fig. 4. PARG activity was determined by thin layer chromatography [16] for the indicated cell lines. Expression of PARP in these cell lines was analyzed by Western blot with a polyclonal antibody raised against PARP.

Western Blot experiment was performed with cell extracts from mouse fibroblasts of different PARP genotypes [20] (Fig. 6), using the antibody raised against PARG. Figure 6 shows that there was no significant difference in the expression of PARG in the genotypes PARP +/+, PARP +/–, or PARP –/–. Furthermore, PARG activity determined for each of these cell extracts confirmed the result (Fig. 6).

Discussion

Until now, the analysis of PARG has been possible only after conventional purification of the enzyme. The availability of specific antibodies against PARG has become a long-awaited molecular tool for further studies of this enzyme, e.g. the distribution within the cell and in different biological samples. In our previous report [16], we described the isolation and characterization of the cDNA encoding bovine PARG allowing us to express the enzyme in heterologous systems such as bacteria or baculovirus. Previously, we have described the construction of a plasmid that expresses the catalytic domain of PARG [16]. This has allowed the preparation of specific antibodies against the catalytic domain of PARG.

An important question debated in the literature concerns the size of PARG *in vivo*. Numerous studies have suggested that different forms of PARG exist in cells, suggesting different roles and locations within the cell. Different sizes for PARG ranging from 50–74 kDa [12, 13, 16, 17, 21–24] have been reported. Some authors associated this heterogeneity with differences in the subcellular distribution of the enzyme and it has been suggested that the multiple forms of PARG result from the expression of multiple genes. The antibodies raised against PARG allowed us to determine the presence and size of the enzyme in different cell and tissue extracts. A major band was found in the bovine tissue extracts that corresponded to a molecular weight of 65 kDa, which is almost identical to the size of the purified enzyme. Other bands of higher MW (74, 84 and 115 kDa) were also detected but the abundance of these forms was much lower. A similar profile of bands was observed within extracts from rat, mouse, and insect cells, showing that PARG is a highly conserved protein. The high degree of conservation includes aa sequence (human, bovine, mouse, *Drosophila*, and *Caenorhabditis elegans*, to be published), the pattern of different size observed, and the low abundance of the enzyme in the cell. In our studies, we demonstrated with the cloning of the bovine PARG cDNA that only one copy of the gene of PARG exists in the bovine genome. Furthermore, our results from Western blotting indicate that all of the observed lower molecular weight forms, including the purified 65 kDa form, are derived from the 111 kDa form of the enzyme.

Previous studies have shown that different forms of PARG differ considerably in their isoelectric point (p*I*). Figure 7 shows the theoretical values of p*I* deduced from the amino acid sequence taken from different sizes of the protein ranging from 35–111 kDa. A significant variation in p*I* is observed in the size range of from 55–80 kDa, which corresponds to the range of sizes described in the literature [17, 24]. This analysis suggests that fragments of different size derived from a single precursor can account for the different isoforms of the enzyme described previously.

The large change in relative amounts of basic and acidic amino acid residues between different regions of the enzyme may also reflect a regulatory function for the amino terminal domain. The N-ter region (aa 1 to 380) contains many more acidic amino acids (27%) than basic amino acids (14%), resulting in p*I* values below 7.5 for isoforms of PARG ranging from 111 to 83 kDa. On the other hand, the 69 kDa fragment of PARG purified previously [16], which has been shown to contain the entire catalytic domain of the enzyme (located at the C-terminus of the protein) has a p*I* of 8.6. The high p*I* value implies a high binding affinity for the negatively charged ADP-ribose polymers. This suggests that when PARG is present as a full length molecule the N-ter acidic region of the protein may regulate PARG activity by interfering with the substrate recognition.

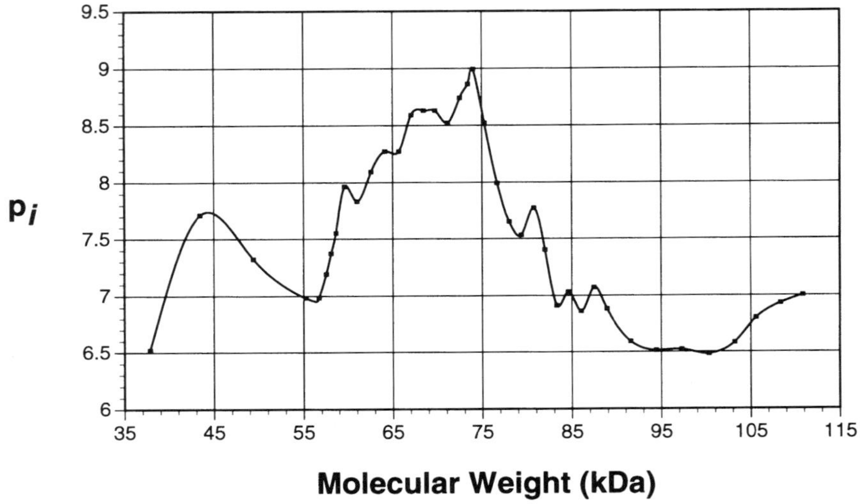

Fig. 7. Variation of the isoelectric point of PARG with the molecular weight of the protein. The p*I* of PARG was calculated from sizes ranging from 35–111 kDa with the program 'Peptidesort' from the GCG package of Bioninformatic Softwares.

The isolation of cDNAs for PARG will facilitate the study of many presently unanswered questions pertaining to the biological significance of PARG and the pathway in which it is involved. It will now be possible to develop molecular genetic approaches to evaluate the biological consequences of PARG depletion and overexpression in cultured cells. The availability of PARG cDNA will allow the isolation and study of the gene encoding PARG, the regulation of PARG gene expression, and the generation of animals with an inactivated PARG gene. The availability of specific antibodies to PARG have already established that the level of expression of PARG is not affected by the presence of a disrupted PARP gene (Fig. 6), but the factors involved in the regulation of PARG expression under different physiological conditions remain to be elucidated. The availability of antibodies also will allow the study of the possible significance of isoforms of PARG previously detected. It is possible that different isoforms of PARG reflect its location in subcellular compartments. The location of a nuclear location signal (NLS) at the junction between the two primary domains raises the possibility that isoforms of PARG less than 65 kDa will have lost the NLS and thus might reside in the cytoplasm, whereas the larger forms may be targeted to the nucleus. However, since there is currently no evidence for extranuclear roles for PARG, the possibility of cross contamination or redistribution of PARG during subcellular fractionation needs to be carefully examined. Even if PARG functions only in the nucleus, the different isoforms may allow differential regulation of PARG in different regions of chromatin by modulation of interactions with other nuclear components. Finally, the availability of amounts of PARG protein that will allow detailed structural studies will allow the evaluation of the potential of PARG as a therapeutic target.

Acknowledgements

We thank Dr. Z-Q Wang who has kindly provided the PARP +/+, +/– and PARP –/– mouse cell lines. We thank Donna Coyle for technical assistance, Melissa Shieh for growth of mouse fibroblast cells and David Koh for testing PARG activity from these cell lines. This work was supported in part by NIH Grant CA43894.

References

1. Thompson CB: Apoptosis in the pathogenesis and treatment of disease. Science 267: 1456–1462, 1995
2. Althaus FR, Richter C: ADP-ribosylation of proteins – enzymology and biological significance. In: M. Solioz (ed). ADP-Ribosylation of Proteins, vol. 37 of Molecular Biology, Biochemistry and Biophysics. Springer-Verlag, Berlin, 1987
3. Lautier D, Lagueux J, Thibodeau J, Ménard L, Poirier GG: Molecular and biochemical features of poly(ADP-ribose) metabolism. Mol Cell Biochem 122: 171–193, 1993
4. Miwa M, Sugimura T: Splitting of the ribose-ribose linkage of poly(adenosine diphosphate-ribose) by a calf thymus extract. J Biol Chem 146: 6362–6364, 1971
5. Ueda K, Oka J, Narumiya S, Miyakawa N, Hayaishi O: Poly(ADP-ribose) glycohydrolase from rat liver nuclei, a novel enzyme degrading the polymer. Biochem Biophys Res Commun 46: 516–523, 1972
6. Oka J, Ueda K, Hayaishi O, Komura H, Nakanishi K: ADP-ribosyl protein lyase. Purification, properties, and identification of the product. J Biol Chem 259: 986–995, 1984

7. Uchida K, Morita T, Sato T, Ogura T, Yamashita R, Nogushi S, Suzuki H, Nyunoya H, Miwa M, Sugimura T: Nucleotide sequence of a full length cDNA for human fibroblast poly(ADP-ribose) polymerase. Biochem Biophys Res Commun 148: 617–622, 1987
8. Cherney BW, McBride OW, Chen D, Alkhatib H, Bhatia K, Hensley P, Smulson ME: cDNA sequence, protein structure and chromosomal location of the human gene for poly(ADP-ribose) polymerase. Proc Natl Acad Sci USA 84: 8370–8374, 1987
9. Kurosaki T, Ushiro H, Mitsuuchi Y, Suzuki S, Matsuda M, Matsuda Y, Katunuma N, Kangawa K, Matsuo H, Hirose T, Inayama S, Shizuta Y: Primary structure of human poly(ADP-ribose) synthetase as deduced from cDNA sequence. J Biol Chem 262: 15990–15997, 1987
10. de Murcia G, Ménissier-de Murcia J, Schreiber V: Poly(ADP-ribose) polymerase: Molecular biological aspects. BioEssays 13: 455–462, 1991
11. de Murcia G, Ménissier-de Murcia J: Poly(ADP-ribose) polymerase: A molecular nick-sensor. Trends Biochem Sci 19: 172–176, 1994
12. Hatakeyama K, Nemoto Y, Ueda K, Hayaishi O: Purification and characterization of poly(ADP-ribose) glycohydrolase. Different modes of action on large and small poly(ADP-ribose). J Biol Chem 261: 14902–14911, 1986
13. Thomassin H, Jacobson MK, Guay J, Verreault A, Aboul-Ela N, Ménard L, Poirier GG: An affinity matrix for the purification of poly(ADP-ribose) glycohydrolase. Nucleic Acids Res 18: 4691–4694, 1990
14. Tanuma S, Kawashima K, Endo H: Purification and properties of an $(ADP\text{-}ribose)_n$ glycohydrolase from guinea pig liver nuclei. J Biol Chem 261: 965–969, 1986
15. Uchida K, Suzuki H, Maruta H, Abe H, Aoki K, Miwa M, Tanuma S: Preferential degradation of protein-bound $(ADP\text{-}ribose)_n$ by nuclear poly(ADP-ribose) glycohydrolase from human placenta. J Biol Chem 268: 3194–3200, 1993
16. Lin W, Amé JC, Aboul-Ela N, Jacobson EL, Jacobson MK: Isolation and characterization of the cDNA encoding bovine poly(ADP-ribose) glycohydrolase. J Biol Chem 272: 11895–11901, 1997
17. Brochu G, Shah GM, Poirier GG: Purification of poly(ADP-ribose) glycohydrolase and detection of its isoforms by a zymogram following one- or two-dimensional electrophoresis. Anal Biochem 218: 265–272, 1994
18. Amé JC, Jacobson MK: Unpublished data, 1997
19. Vaitukaitis JL: Production of antisera with small doses of immunogen: Multiple intradermal injections. Meth Enzymol 73: 46–52, 1981
20. Wang ZQ, Auer B, Stingl L, Berghammer H, Haidacher D, Schweiger M, Wagner EF: Mice lacking ADPRT and poly(ADP-ribosyl)ation develop normally but are susceptible to skin disease. Genes and Dev 9: 509–520, 1995
21. Miwa M, Tanaka M, Matsushima T, Sugimura T: Purification and properties of glycohydrolase from calf thymus splitting ribose-ribose linkages of poly(adenosine diphosphate ribose). J Biol Chem 249: 3475–3482, 1974
22. Tavassoli M, Tavassoli MH, Shall S: Isolation and purification of poly(ADP-ribose) glycohydrolase from pig thymus. Eur J Biochem 135: 449–455, 1983
23. Tanuma S, Endo H: Purification and characterization of an $(ADP\text{-}ribose)_n$ glycohydrolase from human erythrocytes. Eur J Biochem 191: 57–63, 1990
24. Abe H, Tanuma S: Properties of poly(ADP-ribose) glycohydrolase purified from pig testis nuclei. Arch Biochem Biophys 336: 139–146, 1996
25. Towbin H, Staehlin T, Gordon I: Electrophoretic transfer of proteins from polyacrylamide gels to nitrocellulose sheets: Procedure and some applications. Proc Natl Acad Sci USA 76: 4350–4354, 1979

Molecular and Cellular Biochemistry **193:** 83–87, 1999.

Niacin deficiency increases the sensitivity of rats to the short and long term effects of ethylnitrosourea treatment

Ann C. Boyonoski,[1] Lisa M. Gallacher,[1] Michele M. ApSimon,[1] Robert M. Jacobs,[2] Girish M. Shah,[2] Guy G. Poirier[3] and James B. Kirkland[1]

[1]*Department of Human Biology and Nutritional Sciences, University of Guelph, Guelph, Ontario;* [2]*Department of Pathology, Ontario Veterinary College, University of Guelph, Guelph, Ontario;* [3]*Laboratory of Skin Research, Hospital Research Center of University Laval, Ste. Foy, Quebec;* [4]*Unit of Health and Environment, Hospital Research Center of University Laval, Ste. Foy, Quebec, Canada*

Abstract

Most chemotherapy agents function by causing damage to the DNA of rapidly dividing cells, such as those in the bone marrow, leading to anemia and leukopenia during chemotherapy and the development of secondary leukemias in the years following recovery from the original disease. We created an animal model of nitrosourea-based chemotherapy using ethylnitrosourea (ENU) to investigate the effect of niacin deficiency on the side effects of chemotherapy. Weanling Long-Evans rats were fed diets containing various levels of niacin for a period of 4 weeks. ENU treatment started after 1 week of feeding and consisted of 12 doses delivered by gavage, every other day. Cancer incidence was also monitored in the following months. ENU treatment caused many of the acute symptoms seen in human chemotherapy patients, including anemia and neutropenia. Niacin deficiency (ND) had several interesting effects, alone and in combination with ENU. Niacin deficiency alone caused a modest anemia, while in combination with ENU it induced a severe anemia. Niacin deficiency alone caused a 4-fold increase in circulating neutrophil numbers, and this population was drastically reduced by ENU-treatment. In the long term, macin deficiency caused an increased incidence of cancer, especially chronic granulocytic leukemias. (Mol Cell Biochem **193**: 83–87, 1999)

Key words: nitrosourea, chemotherapy, anemia, leukopenia

Introduction

Many cancer chemotherapy agents target the rapidly dividing cells of tumours by causing damage to the DNA, which is poorly repaired during rapid cell division. While this approach kills tumour cells, it also damages other rapidly dividing cells, including the mucosal cells of the gastrointestinal tract, the cells of the hair follicle and the hematopoetic cells within the bone marrow. During replication, DNA is more prone to irreversible DNA damage, hence these cells are at increased risk. This results in hair loss, nausea and diarrhea, anemia and leukopenia, which are frequent acute side-effects of chemotherapy treatment [1]. In fact, bone marrow toxicity is usually the limiting factor in the use of chemotherapy drugs [1], and therefore a serious limitation in the ability to kill enough of the tumor cells to attain a lasting cure. In the longer-term, aggressive chemotherapy treatment places the surviving cancer patient at an increased risk for developing a treatment related (or secondary) malignancy as a result of the DNA damage induced by the chemotherapy treatment. The most frequent type of cancer is leukemia, predominantly non-lymphoblastic in nature [1–3].

Cancer patients are frequently malnourished. Initially this may be due to the response of the body to the disease process, which in its severe form may progress to cancer cachexia. During treatment of the disease, chemotherapy

Address for offprints: J. Kirkland, Department of Human Biology and Nutritional Sciences, University of Guelph, Guelph, Ontario Canada N1G 2W1

exacerbates the problem by causing a loss of appetite, nausea and vomiting. Niacin is one of the nutrients which appears to be deficient in many chemotherapy patients [4] and chemotherapy has been known to induce frank pellagra during treatment [5]. The active form of niacin in the cell is nicotinamide adenine dinucleotide (NAD), which functions in a variety of pathways, including redox reactions, cyclic and mono ADP-ribose reactions, and in the synthesis of poly(ADP-ribose) on nuclear proteins in response to DNA damage [6]. Niacin deficiency (ND) has been shown to dramatically inhibit DNA repair in cell culture models [7, 8], but there is limited knowledge of its effects in the whole animal during exposure to agents which cause damage to DNA. The poorly nourished chemotherapy patient, exposed to maximally tolerable doses of genotoxic chemicals, may require nutritional support to minimize the side effects of treatment.

Our research has focused on developing an *in vivo* model to investigate the effect of niacin deficiency on the side effects of chemotherapy, including the acute leukopenia and anemia which accompanies chemotherapy in essentially all patients, and the secondary cancers which occur in about 5% of surviving patients. We have developed a protocol in young Long-Evans rats based on oral treatment with ENU which produces acute anemia and leukopenia and chronic development of various neoplasms. Niacin deficiency causes a significant increase in the sensitivity of rats to the acute and chronic side effects of ENU.

Materials and methods

Chemicals

Ethylnitrosourea and Modified Wright Giemsa Stain were purchased from Sigma Chemical Co., St. Louis. MO, USA. Halothane, B.P. anesthetic was obtained from MTC Pharmaceuticals, Cambridge, ON, USA. Isotone II and Zap-oglobin II were purchased from Coulter Electronics of Canada, Ltd, Burlington, ON, USA.

Animals

All animal experimentation was approved by the University of Guelph Animal Care Committee, and animal treatment was in accordance with the guidelines of the Canadian Council on Animal Care. Weanling male Long Evans rats 40–50 g (Charles River Canada, St. Constant, PQ, Canada) were housed in suspended wire bottom cages and given free access to water with a 12 h light/dark cycle. Feed intake was determined daily. Rats were placed on a niacin deficient (ND) purified diet or pair-fed (PF) identical quantities of the same diet supplemented with exogenous niacin at 30 mg/kg diet. The diet is based on a mixture of casein and gelatin to lower tryptophan content and has been described previously [9].

Chemotherapy protocol

Using ENU, which is a simple form of the more complex nitrosoureas used in chemotherapy, we simulate a round of chemotherapy in non-tumour bearing rats. Rats were gavaged with twelve doses of either 30 mg/kg ENU in water (pH 4.0), or an equivalent volume of water (pH 4.0) alone (C). Beginning at 28 days of age each rat was gavaged every second day until 50 days of age. All rats were then placed on the high quality AIN-93(Maintenance) diet [10]. This model approximates the situation of a cancer patient going through a round of chemotherapy followed by a recovery period with a return to a higher quality diet.

Blood collection

At 27, 35, 43 and 51 days of age, rats were anesthetized with 3% Halothane under 1.5 L/min O_2 and 0.5 L/min NO and approximately 100 μl of blood were taken from the orbital sinus. Peripheral white cell numbers were determined with a Coulter ZM counter using standard laboratory procedures. Differential counts of 200 cells were made from peripheral blood smears stained with Wright Giemsa stain.

Assessing cancer incidence

At the end of the ENU treatment and niacin deficiency period (8 weeks of age) the rats were placed on a high quality control diet (AIN-93M) and were monitored for the development of cancer. All rats were slightly feed restricted to ensure that they consumed equal amounts of food per unit of body weight. Rat weights were taken weekly and feed intake was monitored daily and adjusted weekly. When rats lost 5% of their maximal body weight they were euthanized and a complete autopsy performed to establish the nature of the cancer(s) involved. While this is not identical to a survival curve, rats are very hardy, and a 5% loss in body weight was usually associated with an advanced state of disease. Mortality is not considered a valid end point in our current animal care regulations.

Results

In the early stages of the study, hematocrit, or packed cell volume (% pcv) was quite variable, as we often see in weanling rats (Fig. 1). There was a developmental increase in hematocrit in the PF-C rats over the first 2 weeks, and a decrease in the variability in all groups. ENU treatment alone caused a small decrease in hematocrit. Niacin deficiency alone (ND-C) resulted in a greater anemia than that caused by ENU treatment. However, ND in combination with ENU (ND-ENU) caused a notable anemia by the end of the treatment period, suggesting a synergistic relationship between these two treatments.

There was a large developmental increase in circulating neutrophils between 4 and 5 weeks of age in the PF-C group (Fig. 2). Unexpectedly, niacin deficiency alone (ND-C) caused a further 3-fold increase in circulating neutrophil numbers relative to the PF-C group. ENU treatment of niacin-replete rats (PF-ENU) caused the expected neutropenia, as seen with most types of chemotherapy. Although the absolute numbers of neutrophils in the ND-ENU remained higher than

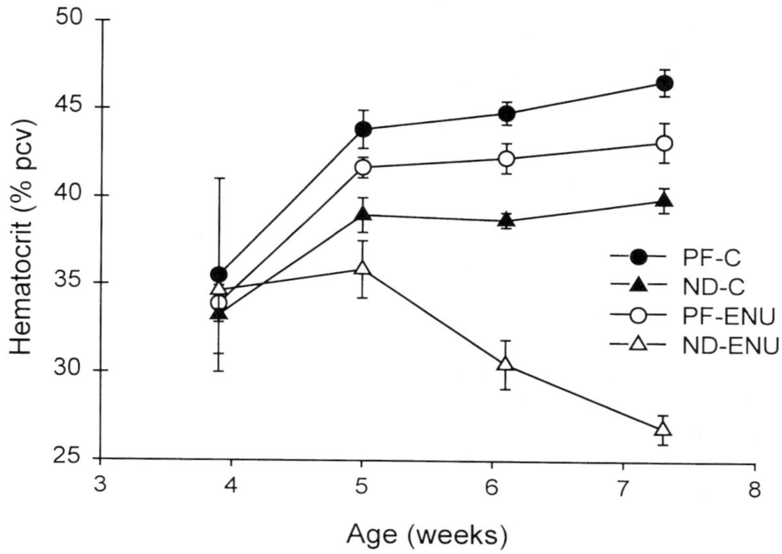

Fig. 1. The effect of niacin deficiency and ethylnitrosourea (ENU) treatment on hematocrit in Long-Evans rats. N = 4–5, bars represent S.E.M. Niacin deficiency began at 3 weeks of age and ENU treatment occurred every other day from 4–7 weeks of age.

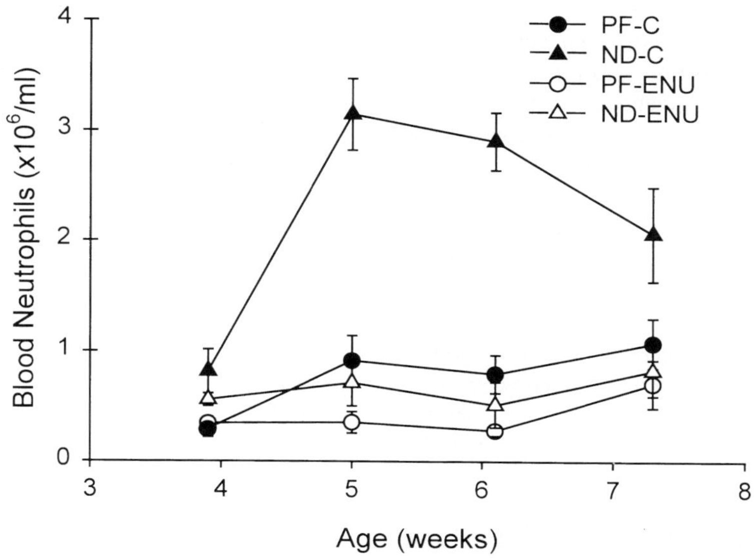

Fig. 2. The effect of niacin deficiency and ethylnitrosourea (ENU) treatment on circulating neutrophils in Long-Evans rats. N = 4–5, bars represent S.E.M. Niacin deficiency began at 3 weeks of age and ENU treatment occurred every other day from 4–7 weeks of age.

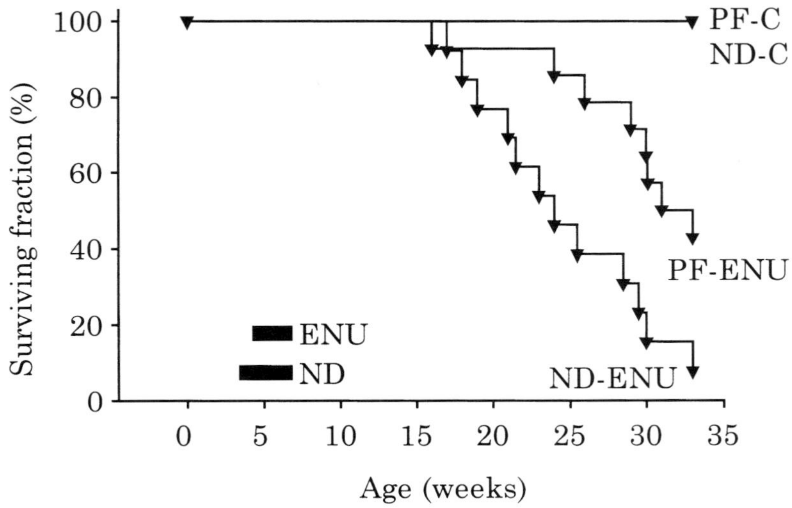

Fig. 3. The effect of niacin deficiency and ethylnitrosourea (ENU) treatment on cancer-induced morbidity in Long-Evans rats. N = 14 for Control rats; N = 13 for ENU-treated rats.

the PF-ENU group, the degree of ENU-induced neutropenia was greater in the ND rats when viewed as a percentage, due to the dramatic increase in neutrophil numbers in ND-C group.

Although the cancer incidence data are incomplete, it is apparent that niacin deficiency decreases the latency period (Fig. 3). The model is quite severe; a large proportion of the population is affected by cancer at a relatively young age. It is probable that at a lower exposure to ENU a portion of the population would have remained disease free, and that the cancer incidence may have been greater over the long term in the niacin deficient group. As in human populations exposed to chemotherapy, we have induced primarily non-lymphoblastic leukemias in the niacin deficient animals. The major type of leukemia is a chronic granulocytic form, occurring in 58% of the niacin deficient rats with cancer. In the PF control rats, only 25% of the animals with cancer had granulocytic leukemias. This is interesting from the perspective of alterations in neutrophillic differentiation induced by niacin deficiency (Fig. 2). Niacin deficiency increased circulating neutrophils 3-fold, and this population was drastically reduced by ENU treatment. If neutrophillic differentiation is more active at the level of bone marrow precursor populations, this lineage may be more susceptible to the side effects of ENU during niacin deficiency.

Discussion

It is well documented that alkylating agents like the nitrosoureas suppress hematopoesis, and that this suppression is often the dose-limiting factor in utilizing these drugs for chemotherapy [1]. Although all lineages are included in this suppression, the response of the marrow differs with drug type, intensity of dose, and duration of treatment. Typically, the depression seen with nitrosourea treatment is delayed and often lasts long after the drugs have been withdrawn [1]. Neutropenia may lead to bacterial and fungal infections, that are a major cause of morbidity and mortality during chemotherapy [11]. However, because alkylating agents have tumour-killing success, modern therapy regimens now incorporate bone marrow transplantation, stem cell reconstitution, and/or growth factor rescue in order to accommodate the bone marrow toxicity of these drugs [1, 12]. Establishing further ways to protect the marrow, not only from short-term toxicity but also from the longer-term effects of alkylating agents, would appear to be valuable.

It is known that niacin deficiency causes anemia in pellagrins [13, 14]. The anemia observed in ND rats indicates that this model mimics the human state. ENU-treatment of ND rats results in a more severe anemia. This may be due to a decrease in production of young red cells by the bone marrow; measurement of circulating red blood cell and reticulocyte numbers and bone marrow precursors will help to determine if this is the case. The group of de Murcia and coworkers have recently reported that disruption of the PARP gene in mice causes an increase in the acute toxicity of methylnitrosourea (MNU) as well as an increase in the incidence of sister chromatid exchanges and chromatid breaks [15]. A single i.p. injection of MNU (75 mg/kg bw) caused 100% mortality in PARP-null mice, and 43% mortality in wild type controls. In addition, the de Murcia PARP-null mouse, and a second PARP-null mouse developed by Wang and coworkers, both show extreme sensitivity to acute radiation injury ([15, 16]; note added in proof). Our data show that niacin deficiency increases the acute toxicity of ENU in bone marrow, and suggest that niacin deficiency may limit the function of PARP (through decreased substrate availability), leading to effects similar to those seen in the PARP-null mouse.

The changes in neutrophil populations, in the absence of chemotherapy and without any apparent infectious disease in the rats, suggest that niacin status influences the differentiation pathways for neutrophils in the whole animal. Recent studies have demonstrated a requirement for decreased expression of poly(ADP-ribose) polymerase during neutrophil differentiation using the NB4 acute promyelocytic leukemia model of differentiation [17, 18]. The effect of niacin deficiency on neutrophil numbers *in vivo* suggests that a decrease in poly(ADP-ribose) synthesis induced by decreased substrate may also encourage neutrophilic differentiation. The large increase in neutrophil numbers in niacin deficiency and their decimation by ENU treatment also suggests that this population may be susceptible to ENU-induced leukemogenesis. Signals causing rapid division of neutrophilic precursors during exposure to ENU could limit effective DNA repair and lead to the accumulation of mutations in these cells, giving rise to granulocytic leukemias. This is observed in the long term, as niacin deficiency during exposure to ENU causes earlier cancer incidence, with a large number of chronic granulocytic leukemias. Taken together, the data show that niacin deficiency increases the sensitivity of bone marrow precursors to ENU, leading to both acute and chronic problems.

Acknowledgments

This work was supported by a Strategic Grant from the Cancer Research Society of Montreal, Canada. The authors appreciate critical input from Dr. Scott Kaufmann, of the Division of Oncology, Mayo Clinic, and the support that Dr. Kirkland received during his sabbatical stay in the laboratory of Dr. Kaufmann.

References

1. Tew KD, Colvin M, Chabner BA: Alkylating Agents. In: B.A. Chabner, D.L. Longo, (eds). Cancer Chemotherapy and Biotherapy, 2nd Edition. Lippencott-Raven Publishers, Philadelphia, 1996, pp 297–332
2. Boffetta P, Kaldor JM: Secondary malignancies following cancer chemotherapy. Acta Oncologica 33: 591–598, 1994
3. Tucker MA: Secondary Cancers. In: V.T. DeVita, S. Hellman, S.A. Rosenberg (eds). Cancer: Principles & Practice of Oncology, 4th Edition. J.B. Lippincott Co, Philadelphia, 1993, pp. 2407–2416
4. Inculet RI, Norton JA, Nichoalds GE, Maher MM, White DE, Brennan MF: Water soluble vitamins in cancer patients on parenteral nutrition: A prospective study. J Parent Ent Nutr 11: 243–249, 1987
5. Stevens HP, Ostlere LS, Begent RHJ, Dooley JS, Rustin MHA: Pellagra secondary to 5-fluorouracil. Br J Dermatol 128: 578–580, 1993
6. Lautier D, Lagueux J, Thibodeau J, Menard L, Poirier GG: Molecular and biochemical features of poly(ADP-ribose) metabolism. Mol Cell Biochem 122: 171–193, 1993
7. Durkacz BW, Omidijii O, Gray DA, Shall S: (ADP-ribose)n participates in DNA excision repair. Nature 283: 593–596, 1980
8. Jacobson EL, Nunbliakdi-Craig V, Smith DG, Chen HY, Wasson BL, Jacobson MK: ADP-ribose polymer metabolism: Implications for human nutrition. In: G.G. Poirier, P. Moreau (eds). ADP-ribosylation reactions. Springer-Verlag, New York, 1992, pp 153–162
9. Rawling JM, Jackson TM, Driscoll E, Kirkland JB: Dietary niacin deficiency lowers tissue poly(ADP-ribose) and NAD^+ concentrations in Fischer-344 rats. J Nutr 124: 1597–1603, 1994
10. Reeves PG, Nielson FH, Fahey GC: AIN-93 Purified diets for laboratory rodents: Final report of the American Institute of Nutrition Ad Hoc Writing Committee on the reformulation of the AIN-76A rodent diet. J Nutr 123: 1939–1951, 1993
11. Gabrilove JL, Golde DW: Hematopoetic Growth Factors. In: V.T. DeVita, S. Hellman and S.A. Rosenberg (eds). Cancer: Principles & Practice of Oncology, 4th Edition, J.B. Lippincott Co, Philadelphia, 1993, pp 2275–2291
12. Peters WP, Rosner G, Ross M, Vredenburgh J, Melsenburg B, Gilbert C, Kurtzberg J: Comparative effects of granulocyte-macrophage colony stimulating factor (GM-CSF) and granulocyte colony-stimulating factor (GCSF) on priming peripheral blood progenitor cells for use with autologous bone marrow after high dose chemotherapy. Blood 81: 1709–1719, 1993
13. Spivak JL, Jackson, DL: Pellagra: An analysis of 18 patients and a review of the literature. In: K.J. Carpenter (ed). Pellagra. Hutchinson Ross Publishing Co, Stroudsburg, 1981, pp 345–359
14. Gillman J, Gillman T: Perspectives in Malnutrition, Grune and Stratton, New York, 1951, pp 326–327
15. De Murcia JM, Niedergang C, Trucco C, Ricoul M, Dutrillaux B, Mark M, Oliver J, Masson M, Dierich A, LeMeur M, Walztinger C, De Murcia G: Requirement of poly(ADP-ribose) polymerase in recovery from DNA damage in mice and in cells. Proc Natl Acad Sci USA 94: 7303–7307, 1997
16. Wang ZQ, Stingl L, Morrison C, Jantsch M, Los M, Schultz-Osthoff K, Wagner EF: PARP is important for genomic stability but dispensable in apoptosis. Genes Dev 11: 2347–2358, 1997
17. Bhatia M, Kirkland JB, Meckling-Gill KA: Modulation of poly(ADP-ribose) polymerase during neutrophilic and monocytic differentiation of promyelocytic (NB4) and myelocytic (HL-60) leukaemia cells. Biochem J 308: 131–137, 1995
18. Bhatia M, Kirkland JB, Meckling-Gill KA: Over-expression of poly(ADP-ribose) polymerase inhibits growth and blocks neutrophillic differentiation of NB4 acute promyelocytic leukaemia cells. Cell Growth Diff 7: 91–100, 1996

Molecular and Cellular Biochemistry **193**: 89–98, 1999.

Structures and activities of cyclic ADP-ribose, NAADP and their metabolic enzymes

Hon Cheung Lee, Cyrus Munshi and Richard Graeff
Department of Physiology, University of Minnesota, Minneapolis, USA

Abstract

ADP-ribosyl cyclase and CD38 are multi-functional enzymes involved in calcium signaling. Both can cyclize NAD and its guanine analog, NGD, at two different sites of the purine ring, N1 and N7, respectively, to produce cyclic ADP-ribose (cADPR) and cyclic GDP-ribose, a fluorescent but inactive analog. Both enzymes can also catalyze the exchange of the nicotinamide group of NADP with nicotinic acid, producing yet another potent activator of Ca^{2+} mobilization, nicotinic acid adenine dinucleotide phosphate (NAADP). The Ca^{2+} release mechanism activated by NAADP is totally independent of cADPR and inositol trisphosphate indicating it is a novel and hitherto unknown Ca^{2+} signaling pathway. This article summarizes the current results on the structures and activities of cADPR, NAADP and the enzymes that catalyze their syntheses. A comprehensive model accounting for the novel multi-functionality of ADP-ribosyl cyclase and CD38 is presented. (Mol Cell Biochem **193**: 89–98, 1999)

Key words: cyclic ADP-ribose (cADPR), nicotinic acid adenine dinucleotide phosphate (NAADP), Ca^{2+}

Introduction

Ca^{2+} mobilization is a general mechanism cells employ to effect changes in their functions in response to signals and stimuli from diverse sources. Two major mechanisms for mobilizing Ca^{2+} have been described. The best known is mediated by inositol trisphosphate (IP_3). Binding of external ligands to surface receptors can activate phospholipase C resulting in cleavage of the head group of phosphatidylinositol bisphosphate and the production of IP_3, which, upon binding to a specific receptor on the endoplasmic reticulum, activates Ca^{2+} release. Ca^{2+}-induced Ca^{2+} release is another major mechanism of Ca^{2+} mobilization which is best described for cardiac myocytes whereby influx of Ca^{2+} can itself activate further Ca^{2+} release from intracellular stores. This mechanism is believed to be mediated by ryanodine receptors.

Two metabolites of pyridine dinucleotides have been shown to be effective activators of release of intracellular Ca^{2+} stores [1]. Cyclic ADP-ribose (cADPR) is a newly discovered cyclic nucleotide derived from NAD [2, 3] and accumulating evidence suggests it is an endogenous modulator of the Ca^{2+}-induced Ca^{2+} release mechanism in cells ([4, 5] and reviewed in [6]). The cADPR-sensitive Ca^{2+} release mechanism specifically requires calmodulin [7–9] and Ca^{2+} acts as a co-agonist [5, 7]. In addition to cADPR, a metabolite of NADP, nicotinic acid adenine dinucleotide phosphate (NAADP) is also a potent mobilizer of intracellular Ca^{2+} stores [1, 10]. The Ca^{2+} release mechanism activated by NAADP is pharmacologically different from those activated by either cADPR or IP_3 [1, 10, 11] and the Ca^{2+} stores NAADP acts on are separable by gradient fractionation from those sensitive to cADPR and IP_3 [1, 10]. These stores stimulated by NAADP are novel in that they are insensitive to thapsigargin [12], a specific inhibitor of the Ca^{2+} pump present in endoplasmic reticulum. The NAADP-sensitive Ca^{2+} release is thus totally independent from that activated by cADPR and IP_3, indicating it is a hitherto unknown Ca^{2+} signaling mechanism.

Although cADPR and NAADP are structurally and functionally distinct, these two Ca^{2+} release activators, nevertheless, can be synthesized by the same enzymes. Both ADP-ribosyl cyclase and CD38, a lymphocyte antigen, not only can cyclize NAD to produce cADPR [13, 14] they can also catalyze the exchange of the nicotinamide group of NADP with nicotinic acid to produce NAADP as well [15].

Address for offprints: H.C. Lee, Department of Physiology, 6-255 Millard Hall, 435 Delaware Street SE, University of Minnesota, Minneapolis, MN 55455, USA

That a single enzyme can produce two signaling molecules is reminiscent of phospholipase C catalyzing the synthesis of both IP_3 and diacylglycerol. This article reviews our current understanding of the catalytic mechanisms of ADP-ribosyl cyclase and CD38 in synthesizing two totally different Ca^{2+} signaling molecules.

Structure and activity of cADPR and NAADP

The chemical structures of cADPR and NAADP are shown in Fig. 1. The cyclic structure of cADPR is formed as a result of linking the adenine ring of NAD to the terminal ribose and displacing the nicotinamide group [2]. X-ray crystallography results confirm the cyclic structure and show that the site of cyclization is at N1 of the adenine ring [3]. Both ribosyl linkages to the adenine are in the β-configuration and C6 is double bonded to N6. The cyclic linkage can be hydrolyzed chemically by heat [16] or enzymatically by CD38 [14, 17] to produce ADP-ribose.

NAADP is not a cyclic molecule but a very simple derivative of NADP in which the nicotinamide group is replaced by nicotinic acid (NA) [10]. It can be formed from NADP chemically by an alkaline treatment [10] or enzymatically via a base exchange reaction with NA [15]. NAADP can be readily degraded by either alkaline phosphatase, which cleaves the 2′-phosphate, or by nucleotide pyrophosphatase, which cleaves the pyrophosphate linkage [18]. Therefore, natural degradation and synthesis pathways exist for cADPR and NAADP, necessary prerequisites for signaling molecules.

That both cADPR and NAADP are effective in mobilizing Ca^{2+} stores in live cells is most convincingly demonstrated by using inactive caged analogs which can be activated by UV-light [18–20]. Covalently attaching a caging group (Fig. 1) to either of the two phosphates in cADPR or to the 2′-phosphate of NAADP render them biologically inactive. Brief UV-exposure of sea urchin eggs loaded with caged cADPR by microinjection releases the caging group and regenerates cADPR, activating Ca^{2+} mobilization [19, 21]. Photolyzing caged NAADP in sea urchin eggs produces not only transient Ca^{2+} release as seen with caged cADPR but a sustained Ca^{2+} oscillation lasting for more than 30 min [18, 20]. The Ca^{2+} oscillation has been proposed to be the result of interactions between separate Ca^{2+} stores respectively sensitive to either NAADP or cADPR [18, 22]. These results show that the Ca^{2+} signaling mechanisms activated by cADPR and NAADP are present and fully operational in live cells.

In addition to caged cADPR and NAADP, various other analogs of both metabolites have been synthesized. The following summarizes the structure-function relationships of cADPR, NAADP and their analogs.

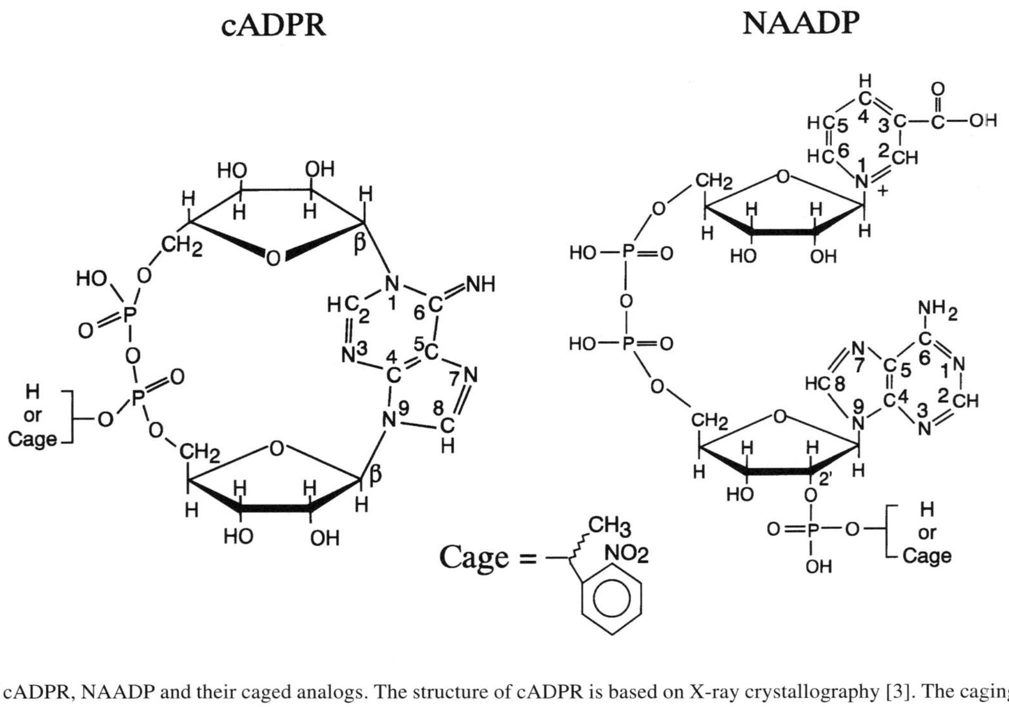

Fig. 1. Structure of cADPR, NAADP and their caged analogs. The structure of cADPR is based on X-ray crystallography [3]. The caging group is attached to either of the phosphates [19]. The structure of NAADP was based on measurements published previously [10] and the caging group is attached to the 2'-phosphate [18].

Structure-activity relationships of cADPR analogs

As described above, the phosphate groups in cADPR are crucial for its Ca^{2+} release activity as attachment of a caging group inactivates it. Caged cADPR does not, however, inhibit cADPR from releasing Ca^{2+} [19], suggesting the modification renders the analog incapable of interacting with the cADPR-receptor. Modifications at the 8-position of the adenine ring of cADPR have quite different effects. Although the position is also critical for the Ca^{2+} release activity and changing it totally eliminates the activity, the modified analogs can still bind to the cADPR-receptor and competitively inhibit cADPR from doing so. In this manner, the 8-derivatives of cADPR behave as effective antagonists [23]. Among these derivatives, 8-amino-cADPR is the most potent antagonist and can block the Ca^{2+} release activity of cADPR in the nanomolar range [23, 24]. These specific antagonists were first demonstrated to be effective in intact sea urchin eggs and egg microsomes [23, 24] and have since been shown to be also effective in a variety of mammalian systems ([25–28] and reviewed in [22]). Another 8-derivative, 8-Br-cADPR, is also an antagonist of cADPR although with less potency [23]. It does, however, possess a novel property that it is cell-permeant [29]. External application of 8-Br-cADPR in the micromolar range inhibits the Ca^{2+} mobilization associated with fertilization in sea urchin eggs preloaded with heparin to block the IP_3-receptor [29]. It appears that the 8-Br-group imparts enough lipophilicity to the molecule to facilitate its movement across the cell membrane. This property of the 8-Br-group is well documented in the cases of 8-Br-derivatives of cAMP and cGMP.

The approach used for synthesizing the series of 8-derivatives of cADPR is to take advantage of the leniency of ADP-ribosyl cyclase toward its substrate. NAD is first prepared from AMP which is derivatized at the 8-position of the adenine ring and then the modified NAD is converted to the corresponding derivative of cADPR using the cyclase [23]. The variety of widely available analogs of NAD and the well established chemistry of synthesizing them make this a particularly versatile approach for producing various analogs of cADPR (reviewed in [22]).

Several of the analogs synthesized by this approach have particularly novel structural and functional properties. Cyclic aristeromycin diphosphate ribose, which has the oxygen in the ribosyl ring of the adeninyl ribose replaced by carbon, is highly resistant to hydrolysis by heat as well as by hydrolytic enzymes of cADPR. The analog retains the Ca^{2+} release activity with a potency comparable to cADPR [30]. This analog should be useful in prolonging the physiological action of cADPR, especially in cells that have a high metabolism for cADPR. Even higher stability can be conferred to cADPR by converting the 7-nitrogen of the adenine ring to carbon as in 7-deaza-cADPR [31]. Perhaps, the most novel analog synthesized so far is 7-deaza-8-Br-cADPR. By converting both the 7-nitrogen to carbon and attaching a bromine to the 8-position of the adenine ring, a derivative is made that is cell-permeant, antagonistic and resistant to hydrolysis [29]. The membrane permeability of the analog appears to be higher than 8-Br-cADPR, which can be due to the additional lipophilicity contributed by the 7-CH-group. As a result, lower external concentrations are needed to produce even greater inhibition of Ca^{2+} mobilization in live sea urchin eggs than 8-Br-cADPR [29]. The availability of this novel derivative of cADPR should greatly facilitate future studies aimed at elucidating the physiological role of cADPR in various cell functions.

Structure-activity relationships of NAADP analogs

A series of analogs of NAADP has been synthesized to assess the structural determinants important for its Ca^{2+} release activity [32]. The synthesis makes use of the base-exchange reaction catalyzed by ADP-ribosyl cyclase [15]. The reaction exchanges the nicotinamide group of NADP with NA or one of its analogs. The large number of pyridine derivatives and NADP analogs available makes this method a versatile and general approach for synthesizing NAADP analogs.

Three structural determinants in NAADP have been shown to be important to its Ca^{2+} release activity [32]. The first one is the 3-carboxyl group of the NA moiety. A series of analogs of pyridine with various substitutions at the 3-position have been used as substrates for the base-exchange reaction with NADP to synthesize the corresponding analogs of NAADP shown in Fig. 2 [32]. Increasing the bulkiness of the group at the 3-position from a carboxyl group to either a sulfonic acid group or an acetic acid group reduces the effectiveness of Ca^{2+} release activity by 2–3 orders of magnitude. More important than the size of the group is its negative charge, since changing the carboxyl group to a neutral methanol group totally eliminates its ability to release Ca^{2+}, even though the methanol group is smaller. NADP, which contains a neutral amide group at the 3-position, also has no Ca^{2+} release activity [10]. Finally, the position of the carboxyl group is as critical as its charge and moving it to the 4-position of the pyridine ring likewise eliminates its Ca^{2+} release activity [32].

The second important determinant of the Ca^{2+} release activity of NAADP is the 2′-phosphate. Changing the position of the phosphate group to the 3′-position of the adenylyl ribose or cyclically linking it to both the 2′- and 3′-positions reduces its activity by about 10-fold [32]. The position of the phosphate group is thus important but not critical. However, removal of the 2′-phosphate by alkaline

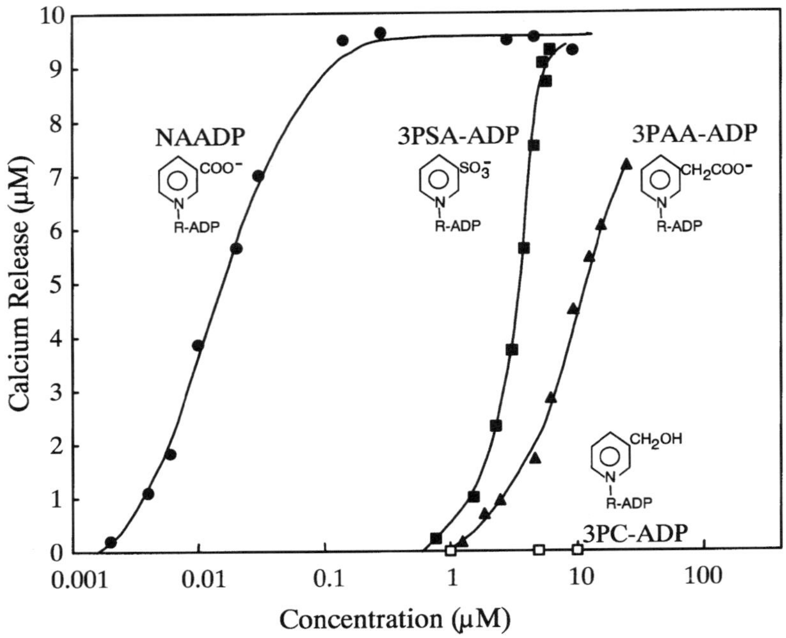

Fig. 2. Relationships between the structure of NAADP analogs and their Ca^{2+} release activity. The analogs were synthesized by exchanging the nicotinamide group of NADP with various pyridine analogs using ADP-ribosyl cyclase and the Ca^{2+} release activity was measured in sea urchin egg homogenates. Experimental conditions were as described previously [32]. (modified from [32]).

phosphatase, as in nicotinic acid adenine dinucleotide (NAAD), does eliminate the Ca^{2+} release activity [10, 18].

The third important determinant is the 6-amino group on the adenine ring. The effectiveness of deamino-NAADP produced from the base-exchange reaction using deamino-NADP and nicotinic acid as substrates is more than two orders of magnitude lower than NAADP [32]. These results demonstrate the exquisite structural specificity of the Ca^{2+} release activity of NAADP and indicate that its action is likely to be mediated by a receptor. That NAADP binds specifically to sea urchin egg microsomes and with high affinity is consistent with a receptor mechanism [20].

All the agonistic analogs described above also activate a novel desensitization mechanism unique to the NAADP-sensitive Ca^{2+} release. Prior treatment of microsomes with sub-threshold concentrations of either NAADP or one of its analogs can inactivate the Ca^{2+} release mechanism such that the microsomes do not respond to subsequent challenges with maximal concentrations [20, 32, 33]. Binding studies done with microsomes show that the desensitization occurs at the receptor level [20]. The effectiveness of the agonistic analogs in inducing desensitization parallels their Ca^{2+} release activities and the analogs that have no Ca^{2+} release activity also do not desensitize, suggesting that this novel regulation of the receptor is obligatorily linked to its Ca^{2+} release function [32]. It is possible that the NAADP-receptor contains two sites with the regulatory site having higher affinity. Occupation of the regulatory site at sub-threshold agonist concentrations may lead to alteration of the Ca^{2+} release site rendering it incapable of binding NAADP and/or activating the Ca^{2+} release channel. In any case, the NAADP-sensitive Ca^{2+} release is novel in having built in and intricate regulatory mechanisms, which is not observed in the simple ligand-gated channel of the IP_3-dependent mechanism.

Structure of ADP-ribosyl cyclase and CD38

The cyclase and CD38, a lymphocyte surface antigen, are homologous proteins sharing 34% identity in amino acid sequence and an additional 42% of the amino acid residues are conservative substitutions [34]. A schematic comparing the primary structure of the two proteins is shown in Fig. 3. The cyclase is a 30,000 kDa soluble protein first purified from *Aplysia* ovotestis and contains 258 amino acid residues [13, 35, 36]. A sequence of 19 amino acids in the middle of the protein is highly conserved among two different species of *Aplysia* as well as in CD38 from three species (Fig. 3). The cyclase also contains ten cysteine residues and all can be aligned with those in CD38. The crystal structure of the cyclase shows that all ten cysteines are paired to form five disulfide linkages [37] as diagrammed in Fig. 3. An extra stretch of amino acid residues in the N-terminal of CD38,

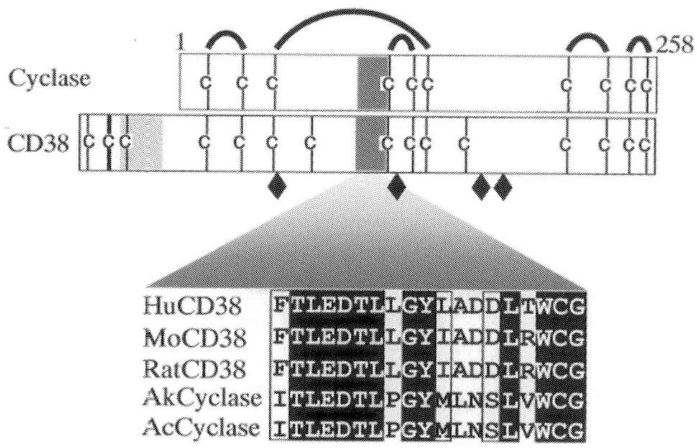

Fig. 3. Schematic comparison of the primary structures of *Aplysia* ADP-ribosyl cyclase and human CD38. The lines labeled C represent the positions of cysteine in the sequences. The arches indicate disulfide linkage based on X-ray crystallography results [37]. The most highly conserved sequence and the transmembrane domain are respectively represented by gray and hatched boxes. The amino acid residues in the conserved sequence are listed for three species of CD38; human (Hu) [38], mouse (Mo) [58] and rat [59], and compared with two species of cyclase *Aplysia kurodai* (Ak) [60] and *Aplysia california* (Ac) [35]. The diamonds represent putative gylcosylation sites.

which is absent in the cyclase, contains a hydrophobic sequence representing the transmembrane domain [38]. Also different from the cyclase is the presence of four putative glycosylation sites in the sequence of CD38. These features are consistent with CD38 being a surface antigen (reviewed in [39]).

The crystal structure of the cyclase has been solved at 2.4 Å [37]. It is a bean shaped molecule containing a cleft that separates the N- and C-terminal domains (Fig. 4A). The sequence that is highly conserved with CD38 is present right at the cleft and is shown in dark grey in Fig. 4A. The secondary structures of the cyclase are made up of five β-sheets and ten helices [37]. All the β-structures are in the C-terminal portion of the molecule, while the N-domain is composed of only α-helices (cf. Fig. 4B). The molecule resembles a hand with the palm representing the β-structures and the fingers representing the α-helices. The ten cysteine residues of the molecule are all paired up and form disulfide bridges, three of them in the N-terminal domain and two in the C-terminal domain (cf. Fig. 4B). Perhaps, the most novel feature of the molecule is that, in three different crystal forms of the cyclase, it is crystallized as a dimer in a head to head fashion as shown in Fig. 4B [37, 40]. Three (α1, α4 and α10) of the ten helices of the molecule are involved in the formation of the dimer (Fig. 4). The dimeric structure is likely to be highly stable since the sequences of the interacting helixes suggest the involvement of hydrogen bonding, salt bridges as well as hydrophobic interaction [37]. The central cavity in the dimer has a dimension comparable to a molecule of cADPR [37]. Since the highly conserved sequence is also part of the amino acid residues that line this

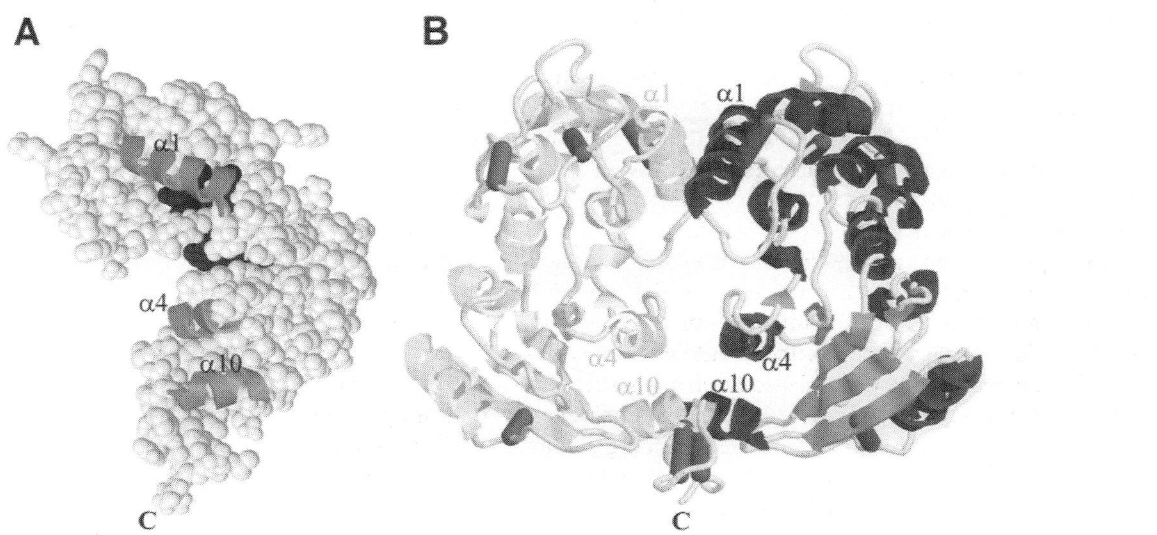

Fig. 4. The crystal structure of *Aplysia* ADP-ribosyl cyclase. A. A space-filling view of the monomeric cyclase. The black region represents the highly conserved sequence shown in Fig. 3. The three helices (α) shown are involved in formation of a cyclase dimer. B. Secondary structure of dimeric cyclase. The central cavity of the dimer has the dimension similar to a molecule of cADPR. Helixes are shown in dark gray. β-structures are represented as flat ribbons in light gray and the C-termini are displayed as arrow heads. The cylindrical rods in gray represent disulfide linkages. The carboxyl end of the protein is labeled C. The crystallographic coordinates of the structures are from reference [37] and displayed using the RasMol program [61].

central cavity, these structural features suggest that the active site of the cyclase may be located at the central cavity. Indeed, it has been speculated that the functional unit of the cyclase is a dimer and that the central cavity may function to bind and mold the substrate NAD, a long linear molecule, into a folded conformation such that the two ends of the molecule can be linked [22, 41].

Computer modeling based on the crystal structure of the cyclase and the homologous sequence of CD38 suggests that it is also a dimer [37]. That CD38 molecules are prone to form multimers in solution as well as in cell membranes is consistent with such a notion [42, 43]. Questions concerning the structure-function relationship of CD38 can best be addressed by solving its crystal structure. A highly efficient method for expressing CD38 and the *Aplysia* cyclase in yeast has been developed for this purpose [44–46].

Figure 5 shows the schematic of the expression constructs. The sequences of the enzymes are spliced behind an alcohol oxidase promotor and a secretion signal sequence derived from the α-factor, a protein secreted by yeast during mating. When the yeast is cultured under the condition that methanol is the sole carbon source, the promotor is activated and the synthesis of the cyclase is induced. The signal sequence of the α-factor, which directs the induced cyclase to be secreted, is endogenously removed by the yeast before secretion. In the case of CD38, the sequence used for expression is modified by site directed mutagenesis to remove the four glycosylation sites and the N-terminal sequence containing the transmembrane domain [46]. The expressed protein is soluble and without heterogeneous glycosylation that may hinder the crystallization of the protein. Using these constructs in conjunction with large scale fermentation, as much as 0.3 g/L of the cyclase [44] and 0.5 g/L of the catalytic domain of human CD38 [45] have been produced. Moreover, more than 90% of the secreted proteins present in the fermentation media are the expressed enzymes, allowing simple purification of the proteins. The purified proteins are shown to retain full enzymatic activity with characteristics similar to those of the native proteins [44, 45]. The stage is thus set for detailed analysis of the structural and functional relationships of these enzymes.

Enzymatic activities of ADP-ribosyl cyclase and CD38

In addition to the structural homology described above, the cyclase and CD38 are also functionally similar. Both are multifunctional enzymes capable of catalyzing the synthesis of cADPR and NAADP, two structurally distinct Ca^{2+} signaling molecules (reviewed in [47, 48]). Both enzymes cyclize NAD to produce cADPR by linking the N1 of the adenine with the anomeric carbon of the terminal ribose [13, 14, 17]. They can also cyclize nicotinamide guanosine

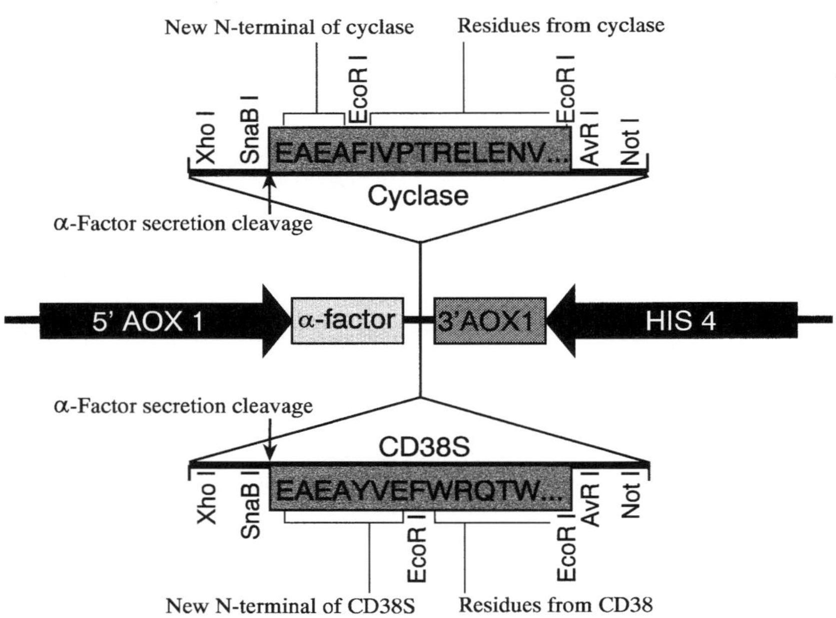

Fig. 5. Schematics of the yeast expression constructs for the cyclase and CD38. The modified sequence of human CD38 (CD38S) or the *Aplysia* cyclase is inserted into the expression vector after the secretion signal of the yeast a-mating factor and is under the control of the alcohol oxidase promotor (AOXI). The secreted proteins are cleaved by the transfected yeast at the indicated sites and contain 5–8 extra amino acids at the N-terminal as listed. The presence of HIS4 or the dominant selection marker, Zeocin in the constructs allow for histidine and Zeocin selection respectively of the transfectants.

dinucleotide (NGD), an analog of NAD, at an alternative site (N7) to produce cyclic GDP-ribose (cGDPR), a fluorescent analog of cADPR [49, 50]. The main catalytic difference between the two enzymes is that CD38 can readily hydrolyze cADPR to ADP-ribose while the cyclase cannot [13, 14, 17]. The two enzymes also differ in the amount of cADPR produced from NAD. While the only product of the cyclase is cADPR [13], the amount synthesized by CD38 is relatively small [14, 17], with the major product produced being ADP-ribose. The overall reaction catalyzed by CD38 thus resembles that of NADase [51]. However, CD38 can easily be distinguished from classical NADase such as that from *Neurospora* since the latter can cyclize neither NAD nor NGD, nor can it hydrolyze cADPR [52, 53]. Strictly speaking, the term NADase refers only to an enzymatic reaction and not to a defined enzyme, since many structurally and functionally distinct enzymes, including ADP-ribosyl transferases and poly(ADP-ribose) polymerases, all possess NADase activity (reviewed in [54, 55]). On the other hand, synthesis and/or hydrolysis of cADPR is only catalyzed by the cyclase and its related enzymes (reviewed in [22, 47]). It is thus clear that ADP-ribosyl cyclases are a functionally and structurally defined class of enzymes and should not be confused with NADases. Indeed, the simplest diagnostic test that distinguishes between the two classes of enzymes is to use NGD as substrate. CD38 and other enzymes related to the cyclase produce a fluorescent product, cGDPR, whereas, NADases can only hydrolyze it to non-fluorescent GDP-ribose [49, 50, 56].

It should be noted that some NADases have been documented in the literature to also have the ability of metabolizing cADPR. Among these, the best characterized is the one purified from spleen [51, 57]. It is a 39 kD protein capable of synthesizing cADPR as well as hydrolyzing it [51, 57]. However, both the size and its enzymatic properties are very similar to CD38. The spleen, of course, contains a large number of lymphocytes, and is a rich source of CD38, making it likely that the spleen NADase is in fact CD38. It is thus imperative to obtain sequence data so that these NADases can be definitively identified.

In addition to cyclizing NAD to produce cADPR, both the cyclase and CD38 can also use NADP as substrate and catalyze the exchange of its nicotinamide group with nicotinic acid (NA) [15]. Which of these two catalytic pathways the enzymes take is critically regulated by pH [15]. At acidic pH and in the presence of NA, the base-exchange reaction dominates and NAADP is produced. The optimum of the exchange reaction is at about pH 4.5. Above neutrality, the cyclization of the substrate is the main reaction even in the presence of NA. Therefore, depending on the pH and the presence of particular substrates different Ca^{2+} messengers can be produced, suggesting a central role for the cyclase in Ca^{2+} signaling.

The base exchange reaction also shows a high degree of specificity with respect to NA [32]. Figure 6 shows HPLC analyses of the products of the base-exchange reaction catalyzed by the cyclase [32]. With NADP and NA as substrates (Fig. 6A) and at pH 4.5, the major product is NAADP. A small amount of cyclized NADP, cADPR-phosphate (cADPRP), is also produced under this condition. The product profile is similar if pyridine 4-carboxylate is used instead of NA, which has a carboxyl group at the 4-position (Fig. 6B). The situation is totally different if the carboxyl group is moved to the 2-position. Pyridine 2-carboxylate does not support the base-exchange reaction. In its presence, the cyclase mainly cyclizes NADP to cADPRP and no exchange product is produced (Fig. 6C). In fact, any substitution at the 2-position of pyridine ring appears to

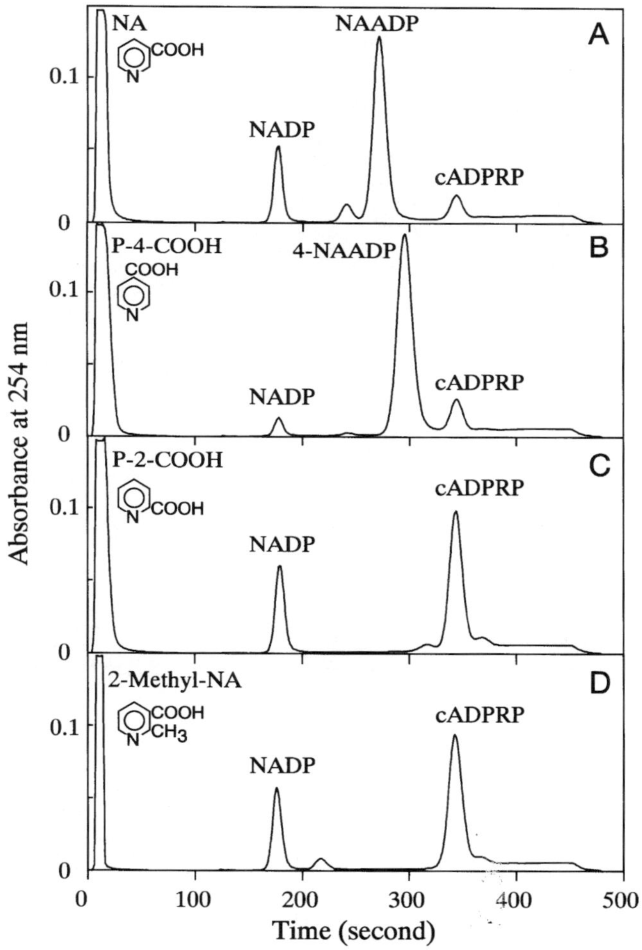

Fig. 6. Structural determinants of nicotinic acid analogs important for the base-exchange reaction. Nicotinic acid (NA) and its analogs were incubated with the *Aplysia* cyclase together with NADP. The reactants and products after the reaction were analyzed by HPLC. Experimental conditions are as described in [32]. (modified from [32]).

block the exchange reaction. As shown in Fig. 6D, addition of an extra methyl group at the 2-position of NA makes it incapable of supporting the exchange reaction. Again, like the 2-carboxylate analog, the main reaction product is through the cyclization reaction producing cADPRP.

The negative charge of NA is, however, not essential for supporting the base-exchange reaction. Pyridinyl 3-carbinol, an analog of NA with a neutral -CH_2OH group at the 3-position, effectively supports the base-exchange reaction, as do NA analogs with various negatively charged group substitutions at the 3-position [32]. Indeed, the neutral analogs of NA are so effective in supporting the base-exchange reaction that the cyclase fails to cyclize NAD or NADP in their presence [32]. This can also be seen in Fig. 6 by comparing the extent of cADPRP produced (i.e. cyclization) in the reactions shown in Figs 6C and 6D with the control shown in Fig. 6A. These results suggest that the exchange and the cyclization reactions are in competition. In the presence of an effective nucleophile such as NA or its carbinol analog, the exchange reaction dominates. Otherwise, the cyclase mainly cyclizes the substrate.

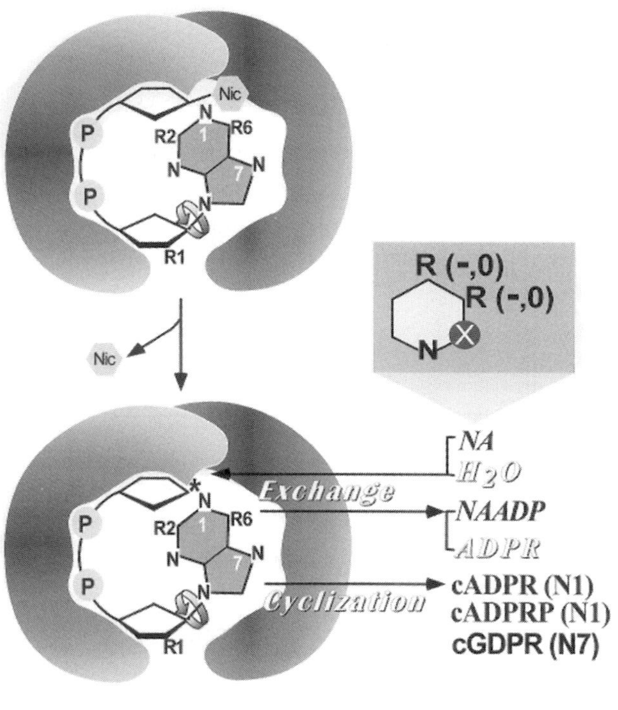

Fig. 7. A model for the catalytic mechanisms of ADP-ribosyl cyclase and CD38. Details of the model are described in the text. The main features include the proposal that the functional form of the cyclase is a dimer and the central cavity is involved in molding the substrate into a folded conformation. An ADP-ribosyl intermediate is then formed with the release of nicotinamide. The subsequent catalysis is the result of the competition between intramolecular cyclization and base-exchange with an available nucleophile (producing NAADP with nicotinic acid (NA) and ADP-ribose (ADPR) with water). NA analogs with neutral or negative charged (0,–) groups at the 3- or 4-position can support the base-exchange reaction, while any substitutions at the 2-position inhibits it. The free rotation of purine ring around the ribosyl bond allows the cyclization to occur at either N1-(cADPR, cADPRP) or N7-position (cGDPR), depending on which substrate is used.

A model for the catalytic mechanisms of ADP-ribosyl cyclase and CD38

The catalytic model shown in Fig. 7 is designed to account for the multi-functionality of the cyclase and CD38. It incorporates many of the features proposed previously [22, 47, 49, 51, 57]. Recent information on the crystal structure of the cyclase [37] indicates that it is a dimer with a central cavity. The conserved sequence is around this cavity suggesting that it may be the active site. In order to cyclize the substrate, the cyclase must bind it in a folded conformation such that the two ends of the molecule (adenine and ribose) can be joined. The central cavity of the dimeric cyclase can conceivably function to mold the substrate into such a conformation.

Since the cyclase can use either NAD or NADP, the binding site must not recognize the 2′-position of the ribose. The R1-group can be either OH for NAD or a phosphate for NADP and is shown as directing out of the active site (Fig. 7). The next step in the catalysis is likely to be the release of the nicotinamide group (Nic) and the formation of the an enzyme intermediate with the anomeric carbon of the terminal ribose being in an activated state. At an acidic pH, nucleophiles such as NA have ready access to the intermediate and the dominant reaction is the base-exchange leading to the formation of NAADP. There are specific structural requirements for the nucleophile. Either neutral or negatively charged groups can be present at the 3- or 4-position of the pyridine ring, but any substitutions at the 2-position make them ineffective (cf. Figs 6C and 6D).

In the absence of an attacking group, the activated carbon can react with the N1 of the adenine ring; the result is the formation of either cADPR or cADPRP, respectively, depending on whether the substrate is NAD or NADP. This mode of catalysis is favored at neutral and alkaline pH. To account for cyclizing NGD at the N7 instead of the N1 position of NAD, it is proposed that the active site of the cyclase allows free rotation of the purine ring of the bound substrate with respect to the N9-ribose bond. Since N7 of guanine is known to be more reactive than N1, as the purine base is rotated into position, N7 would preferentially react with the activated C1 of the ribose.

All the features of the model also apply to CD38. To account for the observed catalytic differences between CD38 and the cyclase, however, it is further proposed that the active site of CD38 may have structural features that favor cyclization through the N7-linkage at the purine ring, perhaps through restraining free rotation of the ring and stabilizing a close apposition of N7 of the purine to the C1 of the ribosyl intermediate. With NAD as the substrate, this positioning of the bound substrate would reduce its chance of cyclization. A necessary consequence is the increased chance of the intermediate being attacked by water, a ubiquitous nucleophile. The end result is the formation of the hydrolysis product ADP-ribose instead of cADPR. In contrast, with NGD as substrate, the positioning would enhance cyclization at the N7-position to produce cGDPR and reduce its chance of hydrolysis to GDP-ribose.

In summary, the model proposes that the catalysis of the cyclase and CD38 is determined by the competition between the intra-molecular cyclization and the exchange of the nicotinamide group with the available nucleophile, including water. Factors that enhance cyclization, such as neutral pH and optimal substrate positioning as in the cyclase, would reduce hydrolysis (exchange). The converse is also true in the case of CD38, where exchange (hydrolysis) dominates over cyclization. The proposed effects of substrate positioning can best be tested by crystallography. The success of crystallizing the cyclase and the development of an efficient expression system in yeast should set the stage for the eventual understanding of the catalytic mechanisms of these novel Ca^{2+} signaling enzymes.

References

1. Clapper DL, Walseth TF, Dargie PJ, Lee HC: Pyridine nucleotide metabolites stimulate calcium release from sea urchin egg microsomes desensitized to inositol trisphosphate. J Biol Chem 262: 9561–9568, 1987
2. Lee HC, Walseth TF, Bratt GT, Hayes RN, Clapper DL: Structural determination of a cyclic metabolite of NAD^+ with intracellular Ca^{2+}-mobilizing activity. J Biol Chem 264: 1608–1615, 1989
3. Lee HC, Aarhus R, Levitt D: The crystal structure of cyclic ADP-ribose. Nature Struct Biol 1: 143–144, 1994
4. Galione A, Lee HC, Busa WB: Ca^{2+}-induced Ca^{2+} release in sea urchin egg homogenates: Modulation by cyclic ADP-ribose. Science 253: 1143–1146, 1991
5. Lee HC: Potentiation of calcium- and caffeine-induced calcium release by cyclic ADP-ribose. J Biol Chem 268: 293–299, 1993
6. Lee HC: Modulator and messenger functions of cyclic ADP-ribose in calcium signaling. Re Prog Horm Res 51: 355–388, 1996
7. Lee HC, Aarhus R, Graeff RM: Sensitization of calcium-induced calcium release by cyclic ADP-ribose and calmodulin. J Biol Chem 270: 9060–9066, 1995
8. Lee HC, Aarhus R, Graeff R, Gurnack ME, Walseth TR: Cyclic ADP ribose activation of the ryanodine receptor is mediated by calmodulin. Nature 370: 307–309, 1994
9. Tanaka Y, Tashjian AH Jr: Calmodulin is a selective mediator of Ca^{2+}-induced Ca^{2+} release via the ryanodine receptor-like Ca^{2+} channel triggered by cyclic ADP-ribose. Proc Natl Acad Sci USA 92: 3244–3248, 1995
10. Lee HC, Aarhus R: A derivative of NADP mobilizes calcium stores insensitive to inositol trisphosphate and cyclic ADP-ribose. J Biol Chem 270: 2152–2157, 1995
11. Chini EN, Dousa TP: Nicotinate-adenine dinucleotide phosphate-induced Ca^{2+}-release does not behave as a Ca^{2+}-induced Ca^{2+}-release system. Biochem J 316: 709–711, 1996
12. Genazzani AA, Galione A: Nicotinic acid-adenine dinucleotide phosphate mobilizes Ca^{2+} from a thapsigargin-insensitive pool. Biochem J 315: 721–725, 1996
13. Lee HC, Aarhus R: ADP-ribosyl cyclase: An enzyme that cyclizes NAD^+ into a calcium-mobilizing metabolite. Cell Reg 2: 203–209, 1991
14. Howard M, Grimaldi JC, Bazan JF, Lund FE, Santos-Argumedo L, Parkhouse RM, Walseth TF, Lee HC: Formation and hydrolysis of cyclic ADP-ribose catalyzed by lymphocyte antigen CD38. Science 262: 1056–1059, 1993
15. Aarhus R, Graeff RM, Dickey DM, Walseth TF, Lee HC: ADP-ribosyl cyclase and CD38 catalyze the synthesis of a calcium-mobilizing metabolite from NADP. J Biol Chem 270: 30327–30333, 1995
16. Lee HC, Aarhus R: Wide distribution of an enzyme that catalyzes the hydrolysis of cyclic ADP-ribose. Biochim Biophys Acta 1164: 68–74, 1993
17. Lee HC, Zocchi E, Guida L, Franco L, Benatti U, De Flora A: Production and hydrolysis of cyclic ADP-ribose at the outer surface of human erythrocytes. Biochem Biophys Res Commun 191: 639–645, 1993
18. Lee HC, Aarhus R, Gee KR, Kestner T: Caged nicotinic acid adenine dinucleotide phosphate-synthesis and use. J Biol Chem 272: 4172–4178, 1997
19. Aarhus R, Gee K, Lee HC: Caged cyclic ADP-ribose-synthesis and use. J Biol Chem 270: 7745–7749, 1995
20. Aarhus R, Dickey DM, Graeff RM, Gee KR, Walseth TF, Lee HC: Activation and inactivation of Ca^{2+} release by $NAADP^+$. J Biol Chem 271: 8513–8516, 1996
21. Guo X, Becker PL: Cyclic ADP-ribose-gated Ca^{2+} release in sea urchin eggs requires an elevated $[Ca^{2+}]$. J Biol Chem 272: 16984–16989, 1997
22. Lee HC: Mechanisms of calcium signaling by cyclic ADP-ribose and NAADP. Physiological Reviews: 77: 1133–1164, 1997
23. Walseth TF, Lee HC: Synthesis and characterization of antagonists of cyclic-ADP-ribose induced Ca^{2+} release. Biochim Biophys Acta 1178: 235–242, 1993
24. Lee HC, Aarhus R, Walseth TF: Calcium mobilization by dual receptors during fertilization of sea urchin eggs. Science 261: 352–355, 1993
25. Clementi E, Riccio M, Sciorati C, Nistico G, Meldolesi J: The type 2 ryanodine receptor of neurosecretory PC12 cells is activated by cyclic ADP-ribose. Role of the nitric oxide/cGMP pathway. J Biol Chem 271: 17739–17745, 1996
26. Guse AH, Dasilva CP, Emmrich F, Ashamu GA, Potter BVL, Mayr GW: Characterization of cyclic adenosine diphosphate-ribose-induced Ca^{2+} release in T lymphocyte cell lines. J Immunol 155: 3353–3359, 1995
27. Kuemmerle JF, Makhlouf GM: Agonist-stimulated cyclic ADP ribose. Endogenous modulator of Ca^{2+}-induced Ca^{2+} release in intestinal longitudinal muscle. J Biol Chem 270: 25488–25494, 1995
28. Rakovic S, Galione A, Ashamu GA, Potter BVL, Terrar DA: A specific cyclic ADP ribose antagonist inhibits cardiac excitation-contraction coupling. Curr Biol 6: 989–996, 1996
29. Sethi JK, Empson RM, Bailey VC, Potter BVL, Galione A: 7-Deaza-8-bromo-cyclic ADP-ribose, the first membrane-permeant, hydrolysis-resistant cyclic ADP-ribose antagonist. J Biol Chem 272: 16358–16363, 1997
30. Bailey VC, Fortt SM, Summerhill RJ, Galione A, Potter BVL: Cyclic aristeromycin diphosphate ribose: A potent and poorly hydrolysable Ca^{2+} minic of cyclic adenosine diphosphate ribose. FEBS Lett 379: 227–228, 1996
31. Bailey VC, Sethi JK, Font SM, Galione A, Potter BVL: 7-Deaza cyclic adenosine 5′-diphosphate ribose – first example of a Ca^{2+}-mobilizing partial agonist related to cyclic adenosine 5′-diphosphate ribose. Chem Biol 4: 51–61, 1997

32. Lee HC, Aarhus, R.: Structural determinants of nicotinic acid adenine dinucleotide phosphate important for its calcium-mobilizing activity. J Biol Chem 272: 20378–20383, 1997
33. Genazzani AA, Empson RM, Galione A: Unique inactivation properties of NAADP-sensitive Ca^{2+} release. J Biol Chem 271: 11599–11602, 1996
34. States DJ, Walseth TF, Lee HC: Similarities in amino acid sequences of *Aplysia* ADP-ribosyl cyclase and human lymphocyte antigen CD38. Trends Biochem Sci 17: 495, 1992
35. Glick DL, Hellmich MR, Beushausen S, Tempst P, Bayley H, Strumwasser F: Primary structure of a molluscan egg-specific NADase, a second-messenger enzyme. Cell Reg 2: 211–218, 1991
36. Hellmich MR, Strumwasser F: Purification and characterization of a molluscan egg specific NADase, a second-messenger enzyme. Cell Reg 2: 193–202, 1991
37. Prasad GS, McRee DE, Stura EA, Levitt DG, Lee HC, Stout CD: Crystal structure of *Aplysia* ADP ribosyl cyclase, a homologue of the bifunctional ectozyme CD38. Nature Struct Biol 3: 957–964, 1996
38. Jackson DG, Bell JI: Isolation of a cDNA encoding the human CD38 (T10) molecule, a cell surface glycoprotein with an unusual discontinuous pattern of expression during lymphocyte differentiation. J Immunol 144: 2811–2815, 1990
39. Malavasi F, Funaro A, Roggero S, Horenstein A, Calosso L, Mehta K: Human CD38: A glycoprotein in search of a function. Immunol Today 15: 95–97, 1994
40. Prasad GS, Levitt DG, Lee HC, Stout CD: Crystallization of ADP-ribosyl cyclase from *Aplysia californica*. Proteins 24: 138–140, 1996
41. Lee HC: Calcium signaling by cyclic ADP-ribose and NAADP – A decade of exploration. Cell Biochem Biophys 28: 1–17, 1998
42. Zocchi E, Franco L, Guida L, Calder L, De Flora A: Self-aggregation of purified and membrane-bound erythrocyte CD38 induces extensive decrease of its ADP-ribosyl cyclase activity. FEBS Lett 359: 35–40, 1995
43. Umar S, Malavasi F, Mehta K: Post-translational modification of CD38 protein into a high molecular weight form alters its catalytic properties. J Biol Chem 271: 15922–15927, 1996
44. Munshi C, Lee HC: High-level expression of recombinant *Aplysia* ADP-ribosyl cyclase in *pichia pastoris* by fermentation. Prot Exp Pur, 1997 (in press)
45. Munshi CB, Fryxell KB, Lee HC, Branton WD: Large scale production of human CD38 in yeast by fermentation. Meth Enzymol 280: 318–330, 1997
46. Fryxell KB, O'Donoghue K, Graeff RM, Lee HC, Branton WD: Functional expression of soluble forms of human CD38 in *Escherichia coli* and *Pichia pastoris*. Prot Exp Pur 6: 329–336, 1995
47. Lee HC, Graeff, RM, Walseth TF: ADP-ribosyl cyclase and CD38. Multifunctional enzymes in Ca^{2+} signaling. Adv Exper Med Biol 419: 411–419, 1997
48. Lee HC, Galione A, Walseth TF: Cyclic ADP-ribose: metabolism and calcium mobilizing function. Vit Horm 48: 199–257, 1994
49. Graeff RM, Walseth TF, Hill HK, Lee HC: Fluorescent analogs of cyclic ADP-ribose: Synthesis, spectral characterization, and use. Biochemistry 35: 379–386, 1996
50. Graeff RM, Walseth TF, Fryxell K, Branton WD, Lee HC: Enzymatic synthesis and characterizations of cyclic GDP-ribose. A procedure for distinguishing enzymes with ADP-ribosyl cyclase activity. J Biol Chem 269: 30260–30267, 1994
51. Kim H, Jacobson EL, Jacobson MK: Synthesis and degradation of cyclic ADP-ribose by NAD glycohydrolases. Science 261: 1330–1333, 1993
52. Lee HC, Graeff R, Walseth TR: Cyclic ADP-ribose and its metabolic enzymes. Biochimie 77: 345–355, 1995
53. Graeff RM, Walseth TF, Lee HC: A radio-immunoassay for measuring endogenous levels of cyclic ADP-ribose in tissues. Meth Enzymol 280: 230–241, 1997
54. Hilz H: ADP-ribose: A historical overview. Adv Exp Med Biol 419: 15–24, 1997
55. Koch-Nolte FaH F: Mono(ADP-ribosyl)transferases and related enzymes in animal tissues: Emerging gene families. Adv Exp Med Biol 419: 1–13, 1997
56. Graeff RM, Mehta K, Lee HC: GDP-ribosyl cyclase activity as a measure of CD38 induction by retinoic acid in HL-60 cells. Biochem Biophys Res Commun 205: 722–727, 1994.
57. Mullersteffner HM, Augustin A, Schuber F: Mechanism of cyclization of pyridine nucleotides by bovine spleen NAD^+ glycohydrolase. J Biol Chem 271: 23967–23972, 1996
58. Harada N, Santos-Argumedo L, Chang R, Grimaldi JC, Lund FE, Brannan CI, Copeland NG, Jenkins NA, Heath AW, Parkhouse RME, Howard M: Expression and cloning of a cDNA encoding a novel murine B cell activation marker. J Immunol 151: 3111–3118, 1993
59. Koguma T, Takasawa S, Tohgo A, Karasawa T, Furuya Y, Yonekura H, Okamoto H: Cloning and characterization of cDNA encoding rat ADP-ribosyl cyclase/cyclic ADP-ribose hydrolase (homologue to human CD38) from islets of Langerhans. Biochim Biophys Acta 1223: 160–162, 1994
60. Nata K, Sugimoto T, Tohgo A, Takamura T, Noguchi N, Matsuoka A, Numakunai T, Shikaina K, Yonekura H, Takasawa S *et al.*: The structure of the *Aplysia kurodai* gene encoding ADP-ribosyl cyclase, a second-messenger enzyme. Gene 158: 213–218, 1995
61. Sayle R: RasMol v2.5. Glaxo Research and Development, Greenford, Middlesex, UK, 1994

Molecular and Cellular Biochemistry **193**: 99–102, 1999.

Identification of the archaeal NMN adenylyltransferase gene

Nadia Raffaelli,[1] Monica Emanuelli,[1] Francesca M. Pisani,[2] Adolfo Amici,[1] Teresa Lorenzi,[1] Silverio Ruggieri[3] and Giulio Magni[1]

[1]*Istituto di Biochimica, Facoltà di Medicina, Università di Ancona;* [2]*Istituto di Biochimica delle Proteine ed Enzimologia, Consiglio Nazionale delle Ricerche, Napoli;* [3]*Dipartimento di Biotecnologie Agrarie ed Ambientali, Università di Ancona, Italy*

Abstract

Increasing evidence on the importance of fluctuations in NAD^+ levels in the living cell is accumulating. Therefore a deeper knowledge on the regulation of coenzyme synthesis and recycling is required. In this context the study of NMN adenylyltransferase (EC 2.7.7. 1), a key enzyme in the NAD^+ biosynthetic pathway, assumes a remarkable relevance. We have previously purified to homogeneity and characterized the protein from the thermophilic archaeon *Sulfolobus solfataricus*. The determination of partial sequence of the *S. solfataricus* enzyme, together with the recent availability of the genome sequence of the archaeon *Methanococcus jannaschii,* allowed us, based on sequence similarity, to identify the *M. jannaschii* NMN adenylyltransferase gene. As far as we know from literature, this is the first report on the NMN adenylyltransferase gene. (Mol Cell Biochem **193**: 99–102, 1999)

Key words: NMN adenylyltransferase, archaea, NAD^+ metabolism

Introduction

The maintenance of a proper NAD^+ level is of utmost importance for living cells. Interest in the metabolic reactions, which use the pyridine dinucleotide as substrate, has been growing, with particular attention to ADP-ribosyltransferase and NAD^+ glycohydrolase-catalyzed reactions [1]. NAD^+ depletion due to these metabolic events requires strict control of dinucleotide synthesis. In the biosynthetic pathway, the enzyme NMN adenylyltransferase catalyzes the reaction common to both the *de novo* synthesis and the salvage routes (NMN or deamido mononucleotide + ATP <=> NAD^+ or deamido adenine dinucleotide + PPi), and it is the only enzyme in the pathway to be located in the cell nucleus [2]. These findings suggest that NMN adenylyltransferase occupies a key position in pyridine nucleotides metabolism. For this reason, for several years, we have been focusing our attention on the study of the enzyme. Indeed, a homogeneous preparation and a comprehensive characterization of the protein from several sources, have been obtained in our laboratory [3–5]. Contrary to most of the enzymes involved in the NAD^+ biosynthetic pathway, which have been already cloned and sequenced, the study on the NMN adenylyltransferase gene has been limited to the identification of its locus in *Salmonella thyphimurium* [6]. The present report deals with experiments leading to the identification of the polypeptide encoded by the thermophilic archaeon *Methanococcus jannaschii* MJ0541*orf* [7] as the enzyme NMN adenylyltransferase. To our knowledge, the archaeal gene is the first and only gene so far known for the enzyme NMN adenylyltransferase.

Materials and methods

Materials

Sulfolobus solfataricus cells, strain MT-4, were kindly provided by Prof A. Gambacorta (Istituto per la Chimica di Molecole di Interesse Biologico, CNR, Napoli). NMN

Address for offprints: G. Magni, Istituto di Biochimica, Facoltá di Medicina e Chirurgia, Via Ranieri, 60131 Ancona, Italy

adenylyltransferase was purified to homogeneity from *S. solfataricus* (unpublished results).

NMN adenylyltransferase sequencing

For direct N-terminal sequencing, about 4 µg of the final preparation (0.004 mg/ml, in 50 mM TRIS/HCl, pH 7.4, 0.5 mM EDTA, 1 mM $MgCl_2$, 0.25 mM 3[(3-cholamidopropyl) dimethylammonio]2-hydroxy-1-propanesulfonate (CHAPSO), 1.6 M NaCl) were diluted 10-fold with water prior to direct spotting on a polyvinylidene difluoride membrane. Dilution was required to decrease the concentration of both NaCl and CHAPSO, therefore increasing the binding of the protein to the membrane.

For the enzymatic digestion, the NMN adenylyltransferase preparation was freed from salts and detergents by using reversed-phase chromatography in fast protein liquid chromatography (FPLC) on a Pharmacia RESOURCE RPC column, equilibrated with 0.1% trifluoroacetic acid. The enzyme was eluted with a 1-min linear gradient of 0–90% acetonitrile in 0.1% trifluoroacetic acid, at a flow rate of 2.5 ml/min. After centrifugal evaporation, the protein (about 5 µg) was digested in 25 µl of 50 mM ammonium bicarbonate, pH 7.9, containing 1.65 µg of sequencing grade endoproteinase Glu-C (Boehringer Mannheim). After overnight incubation at 30°C, peptides were separated by sodium dodecyl sulfate-polyacrylamide gel electrophoresis (SDS-PAGE), and then electroblotted onto polyvinylidene difluoride membrane. Transfer was performed at 4°C, for 4 h at 250 mA, in 10 mM 3-[cyclohexylamino]-1-propanesulfonic acid, pH 11, 10% methanol. The membrane was stained with Coomassie Brilliant Blue R-250 and destained following manufacturer's instructions.

Sequencing was performed on an automated Applied-Biosystems gas-phase sequencer.

Electrophoretic analysis

SDS-PAGE was carried out according to Schagger and Von Jagow [8]. To assess the purity of the enzyme preparation, a separating gel with a 10% total acrylamide concentration (T) was used. For electroblotting, peptides were separated on a 16.5% T gel, overlaid with a 10% T spacer gel. The cross-linker concentration was maintained at 3%. To detect proteins, gels were silver stained [9].

Computer-aided sequence analysis

Protein similarity searches were performed using the BLAST [10] network service at the National Center for Biotechnology Information (NIH, Bethesda, MD, USA).

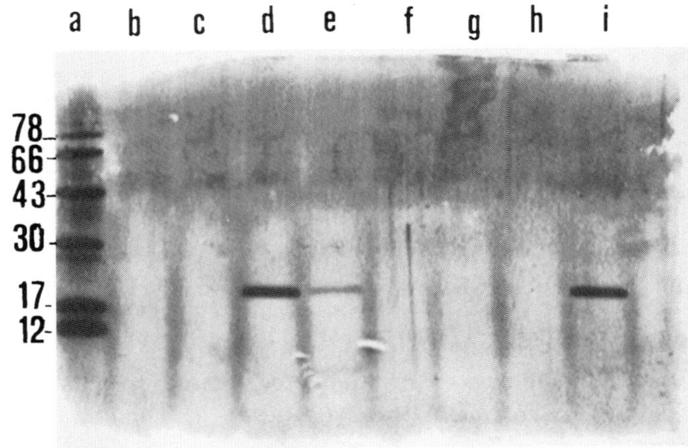

Fig. 1. SDS-PAGE of homogeneous *S. solfataricus* NMN adenylyltransferase. Lane a: standard protein markers, molecular mass values (Mr $\times 10^{-3}$) are indicated on the left; lane i: 1 µg of the enzyme from the final step of the purification procedure; lanes b-h: selected fractions from the gradient of the reversed-phase chromatography column; lanes d and e, comprising the desalted protein peak, correspond to the fractions eluted at about 60% acetonitrile.

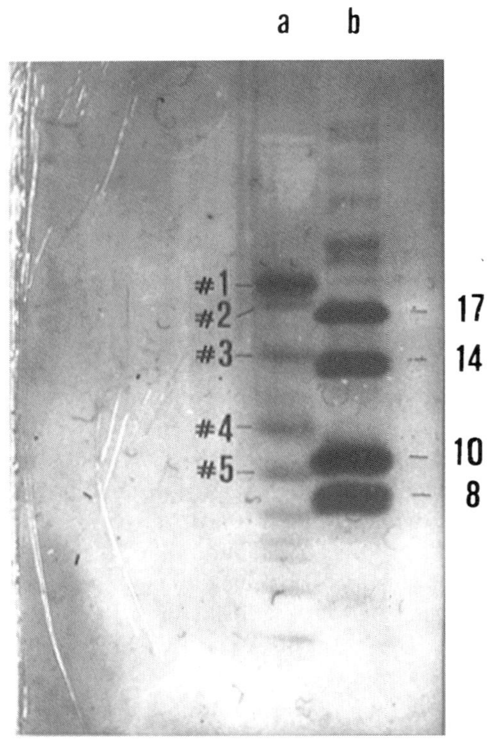

Fig. 2. SDS-PAGE of *S. solfataricus* NMN adenylyltransferase after endoproteinase Glu-C digestion. Lane a: one twentyfifth of the digestion mixture prepared as described in Materials and methods, corresponding to about 0.2 µg protein; lane b: standard protein markers, with molecular mass values (Mr $\times 10^{-3}$) indicated on the right.

```
a)   MRGLYPGRFQPFHLGHLNVIKIKLERVDDPII-------VHTYHNPFTAMEHN
     :||     ||||||| ||| |||    | ||: ||         ||  ||||| |
b)   LRGFIIGRFQPFHKGHLEVIKKIAEEVDEIIIGIGSAQKSHTLENPFTAGERI
```

Fig. 3. Comparison of the partial primary sequence of *S. solfataricus* NMN adenylyltransferase (a), with the MJ0541*orf*-encoded protein (b). Identical and similar residues are marked with vertical and dotted lines, respectively; the gap that was introduced to optimize the alignment is given as horizontal dashes. Amino acids with similar properties are grouped according to Dayhoff classification [11].

Results and discussions

S. solfataricus NMN adenylyltransferase has been purified to homogeneity by classical chromatographic steps yielding a final preparation enriched about 4,000-fold (unpublished results). The purified enzyme migrated on SDS-PAGE as a single band of Mr 18,600 (Fig. 1, lane i). While attempts to determine the N-terminal sequence of the NMN adenylyltransferase purified from eukaryotes failed, probably because of the presence of a modified amino acid residue at the N-terminus, the sequence of the first 32 aminoacids at the N-terminus of the archaeal protein could be easily determined. It was: MRGLYPGRFQPFHLGHLNVIKI KLERVDDPII.

To obtain internal amino acid sequences, the pure enzyme was submitted to reversed-phase chromatography, as described in Materials and methods (Fig. 1, lanes b-h). This step was required to remove both salts and detergent present in the final enzyme preparation, thus eliminating any interference in the subsequent enzymatic digestion. The protein was digested with endoproteinase Glu-C and the resulting peptides were separated by SDS-PAGE, as described in Materials and methods (Fig. 2). Peptides were blotted onto polyvinylidene difluoride membrane, excised with a sterile razor blade, and directly sequenced. As depicted in Fig. 2, a number of different fragments was generated: sequence of band #1 confirmed the N-terminal sequence of the intact protein; bands #2, 4 and 5 were found to correspond to endoproteinase Glu-C peptides; sequence of band #3 was the following: VHTYHNPFTAMEHN. The other peptides of lower size were recovered in too low quantity to be sequenced.

Through computer-assisted comparison of the determined amino acid sequences with those stored in data banks, we found that the N-terminal sequence of the *S. solfataricus* enzyme shared a 65% identity and a 68% similarity with the N-terminal sequence of a hypothetical protein coded by the archaeon *Methanococcus jannaschii* MJ0541 gene (Fig. 3). Interestingly, a sequence significantly similar to peptide #3, generated by the NMN adenylyltransferase enzymatic digestion, was also present in the hypothetical protein (Fig. 3). In addition, the polypeptide encoded by the MJ0541 gene comprises 168 amino acid residues, with a predicted Mr of 19,600, which closely corresponds to the subunit Mr of about 18,600 obtained for the *S. solfataricus* NMN adenylyltransferase. The high similarity observed both in the primary structure and the molecular size of the *S. solfataricus* enzyme and the MJ0541*orf*-encoded protein, strongly suggest that the MJ0541*orf*-encoded protein could be the *M. jannaschii* NMN adenylyltransferase. To confirm the identity of the hypothetical protein, MJ0541*orf* was isolated and cloned into a T7-based vector. Its subsequent expression in *E. coli* resulted in the production of high levels of thermophilic NMN adenylyltransferase activity. The details of the gene cloning and expression, as well as the purification to homogeneity of the recombinant enzyme, will be published elsewhere. In this report we wished to highlight the value of the sequence information on the *S. solfataricus* enzyme in aiding the identification of the NMN adenylyltransferase gene.

Acknowledgements

We wish to thank Prof. Donatella Barra and Dr. Bruno Maras, Dipartimento di Scienze Biochimiche, Università 'La Sapienza', Roma, for their help in enzyme sequencing. We are grateful to Prof. Agata Gambacorta, Laboratorio di Molecole di Interesse Biologico, CNR, Napoli, for kindly providing the *S. solfataricus* cells.

References

1. The 12th International Symposium on ADP-ribosylation Reactions: From Bacterial Pathogenesis to Cancer. Cancun, Mexico, May 10–14, 1997
2. Hogeboom GH, Schneider WC: The synthesis of diphosphopyridine nucleotide by liver cell nuclei. J Biol Chem 197: 611–6203, 1952
3. Natalini P, Ruggieri S, Raffaelli N, Magni G: Nicotinamide mononucleotide adenylyltransferase. Molecular and enzymatic properties of the homogeneous enzyme from baker's yeast. Biochemistry 25: 3725–3729, 1986
4. Emanuelli M., Natalini P., Raffaelli N, Ruggieri S, Vita A, Magni G: NAD biosynthesis in human placenta: Purification and characterization of homogeneous NMN adenylyltransferase. Arch Biochem Biophys 298: 29–34, 1992

5. Balducci E, Orsomando G, Poizonetti V, Vita A, Emanuelli M, Raffaelli N, Ruggieri S, Magni G, Natalini P: NMN adenylyltransferase from bull testis: Purification and properties. Biochem J 310: 395–400, 1995
6. Hughes KT, Lodika D, Roth JR, Olivera BM: An indispensable gene for NAD biosynthesis in *Salmonella thyphimurium.* J Bacteriol 155: 213–221, 1983
7. Bult CJ, Venter JC: Complete genome sequence of the methanogenic archaeon *Methanococcus jannaschii.* Science 273: 1058–1073, 1996
8. Schagger H, Von Jagow G: Tricine-sodium dodecyl sulfate-polyacrylamide gel electrophoresis for the separation of proteins in the range from 1–100 kDa. Anal Biochem 166: 368–379, 1987
9. Oakley BR, Kirch DR, Morris NR: A simplified ultrasensitive silver stain for detecting proteins in polyacrylamide gels. Anal Biochem 105: 361–363, 1980
10. Altschul SF, Warren G, Webb M, Myers EW, Lipman DJ: Basic local alignment search tool. J Mol Biol 215: 403–410, 1990
11. Dayhoff MO, Schwartz RM, Ovcutt BC: Atlas of protein sequence and structure. Natl Biomed Res Found Silver Spring, MD, 1978, p 345–352

Molecular and Cellular Biochemistry **193:** 103–108, 1999.

Functional analysis of poly(ADP-ribose) polymerase in *Drosophila melanogaster*

Masanao Miwa, Shuji Hanai, Palmiro Poltronieri, Masahiro Uchida and Kazuhiko Uchida

Institute of Basic Medical Sciences and Center for Tsukuba Advanced Research Alliance, University of Tsukuba, Tsukuba, Japan

Abstract

Poly(ADP-ribose) polymerase (PARP) is conserved in eukaryotes. To analyze the function of PARP, we isolated and characterized the gene for PARP in *Drosophila melanogaster*. The *PARP* gene consisted of six translatable exons and spanned more than 50 kb. The DNA binding domain is encoded by exons 1–4. Although the consensus cleavage site of CED-3 like protease during apoptosis is conserved from human to *Xenopus laevis* PARPs, it is neither conserved in the corresponding region of *Drosophila* nor *Sarcophaga peregrina*. There are two cDNAs species in *Drosophila*. One cDNA could encode the full length PARP protein (PARP I), while the other is a truncated cDNA which could encode a partial-length PARP protein (PARP II), which lacks the automodification domain and is possibly produced by alternative splicing. The expression of these two forms of PARP in *E. coli* demonstrated that while PARP II has the catalytic NAD-binding domain and DNA-binding domain it is enzymatically inactive. On the other hand PARP I is active. A deletion mutant of *PARP* gene could grow to the end of embryogenesis but did not grow to the adult fly. These results suggest that the *PARP* gene plays an important function during the development of *Drosophila*. (Mol Cell Biochem **193**: 103–108, 1999)

Key words: Poly(ADP-ribose) polymerase, *Drosophila melanogaster*, alternative splicing, apoptosis, DNA repair, development

Introduction

Poly(ADP-ribose) polymerase [PARP; NAD$^+$ ADP-ribosyltransferase; NAD$^+$: poly(adenosine diphosphate-D-ribosyl)-acceptor ADP-D-ribosyltransferase, EC 2.4.2.30] is an eukaryotic DNA binding enzyme and requires DNA strand breaks for its enzymatic activity (Fig. 1) [1–4]. By analysis of limited proteolysis, PARP is known to be cleaved into the three functional domains of; DNA-binding (46 kD), automodification (22 kD), and NAD-binding and catalysis (40 kD), from its N-terminus (Fig. 2) [5]. These domains are conserved phylogenetically from human to *Drosophila* [6].

To analyze the biological function of poly(ADP-ribose) polymerase, we have isolated and characterized the gene for PARP from *Drosophila*. Recently PARP was reported to be cleaved at the specific site in the DNA-binding domain, during the process of apoptosis by a CED-3-like cysteine protease, named caspase-3 [7–9].

We have determined the locus for the *PARP* gene in *Drosophila,* isolated the mutants, and characterized the homozygous deletion mutant of the *PARP* gene. The function of PARP in the development is discussed.

Gene structure of poly(ADP-ribose) polymerase from Drosophila

To analyze the biological function of poly(ADP-ribose) polymerase, we have isolated and characterized the gene for PARP from *Drosophila*. One copy of the *Drosophila PARP* gene was present per haploid genome. The gene was 50 kb or more in length and consisted of six translatable exons (Fig. 2). A portion of the intron 4 sequence is not yet fully mapped. The DNA-binding domain is encoded by exons 1–4. The automodification domain is encoded by a portion of exon 5. The catalytic NAD-binding domain is encoded by a portion

Address for offprints: M. Miwa, Institute of Basic Medical Sciences and Center for Tsukuba Advanced Research Alliance, University of Tsukuba, Tsukuba 305, Japan

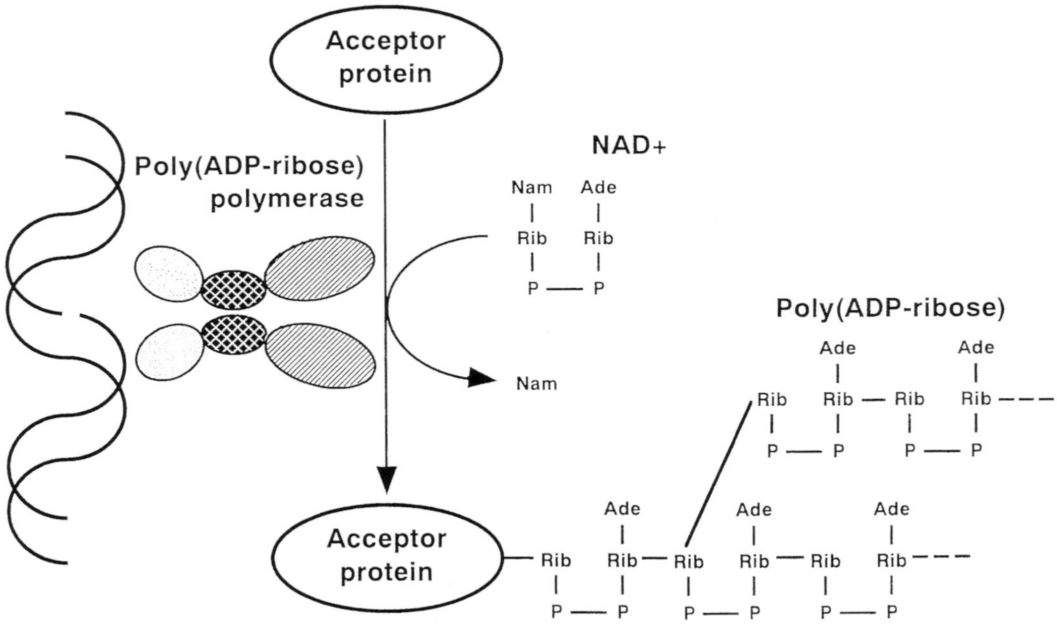

Fig. 1. Poly(ADP-ribosyl)ation of nuclear proteins.

of exon 5 and exon 6. The number of exons in *Drosophila, is* smaller than that of the human *PARP* gene, in which there are 23 exons. It is remarkable that the size of *PARP* gene is possibly larger than that of the human *PARP* gene. Considering that the genome size of *Drosophila* is 5% of that of human or mouse, it is indicated that the relative space of the *PARP* gene occupying the *Drosophila* genome is 20 times more as compared to those in the human or mouse genomes.

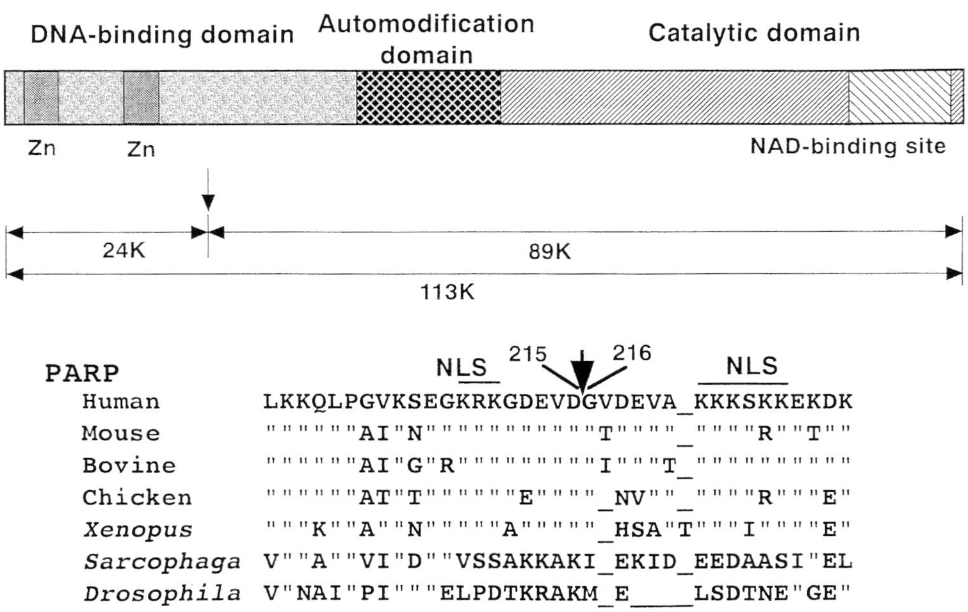

Fig. 2. Cleavage site of PARP by ICE/CED-3-like proteases during apoptosis. The 24 kD peptide is cleaved from the N-terminus of PARP. The corresponding cleavage site is aligned from 7 species. NLS – nuclear location signal.

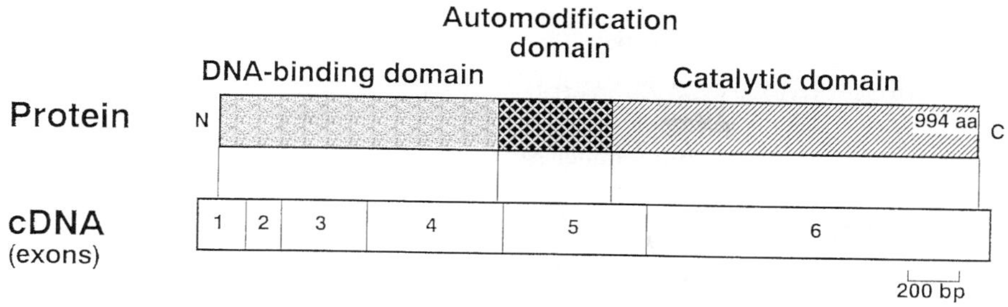

Fig. 3. Domain structures of poly(ADP-ribose) polymerase of *Drosophila* and its coding exons.

Comparison of the PARP cleavage site in Drosophila

Recently PARP was reported to be cleaved at a specific site, the linkage between the 215th aspartic acid and the 216th glycine residue during the process of apoptosis by caspase-3 [7–9] (Fig. 3). This cleavage can produce a 24 kD N-terminal peptide and an 89 kD C-terminal peptide. The physiological significance of this cleavage products in execution of apoptosis is unclear. However, it might be possible that the cleaved 24 kD peptide, which contains two Zinc fingers, can bind to the breaks of DNA, although its release from DNA by automodification is impossible, since the 24 kD peptide has no catalytic activity. Thus continuous binding of this 24 kD peptide to the DNA without automodification might result in deleterious situations, resulting in the inhibition of DNA repair and cell death. This hypothesis is consistent with the report that over expression of the DNA-binding domain blocked alkylation-induced DNA repair synthesis in mammalian cells [10]. We speculate that the DNA-binding domain, thus released by caspases, might bind to the nicked portion of DNA and would not be released from DNA, by the lack of automodification. Although caspases involved in apoptosis are known among species including *Drosophila* [11] and the sequence of the cleavage site of PARP, Asp-Glu-Val-Asp, is conserved from human to *Xenopus laevis,* the sequence is not conserved in *Drosophila* or in another species of fly, *Sarcophaga peregrina* [12]. Therefore, it is not clear whether another uncharacterized caspase is involved in the proteolysis of *Drosophila* PARP or other events comparable to the proteolysis and production of the DNA-binding domain are occurring during apoptosis.

Two forms of PARP and their enzymatic activities

We have analyzed the cDNA species of PARP from the cDNA libraries constructed from the *Drosophila* embryo. We found that there are two kinds of cDNA species for PARP. One is a full length cDNA encoding for the entire length of *Drosophila* PARP and the other is a shorter one. We found that the full length cDNA could encode the complete PARP protein (which we call 'PARP I' in this work) and the shorter cDNA could encode the truncated form of PARP protein (which we call 'PARP II') lacking the automodification domain (Fig. 3). The PARP II cDNA should be derived from alternative splicing, since the exon-intron boundaries of the truncated portion of PARP II are consistent with the splicing consensus rule.

To analyze further the function of the PARP I and PARP II, we introduced their expression vectors in *E. coli* to produce them as GST-fusion proteins. The fusion protein of PARP I has enzymatic activity, while that of PARP II has essentially no activity in the presence of histone H1 (Fig. 4) or in the presence of the recombinant automodification domain (data not shown).

Hypothesis of dominant negative function of PARP II in Drosophila

The following hypothesis would be presented (Fig. 5). PARP I should be activated when it finds breaks of DNA, and binds to them, making the micro environment for DNA repair to

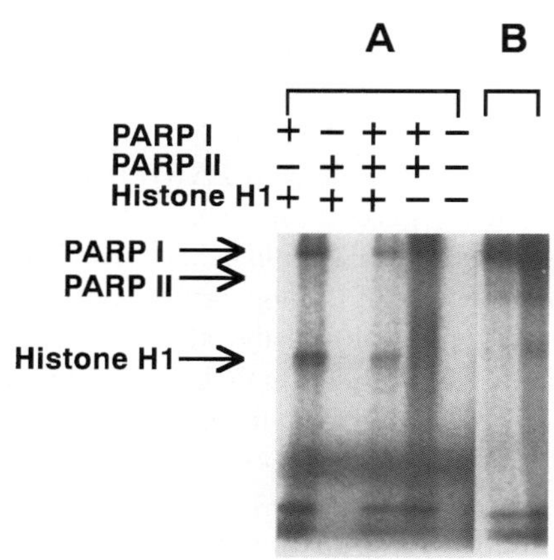

Fig. 4. Enzyme activities of GST-fusion products of PARP I and PARP II revealed by autoradiography. GST-fusion proteins are incubated in the presence or absence of histone H1 with [^{32}P]NAD and separated by SDS polyacrylamide gel electrophoresis (A). HeLa cell extract was used as the positive control (B).

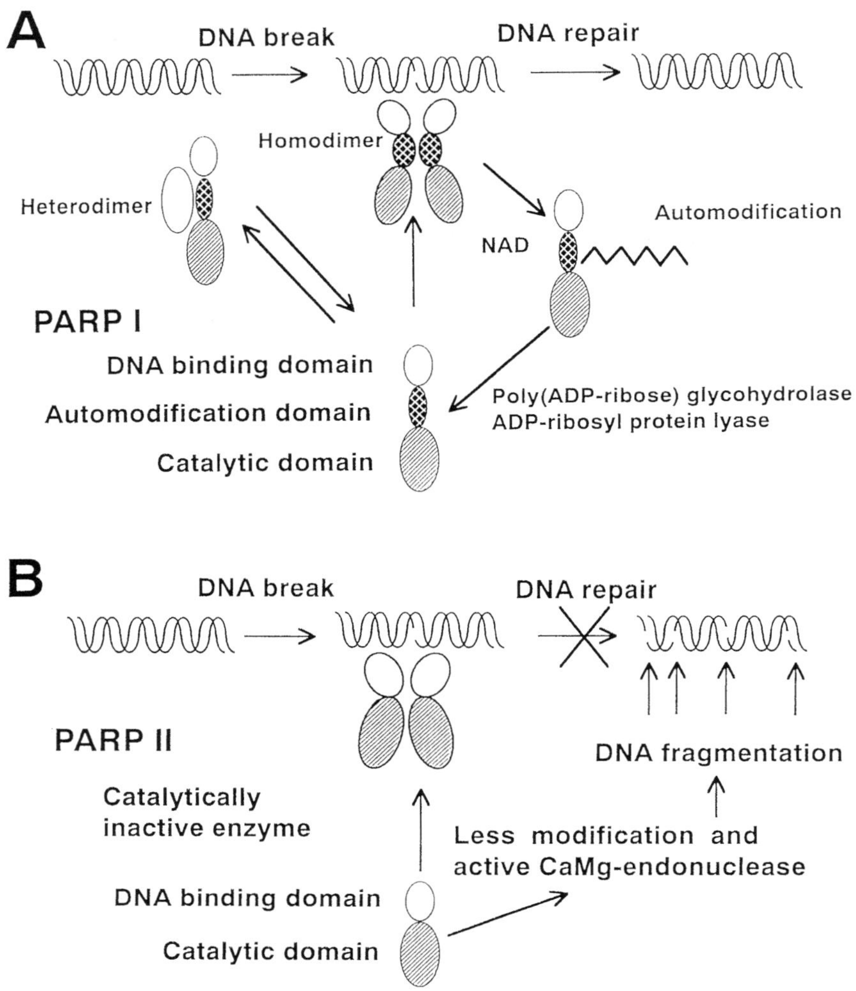

Fig. 5. Hypothetical functions of PARP I (panel A) and PARP II (panel B).

occur. Poly(ADP-ribosyl)ation of PARP I itself (automodification) or other proteins nearby will facilitate the DNA repair then the release of PARP I from the DNA. Then poly(ADP-ribose) attached to PARP is hydrolyzed by poly(ADP-ribose) glycohydrolase and by ADP-ribosyl protein lyase. Thus PARP I could be recycled. PARP I would also bind to other proteins through protein-protein interaction (manuscript in preparation). On the other hand, PARP II might bind to the breaks of DNA through its DNA-binding domain, but might not be released from the DNA.

Deletion mutants of the Drosophila PARP *gene*

The *PARP* gene was localized in chromosome 3 by *in situ* hybridization. Two recessive lethal deletion mutants covering the *PARP* region were obtained. These mutants lack all the exons of the *PARP* gene. These mutant flies are maintained in the heterozygous state, one allele being defective of *PARP* gene and the other allele being a balancer. A cross of the above two mutants would produce a null mutant of the *PARP* gene in 25% of the offspring as Mendelian rule predicts. This homozygous deletion mutant of *PARP* gene yielded a mutant that could grow to the end of embryogenesis, but we could not recover the adult flies. These data suggest that *PARP* gene is an essential gene for development in *Drosophila.* The implications of the above results on cell differentiation and cell death are under investigation.

Future prospect

Genes involved in apoptosis and DNA repair are also found in *Drosophila.* Therefore, it is useful to analyze the function of human genes through the analysis of those homologues in *Drosophila* PARP which is conserved in eukaryotes and its function is now getting to be clearer using knockout mouse systems [13, 14]. However, mouse deficient in *PARP* gene function show little abnormalities as compared to the deletion mutant of *Drosophila*. The different phenotypes of PARP deletion mutants between mouse and *Drosophila* could not be explained at present. This might be due to the low redundancy of gene in *Drosophila* or due to different signal transduction pathways. The elucidation of the function of universally distributed PARP among eukaryotes will further contribute to clarify the basic molecular pathways involved in apoptosis, DNA repair and carcinogenesis.

Acknowledgement

The authors wish to thank Dr. S. Kobayashi for discussion and Mr. K. Maeshima, Ms. Y. Koyama, Mr. S. Katagiri, Ms. T. Yokota, Ms. Y. Yoshida, Mr. T. Hayashi and Mr. O. Nakagawa for assistance with the experiments. This work was supported in part by the Grants-in-Aids for Cancer Research from the Ministry of Education, Science, Sports and Culture, and for the 2nd Term Comprehensive 10-year Strategy for Cancer Control and Cancer Research from the Ministry of Health and Welfare of Japan.

References

1. Miwa M, Sugimura T: ADP ribosylation and carcinogenesis. In: J. Moss, M. Vaughan (eds). ADP Ribosylating Toxins and G Proteins: Insights into Signal Transduction. American Society for Microbiology, Washington, D.C., 1990, pp 543–560
2. Lautier D, Lagueux J, Thibodeau J, Menard L, Poirier GG: Molecular and biochemical features of poly(ADP-ribose) metabolism. Mol Cell Biochem 122: 171–193, 1993
3. de Murcia G, de Murcia JM: Poly(ADP-ribose) polymerase: A molecular nick-sensor. TIBS 19: 172–176, 1994
4. Sugimura T, Miwa M: Poly(ADP-ribose): Historical perspective. Mol Cell Biochem 138: 5–12, 1994
5. Kameshita I, Matsuda M, Nishikimi M, Ushiro H, Shizuta Y: Reconstitution and poly(ADP-ribosyl)ation of proteolytically fragmented poly(ADP-ribose) sythetase. J Biol Chem 261: 3863–3868, 1986
6. Uchida K, Miwa M. Poly(ADP-ribose) polymerase: Structural conservation among different classes of animals and its implications. Mol Cell Biochem 138: 25–32, 1994
7. Lazebnik YA, Kaufmann SH, Desnoyers S, Poirier GG, Earnshaw WC: Cleavage of poly(ADP-ribose) polymerase by a proteinase with properties like ICE. Nature 371: 346–347, 1994
8. Tewari M, Quan LT, O'Rourke K, Desnoyers S, Zeng Z, Beidler DR, Poirier GG, Salvesen GS, Dixit VM: Yama/CPP32beta, a mammalian homolog of CED-3, is a CrmA-inhibitable protease that cleaves the death substrate poly(ADP-ribose) polymerase. Cell 81: 801–809, 1995
9. Nicholson DW, Ali A, Thornberry NA, Vaillancourt JP, Ding CK, Gallant M, Gareau Y, Griffin PR, Labelle M, Lazebnik YA, Munday NA, Raju SM, Smulson ME, Yamin TT, Yu VL, Miller DK: Identification and inhibition of the ICE/CED-3 protease necessary for mammalian apoptosis. Nature 376: 37–43, 1995
10. Molinete M, Vermeulen W, Buerkle A, Menissier-de Murcia J, Kuepper JH, Hoeijmakers JHJ, de Murcia G: Overproduction of the poly(ADP-ribose) polymerase DNA-binding domain blocks alkylation-induced DNA repair synthesis in mammalian cells. EMBO J 12: 2109–2117, 1993
11. Song Z, McCall K, Steller H. DCP-1, a *Drosophila* cell death protease essential for development. Science 275: 536–540, 1997
12. Miwa M, Hanai S, Masuda H, Koyama Y, Hayashi T, Yoshida Y, Poltronieri P, Maeshima. K, Kobayashi S, Okada M, Uchida K: Analysis of the biological function of poly(ADP-ribosyl)ation in *Drosophila melanogaster*. Biochimie 77: 466–471, 1995
13. Wang Z-Q, Auer B, Stingl L, Berghammer H, Haidacher D, Schweiger M, Wagner EF: Mice lacking ADPRT and poly(ADP-ribosyl)ation develop normally but are susceptible to skin disease. Genes Dev 9: 509–520, 1995
14. de Murcia J, Niedergang C, Trucco C, Ricoul M, Dutrillaux B, Mark M, Oliver W, Masson M, Dierich A, LeMeur M, Walztinger C, Chambon P, de Murcia G: Requirement of poly(ADP-ribose) polymerase in recovery from DNA damage in mice and in cells. Proc Natl Acad Sci USA 94: 7303–7307, 1997

Molecular and Cellular Biochemistry **193:** 109–113, 1999.

Characterization of NAD:arginine ADP-ribosyltransferases

Joel Moss, Enrico Balducci, Eleanor Cavanaugh, Hyun Ju Kim, Piotr Konczalik, Elena A. Lesma, Ian J. Okazaki, Maryann Park, Michael Shoemaker, Linda A. Stevens and Anna Zolkiewska

Pulmonary-Critical Care Medicine Branch, National Heart, Lung, and Blood Institute, National Institutes of Health, Bethesda, MD, USA

Abstract

NAD:arginine mono-ADP-ribosyltransferases catalyze the transfer of ADP-ribose from NAD to the guanidino group of arginine on a target protein. Deduced amino acid sequences of one family (ART1) of mammalian ADP-ribosyltransferases, cloned from muscle and lymphocytes, show hydrophobic amino and carboxyl termini consistent with glycosylphosphatidylinositol (GPI)-anchored proteins. The proteins, overexpressed in mammalian cells transfected with the transferase cDNAs, are released from the cell surface with phosphatidylinositol-specific phospholipase C (PI-PLC), and display immunological and biochemical characteristics consistent with a cell surface, GPI-anchored protein. In contrast, the deduced amino acid sequence of a second family (ART5) of transferases, cloned from murine lymphoma cells and expressed in high abundance in testis, displays a hydrophobic amino terminus, consistent with a signal sequence, but lacks a hydrophobic signal sequence at its carboxyl terminus, suggesting that the protein is destined for export. Consistent with the surface localization of the GPI-linked transferases, multiple surface substrates have been identified in myotubes and activated lymphocytes, and, notably, include integrin α subunits. Similar to the bacterial toxin ADP-ribosyltransferases, the mammalian transferases contain the characteristic domains involved in NAD binding and ADP-ribose transfer, including a highly acidic region near the carboxy terminus, which, when disrupted by *in vitro* mutagenesis, results in a loss of enzymatic activity. The carboxyl half of the protein, synthesized as a fusion protein in *E. coli*, possessed NADase, but not ADP-ribosyltransferase activity. These findings are consistent with the existence at the carboxyl terminus of ART1 of a catalytically active domain, capable of hydrolyzing NAD, but not of transferring ADP-ribose to a guanidino acceptor. (Mol Cell Biochem **193**: 109–113, 1999)

Key words: ADP-ribosylation, bacterial toxins, glycosylphosphatidylinositol-linked proteins, NAD glycohydrolases

A family of vertebrate ADP-ribosyltransferases has been purified that, like cholera toxin, modify arginine and other simple guanidino compounds (NAD:arginine ADP-ribosyltransferases) [1]. Several vertebrate transferases of this type have been cloned and characterized including those from rabbit and human skeletal muscle, chicken heterophils and erythroblasts, and mouse lymphocytes (Fig. 1). Expression of NAD:arginine ADP-ribosyltransferases in mammalian tissues appears to be restricted thus far to muscle, hematopoietic cells (e.g., lymphocytes, polymorphonuclear leukocytes) and epithelial cells. Of the mammalian ADP-ribosyltransferases, those from rabbit and human skeletal muscle were first to be cloned [1, 2]. The nucleotide and deduced amino acid sequences of the human skeletal muscle enzyme were 80.8 and 81.3% identical, respectively, to those of the rabbit.

The hydropathy profiles of ART1 from these species demonstrated hydrophobic regions near the amino and carboxy termini as has been found with glycosylphosphatidylinositol (GPI)-linked proteins [1, 2]. Consistent with the hypothesis that the transferase is GPI-linked is the observation that transferase activity is released from the cell

Address for offprints: J. Moss, Room 6D-03, Building 10, 10 Center Drive MSC 1590, National Institutes of Health, Bethesda, MD 20892-1590, USA

```
hART1    MQMPAMMSLL LVSVGLHEAL QAQSHPITRR DLFSQEIQLD MALATFDDQY    50
Rt6-1    MPSNNFK--- -FFLTWWLTQ QVTGL----- ---AVPFMLD MAPNAFDDQY    38
RT6.2    MPSNICK--- -FFLTWWLIQ QVTGL----- ---TGPLMLD TAPNAFDDQY    38
hART3    MKTGHFE--- -IVTMLLATM ILVDI----- -FQVKAEVLD MADNAFDDEY    40
hART4    MQVA------ ---------- -IK------- ------IDFD FAPGSFDDQY    20
mART5    MILEDLL-MV LSCLSLHALW KVRAVPI--- ------LPLS LVPDTFDDAY    40

hART1    VGCAAAMTAA LPDLNHTEFQ ANQVYADSWT LASSQWQERQ ARWPEWSLSP   100
Rt6-1    EGCVEDMEKK -APQLLQEDF NMNEELKL-E WEK------- AE--IKWKEI    77
RT6.2    EGCVNKMEEK -APLLLQEDF NMNAKLKV-A WEE------- AK--KRWNNI    77
hART3    LKCTDRMEIK YVPQLLKEEK ASHQQLDT-V WEN------- AK--AKWAAR    80
hART4    QGCSKQVMEK LTQ---GDYF TKDIEAQKNY FRMW------ QK--AHLAWL    59
mART5    VGCSEEMEEK -AGLLLKEEM ARHALPAP-I LGSS------ TR--GLGTPA    80

hART1    TRPSPPPLGF RDEHGVALLA YTANS-PL-- -HKEFNAAVR EAGRSRAHYL   146
Rt6-1    KNCMSYPAGF HDFHGTALVA YTGNIHRS-- ----LNEATR EFKINPGN--   119
RT6.2    KPSRSYPKGF NDFHGTALVA YTGSIAVD-- ----FNRAVR EFKENPGQ--   119
hART3    KTQIFLPMNF KDNHGIALMA YISEAQEQTP FYHLFSEAVK MAGQSREDYI   130
hART4    NQGKVLPQNM STTHAVAILF YTLNS-NV-- -HSDFTRAMA SVARTPQQYE   105
mART5    SQATLPP-GF KAQHGVAIMV YTNSSNTL-- -YWELNQAVR TGGGSRELYM   126

hART1    HHFSFKTLHF LLTEALQLL- GSGQ-RPPR- CHQVFRGVHG LRFRPAGPRA   193
Rt6-1    --FHYKAFHY YLTRALQLL- --SDQGCRS- ---VYRGTN- VRFR-YTGKG   158
RT6.2    --FHYKAFHY YLTRALQLL- --SNGDCHS- ---VYRGTK- TRFH-YTGAG   158
hART3    YGFQFKAFHF YLTRALQLLR KPCEASSKTV ---VYRTSQG TSFT-FGGLN   176
hART4    RSFHFKYLHY YLTSAIQLLR KDSIMENGTL CYEVHYRTKD VHFNAYT-GA   154
mART5    RHFPFKALHF YLTRALQLLR GSGGCSRGP- GEVVFRGVGS LHFEPKRLGD   175

hART1    TVRLGGFASA SLKHVAAQQ- ---FGEDTFF GIWTCLGAPI KGYSFFPGEE   239
Rt6-1    SVRFGHFASS SLNRSVATSS PFFNGQGTLF IIKTCLGAHI KHCSYYTHEE   208
RT6.2    SVRFGQFTSS SLSKKVAQSQ EFFSDHGTLF IIKTCLGVYI KEFSFRPDQE   208
hART3    QARFGHFT-- -LAYSAKPQA A--NDQLTVL SIYTCLGVDI ENFLDKESER   221
hART4    TIRFGQFLST SLLKEEAQE- ---FGNQTLF TIFTCLGAPV QYFSLKKE--   198
mART5    SVRLGQFTSS SVDERVARR- ---FGNATFF NLRTCFGAPI QALSVFPEER   221

hART1    EVLIPPFETF QVINASRPAQ GPARIYLRAL GKHSTYNCEY IKD-------   282
Rt6-1    EVLIPGYEVF HKVKTQSVER YIQISLDSPK RKKSNFNCFY SGSTQAAN--   256
RT6.2    EVLIPGYEVY QKVRTQGYNE IFLDS---PK RKKSNYNCLY SSA-------   248
hART3    ITLIPLNEVF QVSQEGAGNN LILQS---IN KTCSHYECAF LGGLKTENCI   268
hART4    -VLIPPYELF KVINMSYHPR GD-WLQLRST GNLSTYNCQL LKA-------   239
mART5    EVLIPPHEVF LVTGFSQDGA QSIVTLSSYD QTCSHFNCAY LGG-------   264

hART1    ---------- ---------- ---------- ----KKYK-- ----------   286
Rt6-1    ---------- ---------- ---------- VSSLGSRE-- ----------   264
RT6.2    ---------- ---------- ---------- ----GARE-- ----------   252
hART3    ENLEYFQPIY VYNPGEKNQK LEDHSEKNWK LEDHGEKNQK LEDHGVKILE   318
hART4    ---------- ---------- ---------- ----SSKK-- ----------   243
mART5    ---------- ---------- ---------- ----EKRH-- ----------   268

hART1    ----SGPCHL DNSAMGQSP- ---LSAVWSL LLLLWFLVVR AFPDGPGLL    327
Rt6-1    ---------- -----SCVP- ----LFLVVL LGLLVQQLTL AEP------    287
RT6.2    ---------- -----SCVS- ----LFLVVL PSLLVQLLCL AE-----PE    276
hART3    PTQIPAPGPV PVPGPKCHPS ASSGKLLLPQ FGMVIILISV SAINLFVAL    367
hART4    -----CI--- ----PDPIA- ---IASLSFL TSVIIFSKSR V--------    268
mART5    ----GCVSSR AVGQPEAPS- ---TEALALQ SGKTLLLDPR KLQLSRAGP    309
```

Fig. 1. Alignment of vertebrate ADP-ribosyltransferases. Deduced amino acid sequences of mammalian ADP-ribosyltransferases were aligned using the protein alignment program of GeneWorks. Dashes were inserted to facilitate alignment. Amino acid position is indicated on the right. hART1, ART1 from human skeletal muscle; Rt6-1, mouse Rt6-1; RT6.2, rat RT6.2; hART3, ART3 from human testis; hART4, ART4 from human spleen; mART5, ART5 from mouse lymphoma cells.

surface with phosphatidylinositol-specific phospholipase C (PI-PLC). In addition, the protein released by PI-PLC reacted with anti-cross reacting determinant (anti-CRD) antibodies which recognize an epitope on the oligosaccharide portion of the GPI-anchor [3]. Further, the transferase released by PI-PLC from cells grown in the presence of [^{3}H]ethanolamine, a constituent of the oligosaccharide, contained radiolabel [3]. Based on these studies it appears that ART1 is glycosylphosphatidylinositol (GPI)-anchored and conserved across species.

The muscle enzyme in C2C12 mouse myotubes modified integrin $\alpha 7$ [4, 5]. Integrin $\alpha 7$, as a dimer with integrin $\beta 1$, is a major integrin receptor for extracellular matrix protein laminin in skeletal muscle [6]. ADP-ribose appeared to be incorporated in the extracellular domain of α7, possibly in an arginine-rich region. The fact that an integrin is ADP-ribosylated suggests that the modification may affect cell interactions with the extracellular matrix. ADP-ribosylated integrin was processed by an extracellular phosphodiesterase, releasing 5′AMP, while ribose remains attached to the protein backbone [5].

The fact that the ART1 transferases are GPI-anchored is of interest in view of the prior finding that ADP-ribosylarginine hydrolase, the enzyme that cleaves the ADP-ribose-arginine bond, appears to be soluble and intracellular [7]. Since ADP-ribose is attached to extracellular domains of integrin, the modified proteins may not come in contact with the ADP-ribosylarginine hydrolase. The biochemical data suggest that the ADP-ribosylated integrins are processed by extracellular phosphodiesterases, releasing 5′AMP and leaving ribose covalently attached to the arginine. The presence of ribose would prevent further modification of the arginine residue.

Binding of integrin α7β1 to laminin in solution did not seem to be influenced by ADP-ribosylation of the α7 subunit ([8], data not shown). Interactions in solution, in the presence of detergent, differ, however, from the interactions at the cell surface, where binding of the two proteins may still be affected by ADP-ribosylation. Moreover, ADP-ribosylation could modulate transmembrane signalling by integrin α7β1 triggered by ligand binding.

To address the question of the role of ADP-ribosylation in intact cells, cultured C2C12 myotubes were used in laminin binding assays. The formation of laminin-integrin complex was followed by immunoprecipitation with either anti-laminin or anti-integrin α7 antibodies. As evidenced by laminin dose-response experiments, by dependence of the co-immunoprecipitation of the two proteins on divalent cations, and by chemical cross-linking, laminin did not bind to integrin α7β1 in intact cells. In addition, it was demonstrated that endogenous laminin, which was produced by C2C12 cells and deposited outside the cells, was unable to bind to integrin α7β1, either at the cell surface, or in solution. The lack of integrin binding was paralleled by the lack of reactivity with 5D3 monoclonal antibody, which was shown to recognize the γ subunit in the heterotrimeric laminin molecule.

To investigate the role of ADP-ribosylation of the extracellular domain of integrin α7 on the affinity of α7β1 dimer for laminin or intracellular signalling, the mechanisms leading to integrin activation in cultured cells will be explored. These mechanisms may be similar to those described for several integrins in leukocytes, where a specific cellular activation is required in order to induce an active conformation of integrins in a process known as 'inside-out' signalling [9]. Once integrin α7β1 is activated, its interactions with exogenous laminin can be studied, with little interference from the endogenous laminin.

ADP-ribosyltransferase activity was also investigated in lymphocytes and related cells. It was detected in mouse lymphoma and thymoma cells [10] and cytotoxic T lymphocytes (CTL) [11]; the murine ART1 cDNA was cloned from lymphoma (Yac-1) cells [12]. The deduced amino acid sequence of the GPI-anchored mouse ART1 transferase was 77 and 75% identical, to those from the human and rabbit skeletal muscle enzymes, respectively. Incubation of mouse CTL in the presence of NAD resulted in the ADP-ribosylation of membrane proteins, inhibition of CTL proliferation, and, to a lesser extent, cytotoxicity [11]. The suppressive effects of NAD on CTL were prevented by treatment of the cells with phosphatidylinositol-specific phospholipase C (PI-PLC), which releases most GPI-linked proteins from the cell surface, before incubation with NAD, consistent with the conclusion that a GPI-linked ADP-ribosyltransferase was responsible for modulating CTL function. In addition, ADP-ribosylation of a 40-kDa CTL membrane protein (p40) resulted in inhibition of $p56^{lck}$, a tyrosine kinase that exists in a complex with the 40-kDa protein [13]. Release of the membrane-bound transferase(s) following treatment with PI-PLC prevented the NAD-induced suppression of kinase activity. The relationship between the inhibition of $p56^{lck}$, following ADP-ribosylation of p40, and the inhibition of CTL proliferation has not been established. In other experiments, CTL transferases modified arginines in the extracellular domain of the lymphocyte function-associated molecule-1 (LFA-1), an adhesion molecule of the integrin family [14]. ADP-ribosylation inhibited LFA-1-mediated generation of inositol phosphates and suppressed homotypic cell adhesion. As in C2C12 cells, 5′-AMP was removed from the modified LFA-1 by extracellular phosphodiesterases. These data are consistent with a role for ADP-ribosylation in lymphocyte function.

A second lymphocyte ADP-ribosyltransferase ART5, was cloned from Yac-1 cells [15]. The nucleotide and deduced amino acid sequences of the ART5 transferase are 58 and 33% identical, respectively, to those of mouse ART 1. Further, the ART 1 and ART5 enzymes were only ~28% identical to the rodent RT6 family of proteins which are NAD glycohydrolases and, in the case of murine Rt6, NAD:arginine ADP-ribosyltransferases. Whereas the ART1 protein possessed hydrophobic amino- and carboxy-terminal signal sequences characteristic of GPI-anchored proteins, ART5 contained an amino-, but not carboxy-terminal hydrophobic signal sequence. The amino-terminal signal sequence should direct export of proteins into the endoplasmic reticulum; in the absence of a carboxy-terminal signal sequence, the protein would not be retained by the GPI anchor. Indeed, when rat mammary adenocarcinoma (NMU) cells were transformed with the ART5 cDNA, arginine-specific ADP-ribosyltransferase activity was not released from the membranes following incubation with PI-PLC.

On Northern analysis, an ART1 cDNA probe hybridized with a 1.6 kb band in poly(A)+RNA from mouse cardiac and skeletal muscle. An ART5 cDNA probe, on the other hand, hybridized with 1.6 and 2.0 kb bands in mouse testis mRNA [15]. There was also weak hybridization of the ART5 probe with a 1.6 kb band in cardiac and skeletal muscle mRNA. ART5-specific oligonucleotide probes hybridized with poly(A)+RNA from mouse skeletal muscle and testis, consistent with the fact that both enzymes ART1 and ART5 are expressed in muscle. Expression in rat testis was examined during development; the ART5 mRNA appeared preferentially at 30–60 days and was localized primarily to the seminiferous duct epithelial cells.

ART1 and ART5 enzymes expressed in *E. coli* as GST-fusion proteins were used to compare ADP-ribosyltransferase and NAD glycohydrolase activities of the purified enzymes [15]. The ADP-ribosyltransferase activity of ART1 was twice that of ART5 while the NAD glycohydrolase activity of ART1 was substantially less than its transferase activity. The ADP-ribosyltransferase and NAD glycohydrolase activities of the ART5 protein, however, were approximately equal. The K_m values for NAD with 20 mM agmatine as ADP-ribose acceptor in the ADP-ribosyltransferase assay were 118 and 142 μM for ART1 and ART5, respectively; the values for agmatine in the presence of 0. 1 mM NAD were 9 and 15 mM respectively.

Based on three-dimensional structure, photoaffinity labeling and site-directed mutagenesis, several of the bacterial toxin ADP-ribosyltransferases possess regions of similarity which together form, in part, the catalytic site [16]. The NAD-binding cleft is composed of an α-helix bent over a β-strand. An arginine or histidine and an active site glutamate, which are critical for enzyme activity, are located on two β-strands flanking the active site (Fig. 2). In the group of toxins, of which cholera toxin is a member, region I contains an arginine located on a β-strand which is essential for positioning NAD in the catalytic cleft, and also to form a hydrogen bond with the backbone carbonyl of serine 61 which is located in region II. In region III, the glutamate-X-glutamate sequence is essential for function. The second glutamate is essential for transferase and NAD glycohydrolase activities. Replacing the first glutamate with glutamine abolishes transferase activity but preserves NAD glycohydrolase activity. To date, all these toxins and their eukaryotic homologues possess NAD:arginine ADP-ribosyltransferase activity. In the second group of toxins exemplified by diphtheria toxin, a histidine in region I is

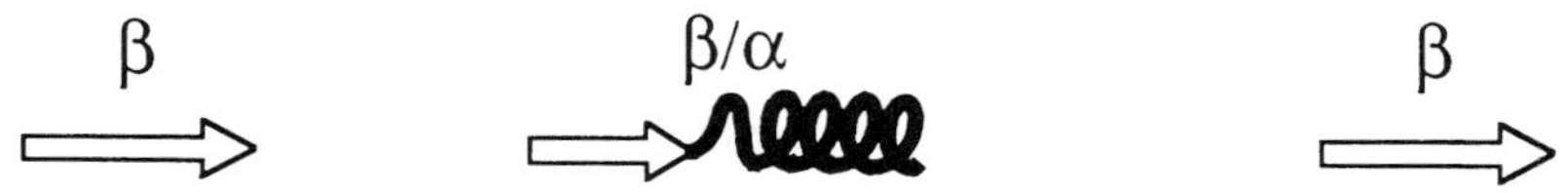

	Region I		Region II		Region III	
		*		*		*
LT	1	NGDKLYRADS	58	GYVSTSLSLR	107	HPYEQEVSAL
CT	1	NDDKLYRADS	58	GYVSTSISLR	107	HPDEQEVSAL
PT	3	PPATVYRYDS	49	AFVSTSSSRR	124	ATYQSEYLAH
iota	289	SNLIVYRRSG	335	NFISTSIGSV	375	YAGEYEVLLN
rART1	173	RCRQVFRGVH	199	GFASASLKNV	235	FPGEEEVLIP
mRt6-1	140	GCRSVYRGTN	164	HFASSSLNRS	204	YTHEEEVLIP
RT6.2	140	DCHSVYRGTK	164	QFTSSSLSKK	204	RPDQEEVLIP
hART4	136	CYEVHYRTKD	160	QFLSTSLLKE	193	FSLKKEVLIP
mART5	155	PGEVVFRGVG	181	QFTSSSVDER	217	FPEEREVLIP
chAT1	158	RCYYVYRGVR	183	QFTSTSLRKE	219	FPSEDEVLIP
		*		* *		*
DT	15	ENFSSYHGTK	50	WKGFYSTDNKYDAAGY	145	SSVEYINNWE
ETA	434	YVFVGYHGTF	466	WRGFYIAGDPALAYGY	550	GRLETILGWP
hART3	88	MNFKDNHGIA	126	REDYIYGFQFKAFHFY	242	KESERITLIP
hPARP	856	NRRLLWHGSR	89	GKGIYFADMVSKSANY	985	LYNEYIVYDI

Fig. 2. Regions of similarity among ADP-ribosyltransferases. Bacterial toxin ADP ribosyltransferases were grouped according to 3-dimensional structure [16]. In the group exemplified by CT, region I contains a critical arginine (asterisk) on a β-strand, region II contains an aromatic amino acid-hydrophobic amino acid-Ser(asterisk)-Thr-Ser-hydrophobic amino acid motif, and region III contains the active site glutamate (asterisk) on a β-strand. In the group exemplified by DT, region I contains a histidine (asterisk) on a β-strand, region II contains a Tyr-X_{10}-Tyr motif within a hydrophobic rich segment of amino acids, and region III possesses the catalytic glutamate (asterisk) on a β-strand. Amino acid position is indicated to the right of each amino acid segment. LT – heat-labile enterotoxin of *Escherichia coli;* CT – cholera toxin; PT – pertussis toxin; iota – *Clostridium perfringens* iota toxin; DT – diphtheria toxin; ETA – *Pseudomonas aeruginosa* exotoxin A; rART1 – rabbit ART1; mRt6-1 – mouse Rt6-1; RT6.2 – rat RT6.2; hART3 – human ART3; hART4 – human ART4; mART5 – mouse ART5; hPARP – human poly(ADP-ribose) polymerase; chAT1 – AT1 from chicken heterophils.

involved in NAD binding, and several closely-spaced aromatic and hydrophobic amino acids in region II are hypothesized to be involved in hydrophobic interactions with the nicotinamide and adenine rings of NAD. Similar to the other group of bacterial toxin transferases, there is an active site glutamate located on a β-strand in region III. The active site glutamate is equivalent to that found in the third position in the glutamate-X-glutamate sequence in the NAD:arginine ADP-ribosyltransferases.

Comparison of the deduced amino acid sequence of the mammalian NAD:arginine ADP-ribosyltransferases and cholera toxin is consistent with the conclusion that the active site amino acid triad is found in the carboxy half of the protein. Further work was done to analyze the catalytic site. Characterization of the active site of ADP-ribosyltransferases was performed by truncations at the amino terminus of ART1. Removal of the amino terminal region resulted in a loss of transferase activity. The enzyme, however, retained NAD glycohydrolase activity. These data support the hypothesis that a catalytic site is present in the carboxy-terminal domain of the transferase [17]

The regions of sequence similarity among bacterial toxin transferases are also apparent in the alignment of the mammalian ADP-ribosyltransferases [18]. Further, site-directed mutagenesis of the rabbit muscle ART1 [18] and computer modeling of mouse ART2 [19] support the notion of a catalytic cleft, similar to that found in cholera toxin, that contains the critical arginine, serine and glutamate-X-glutamate motif in regions I, II, and III, respectively (Fig. 2). The conserved structure of the active site among bacterial toxin and mammalian ADP-ribosyltransferases is consistent with the hypothesis that these enzymes possess a common mechanism of NAD binding and ADP-ribose transfer.

References

1. Okazaki IJ, Moss J: Structure and function of eukaryotic mono-ADP-ribosyltransferases. In: M.P. Blaustein, H. Grunicke, D. Pette, G. Schultz, M. Schweiger, (eds). Reviews of Physiology, Biochemistry and Pharmacology, Springer-Verlag, New York, NY, 1996, pp. 51–104
2. Zolkiewska A, Nightingale MS, Moss J: Molecular characterization of NAD:arginine ADP-ribosyltransferase from rabbit skeletal muscle. Proc Natl Acad Sci USA 89: 11352–11356, 1992
3. Okazaki IJ, Zolkiewska A, Nightingale MS, Moss J: Immunological and structural conservation of mammalian skeletal muscle glycosyl-phosphatidylinositol-linked ADP-ribosyltransferases. Biochemistry 33: 12828–12836, 1994
4. Zolkiewska A, Moss J: Integrin α7 as substrate for a glycosyl-phosphatidylinositol-anchored ADP-ribosyltransferase on the surface of skeletal muscle cells. J Biol Chem 268: 25273–25276, 1993
5. Zolkiewska A, Moss J: Processing of ADP-ribosylated integrin α7 in skeletal muscle myotubes. J Biol Chem 270: 9227–9233, 1995
6. von der Mark H, Durr J, Sonnenberg A, von der Mark K, Deutzmann R, Goodman SL: Skeletal myoblasts utilize a novel β1 -series integrin and not α6β1 for binding to the E8 and T8 fragments of laminin. J Biol Chem 266: 23593–23601, 1991
7. Moss J, Stanley SJ, Nightingale MS, Murtagh JJ Jr., Monaco L, Mishima K, Chen H-C, Williamson KC, Tsai SC: Molecular and immunological characterization of ADP-ribosylarginine hydrolases. J Biol Chem 267: 10481–10488, 1992
8. Zolkiewska A, Thompson WC, Moss J: Interaction of integrin α7β1 in C2C12 myotubes and in solution with laminin. Exp Cell Res 240: 86–94, 1998
9. Ginsberg MH, Du Y, Plow EF: Inside-out integrin signalling. Curr Opin Cell Biol 4: 766–771, 1992
10. Soman G, Haregewoin A, Rom RC, Finberg R: Guanidine group specific ADP-ribosyltransferase in murine cells. Biochem Biophys Res Commun 176: 301–308, 1991
11. Wang J, Nemoto E, Kots AY, Kaslow HR, Dennert G: Regulation of cytotoxic T cells by ecto-nicotinamide adenine dinucleotide (NAD) correlates with cell surface GPI anchored/arginine ADP-ribosyltransferase. J Immunol 153: 4048–4058, 1994
12. Okazaki IJ, Kim H-J, Lesma E, McElvaney G, Moss J: Molecular characterization of a glycosylphosphatidylinositol-linked ADP-ribosyltransferase from lymphocytes. Blood 88: 915–921, 1996
13. Wang J, Nemoto E, Dennert G: Regulation of CTL by ecto-nicotinamide adenine dinucleotide (NAD) involves ADP-ribosylation of a $p56^{lck}$-associated protein. J Immunol 156: 2819–2827, 1996
14. Nemoto E, Yu Y, Dennert G: Cell surface ADP-ribosyltransferase regulates lymphocyte function-associated molecule-1 (LFA-1) function in T-cells. J Immunol 157: 3341–3349, 1996
15. Okazaki IJ, Kim H-J, Moss J: A novel membrane-bound lymphocyte ADP-ribosyltransferase cloned from Yac-1 cells. J Biol Chem 271: 22052–22057, 1996
16. Domenighini M, Rappuoli R: Three conserved consensus sequences identify the NAD-binding site of ADP-ribosylating enzymes, expressed by eukaryotes, bacteria and T-even bacteriophages. Mol Microbiol 21: 667–674, 1996
17. Kim H-J, Okazaki IJ, Takada T, Moss J: An 18-kDa domain of a glycosylphosphatidylinositol-linked NAD:arginine ADP-ribosyltransferase possesses NAD glycohydrolase activity. J Biol Chem 272: 8918–8923, 1997
18. Takada T, Iida K, Moss J: Conservation of a common motif in enzymes catalyzing ADP-ribose transfer. J Biol Chem 270: 541–544, 1995
19. Koch-Nolte F, Petersen D, Balasubramanian S, Haag F, Kahlke D, Willer T, Kastelein R, Bazan F, Thiele H-G: Mouse T cell membrane proteins Rt6-1 and Rt6-2 are arginine/protein mono(ADPribosyl)-transferases and share structure motifs with ADP-ribosylation toxins. J Biol Chem 271: 7686–7693, 1996

Molecular and Cellular Biochemistry **193:** 115–118, 1999.

The CD38-cyclic ADP-ribose signaling system in insulin secretion

Hiroshi Okamoto

Department of Biochemistry, Tohoku University School of Medicine, Sendai, Japan

Abstract

Glucose induces an increase in the intracellular Ca^{2+} concentration in pancreatic β-cells to secrete insulin. CD38 occurs in β-cells and has both ADP-ribosyl cyclase, which catalyzes the formation of cyclic ADP-ribose (cADPR) from NAD^+, and cADPR hydrolase, which converts cADPR to ADP-ribose. ATP, produced by glucose metabolism, competes with cADPR for the binding site, Lys-129, of CD38, resulting in the inhibition of the hydrolysis of cADPR and thereby causing cADPR accumulation in β-cells. Cyclic ADP-ribose then binds to FK506-binding protein 12.6 in the ryanodine receptor Ca^{2+} channel (RyR), dissociating the binding protein from RyR to induce the release of Ca^{2+} from the endoplasmic reticulum. Ca^{2+}/calmodulin-dependent protein kinase II (CaM kinase II) phosphorylates RyR to sensitize and activate the Ca^{2+} channel. Ca^{2+}, released from the RyR, further activates CaM kinase II and amplifies the process. Thus, cADPR acts as a second messenger for Ca^{2+} mobilization to secrete insulin. The novel mechanism of insulin secretion described above is different from the conventional hypothesis in which Ca^{2+} influx from extracellular sources plays a role in insulin secretion by glucose. (Mol Cell Biochem **193**: 115–118, 1999)

Key words: cyclic ADP-ribose, CD38, FK506-binding protein, ryanodine receptor, Ca^{2+}/calmodulin-dependent protein kinase II, insulin, diabetes

Glucose induces an increase in the intracellular Ca^{2+} concentration in pancreatic β-cells of the islets of Langerhans to secrete insulin. Since 1984, this increase in the Ca^{2+} concentration has been explained by the hypothesis of Ashcroft *et al.* in which Ca^{2+} is provided extracellularly (Fig. 1). That is, ATP, produced in the process of glucose metabolism, inhibits the potassium channel, inducing membrane depolarization and the opening of the voltage-dependent Ca^{2+} channels [1]. Recently, we have proposed another model of insulin secretion by glucose via cyclic ADP-ribose (cADPR)-mediated Ca^{2+} mobilization from an intracellular Ca^{2+} pool, the endoplasmic reticulum, (Fig. 1). That is, ATP inhibits the cADPR hydrolase of CD38, causing the accumulation of cADPR, which acts as a second messenger for Ca^{2+} mobilization from the endoplasmic reticulum for insulin secretion [2–6].

Previously, we measured the cADPR content in islets by the cADPR-dependent Ca^{2+} mobilizing activity, and found that the cADPR content in islets was increased by glucose stimulation [2]. Recently, the increase in the cADPR level was confirmed by a radioimmunoassay (RIA) using an anti-cADPR antibody [7, 8], which was kindly provided by Dr. Katada of Tokyo University. Rat pancreatic islets were incubated with low (2.8 mM) and high (20 mM) glucose for 5–40 min, and the cADPR contentof the islet extract was measured by the RIA. The cADPR content in the islets treated with high glucose increased significantly [8].

CD38 has both ADP-ribosyl cyclase and cADPR hydrolase activities [3]. We expressed CD38 cDNA in *E. coli*, purified the CD38 protein and confirmed that the purified CD38 has both ADP-ribosyl cyclase and cADPR hydrolase activities. We found that millimolar concentrations of ATP, the concentration of which is increased in islets by high glucose, markedly inhibited the cADPR hydrolase of CD38, and the inhibition by ATP was in a competitive manner with a Ki value of 4.8 ± 0.53 mM. ATP had almost no effect on the ADP-ribosyl cyclase activity of CD38, and when NAD^+ was used as a substrate, the purified CD38 catalyzed the increased

Address for offprints: H. Okamoto, Department of Biochemistry, Tohoku University School of Medicine, 2-1 Seiryo-machi, Aoba-ku, Sendai 980-8575, Miyagi, Japan

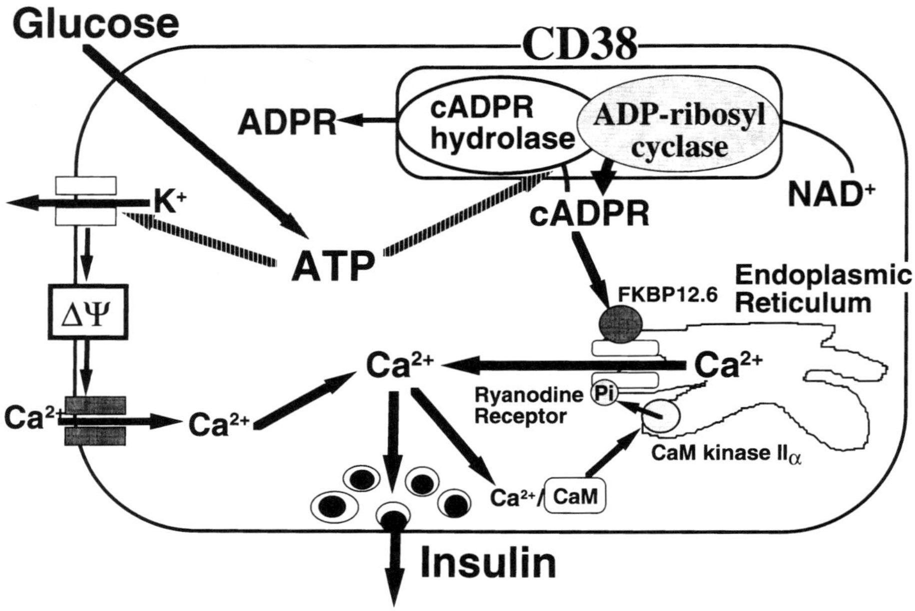

Fig. 1. Insulin secretion by glucose stimulation in β-cells (adapted from [18]). The insulin secretion by the CD38–cyclic ADP-ribose signaling system as described in this paper is shown at right. The conventional insulin secretion mechanism by Ca^{2+}, influx from extracellular sources, proposed by Ashcroft *et al.* [1], is shown at left. In the process of glucose metabolism, millimolar concentrations of ATP are generated, inducing cADPR accumulation by inhibiting the cADPR hydrolase of CD38, and cADPR then acts as a second messenger for intracellular Ca^{2+} mobilization from the endoplasmic reticulum for insulin secretion through RyR. Ca^{2+}/calmodulin (CaM)-dependent protein kinase II (CaM kinase II) phosphorylates RyR to sensitize and activate the Ca^{2+} channel (Pi, phosphorylation of RyR by CaM kinase II) [16]. Ca^{2+}, released from the RyR, further activates CaM kinase II and amplifies the process. In this way, Ca^{2+}-induced Ca^{2+} release (CICR) can be explained. cADPR also binds to FKBP 12.6 to release Ca^{2+}, dissociating FKBP 12.6 from RyR. cADPR, cyclic ADP-ribose. ADPR, ADP-ribose.

formation of cADPR in a manner dependent on the ATP concentration. These results suggest that ATP led to an accumulation of cADPR in islets, inhibiting the cADPR hydrolase of CD38. Competitive inhibition of the cADPR hydrolysis by ATP suggests that ATP and cADPR bind to the same site of CD38 [9].

To identify the binding site for ATP and/or cADPR, we then labeled the purified CD38 with an ATP analogue, 5′-*p*-fluorosulfonylbenzoyladenosine (FSBA). The FSBA labeling of CD38 was completely inhibited by cADPR and by ATP, indicating that the binding site for ATP and/or cADPR was affected by FSBA. We subjected the labeled CD38 to lysylendopeptidase, and the resultant peptide fragments were separated by reverse-phase HPLC. Only the fragment of an 8-amino-acid sequence ILLWSRIK (residues 122–129) was immunoreactive with anti-FSBA, and Lys-129 was suggested as the FSBA-labeled residue, because this residue escaped identification by Edman degradation [9].

We introduced site-directed mutation into the CD38 cDNA by replacing the lys-129 codon with that of alanine and arginine. When the mutant CD38s were incubated with FSBA, the mutant CD38s (K129A- and K129R-CD38s) were not labeled with FSBA, indicating that Lys-129 of CD38 is the actual site of FSBA labeling. We next analyzed the enzymic activities of the mutant CD38s. The mutants did not catalyze the hydrolysis of cADPR to ADPR, whereas their catalytic activity to form cADPR and ADPR from NAD^+ still remained [9]. We next tested the wild and mutant CD38s for their ability to bind to cADPR. cADPR was eluted in a gel filtration chromatography (PD-10, Pharmacia) around fraction 16 (see ref. [9], Fig. 4). After incubation with wild CD38 at 4°C, a new elution peak appeared around fraction 3, in which most of the immunoreactivity against CD38 was recovered, indicating that the elution peak represented the CD38-cADPR complex. On the other hand, no peak around fraction 3 appeared in chromatograms using mutant CD38s (K129A and K129R), whereas the CD38 immunoreactivity was recovered around fraction 3. These results indicate that the wild type CD38 did bind to cADPR, but the mutant CD38s did not, and that Lys-129 participates in the cADPR binding [9].

We have already reported that Cys-119 and Cys-201 of CD38 are essential for the hydrolase reaction [10]. In addition, as described above, when mutations at or around Lys-129 occur in CD38, cADPR metabolism cannot be regulated by ATP, which is generated in the process of glucose metabolism. Such a mutation in CD38 could be a predisposing factor for diabetes mellitus. In fact, most recently, a mutation of Arg-140, which is encoded in exon 3 of CD38 gene with Lys-129 [11], to Trp was found in Japanese diabetic patients [12].

cADPR has been thought to activate the ryanodine receptor (RyR) to release Ca^{2+}, from the intracellular stores [13, 14]. Our recent experiment indicates that the type 2 RyR is expressed in rat and mouse islets [8, 15]. In the presence of calmodulin, rat islet microsomes were sensitized to cADPR at much lower concentrations for Ca^{2+} release, and the Ca^{2+} release was greatly increased. We then examined the effects of calmodulin and Ca^{2+}/calmodulin-dependent protein kinase (CaM kinase) II inhibitors on the insulin secretion from islets as well as on the cADPR/calmodulin-induced Ca^{2+} release from islet microsomes. A calmodulin inhibitor, W-7, and CaM kinase II inhibitors such as KN-62 and KN-93 completely inhibited the Ca^{2+} release and the insulin secretion, but the non-inhibitory analogues, W-5, KN-04 and KN-92 did not. These results strongly suggest that the activation of cADPR-mediated Ca^{2+} mobilization by calmodulin is achieved by activation of CaM kinase II but not by the direct interaction of calmodulin with the Ca^{2+} release machinery [16].

We next examined the existence of CaM kinase II in rat islet microsomes by immunoblot analysis and revealed that the α isoform of CaM kinase II but not the β isoform actually exists in the microsomes. To confirm the essential requirement of CaM kinase II activity for the cADPR-mediated Ca^{2+} mobilization, we added the active 30 kDa chymotryptic fragment of the CaM kinase II, which lacks the autoinhibitory domain and is therefore activated in the absence of calmodulin, to islet microsomes and measured the cADPR-mediated Ca^{2+} release. The cADPR-mediated Ca^{2+} release was as fully and dose-dependently activated by the 30 kDa fragment of CaM kinase II as by the addition of calmodulin to microsomes [16].

These results suggest that the cADPR-mediated Ca^{2+} mobilization for insulin secretion is achieved by the calmodulin-activated α-isoform of CaM kinase II. Possibly, the activated kinase phosphorylates RyR to sensitize the Ca^{2+} channel for the cADPR signal. Furthermore, Ca^{2+}-induced Ca^{2+} release (CICR) has become known as an important regulatory mechanism of the intracellular Ca^{2+} concentration in almost all cells. CICR can be explained by a series of amplification processes in which CaM kinase II is activated by Ca^{2+} released from RyR, and the activated enzyme phosphorylates RyR to further release Ca^{2+}.

Our recent experiments suggest that cADPR does not bind directly to RyR but may act on the receptor through a mediator such as FK506-binding protein (FKBP) 12.6 to release Ca^{2+} [15]. FK506 is one of the most widely used immunosuppressive agents. When rat islet microsomes were treated with FK506, the Ca^{2+} release by cADPR from the microsomes was reduced depending on the concentration of FK506. This result suggests that cADPR and FK506 may act on RyR through a common mediator [15]. The cellular target for FK506 is thought to be FKBP 12 and FKBP 12.6. Therefore, we isolated FKBP 12 and FKBP 12.6 cDNAs from a rat islet cDNA library. Rat FKBP 12 is composed of 108 amino acids and is highly conserved among human, mouse, bovine and rabbit FKBP 12. Rat FKBP 12.6 is also a 108-amino acid protein and completely conserved in human and bovine FKBP 12.6. We then isolated microsomes from rat islets and carried out immunoblot analyses: Although islet microsomes did not contain FKBP 12, the microsomes contained FKBP 12.6 [15].

Next, we examined the binding of FKBP 12.6 to cADPR. cADPR was found to bind to FKBP 12.6 at a M value of 35 nM. The cADPR-binding was inhibited by FK506 and neither structurally nor functionally related analogues of cADPR such as NAD^+, ATP, ADP-ribose, nicotinamide, inositol 1,4,5-trisphosphate (IP_3) and ryanodine inhibited the cADPR binding to FKBP 12.6. These results indicate that FKBP 12.6 acts as a cADPR-binding protein and strongly suggest that cADPR is the actual ligand for FKBP 12.6 since FK506 does not normally exist in mammalian cells [15].

FKBP 12.6 occurs in islet microsomes. However, when islet microsomes were treated with cADPR, FKBP 12.6 dissociated from the microsomes and moved to the supernatant, releasing Ca^{2+} from the intracellular stores. The microsomes that were treated by cADPR and were then devoid of FKBP 12.6 did not show Ca^{2+} release by cADPR. From these results, it is strongly suggested that, when cADPR binds to FKBP 12.6 in the islet microsome RyR and causes the dissociation of FKBP 12.6 from RyR to form FKBP 12.6-cADPR complex, the channel activity of RyR is thereby increased to release Ca^{2+}, from the endoplasmic reticulum [15].

Figure 1 summarizes our hypothesis of cADPR-mediated insulin secretion. In the process of glucose metabolism, millimolar concentrations of ATP are generated, inducing cADPR accumulation by inhibiting the cADPR hydrolase of CD38, and cADPR then acts as a second messenger for intracellular Ca^{2+} mobilization from the endoplasmic reticulum for insulin secretion through RyR. Furthermore, RyR can be activated by CaM kinase II. The novel mechanism of insulin secretion described in this paper is different from the conventional hypotheses such as the IP_3 hypothesis and the ATP-sensitive K^+ channel hypothesis. In the ATP-sensitive K^+ channel hypothesis [1], Ca^{2+} influx from extracellular sources plays a role in insulin secretion by glucose. Insulin secretion by glucose is composed of two phases. Available evidence suggests that the first phase of insulin secretion may be due to the Ca^{2+} influx from extracellular spaces. On the other hand, the novel mechanism of insulin secretion presented here may work in the second phase of glucose-induced insulin secretion. IP_3 has been thought to be a second messenger for Ca^{2+} mobilization from intracellular stores [17]. As described earlier [2], cADPR induces Ca^{2+} release from pancreatic islet microsomes but IP_3 does not. In cerebellum microsomes, both cADPR and IP_3

induced Ca^{2+} release. Our recent study [8, 18] showed that the increase of the cADPR content by glucose did not occur in diabetic β-cells such as *ob/ob* mouse islets and RINm5F cells. Microsomes of these diabetic β-cells released Ca^{2+} in response to IP_3 but not in response to cADPR. In the diabetic β-cells, CD38 (ADP-ribosyl cyclase/cADPR hydrolase) and type 2 ryanodine receptor mRNAs were scarcely detected and, in contrast, an increased expression of IP_3 receptor mRNAs was observed. Cells can utilize two second messengers, cADPR and IP_3, for Ca^{2+} mobilization depending on the species of cells and differences in cellular conditions, physiological or pathological, performing a variety of cellular functions.

References

1. Ashcroft FM, Ashcroft SJH: Insulin – Molecular Biology to Pathology. Oxford University Press, Oxford, 1992, pp 97–150
2. Takasawa S, Nata K, Yonekura H, Okamoto H: Cyclic ADP-ribose in insulin secretion from pancreatic β cells. Science 259: 370–373, 1993
3. Takasawa S, Tohgo A, Noguchi N, Koguma T, Nata K, Sugimoto T, Yonekura H, Okamoto H: Synthesis and hydrolysis of cyclic ADP-ribose by human leukocyte antigen CD38 and inhibition of the hydrolysis by ATP. J Biol Chem 268: 26052–26054, 1993
4. Okamoto H, Takasawa S, Tohgo A: New aspects of the physiological significance of NAD, poly ADP-ribose and cyclic ADP-ribose. Biochimie 77: 356–363, 1995
5. Kato I, Takasawa S, Akabane A, Tanaka O, Abe H, Takamura T, Suzuki Y, Nata K, Yonekura H, Yoshimoto T, Okamoto H: Regulatory role of CD38 (ADP-ribosyl cyclase/cyclic ADP-ribose hydrolase) in insulin secretion by glucose in pancreatic β cells: Enhanced insulin secretion in CD38 transgenic mice. J Biol Chem 270: 30045–30050, 1995
6. Okamoto H, Takasawa S, Tohgo A, Nata K, Kato I, Noguchi N: Synthesis and hydrolysis of cyclic ADP-ribose by human leukocyte antigen CD38: Inhibition of hydrolysis by ATP and the physiological significance. Meth Enzymol 280: 306–318, 1997
7. Takahashi K, Kukimoto I, Tokita K, Inageda K, Inoue S, Kontani K, Hoshino S, Nishina H, Kanaho Y, Katada T: Accumulation of cyclic ADP-ribose measured by a specific radioimmunoassay in differentiated human leukemic HL-60 cells with all-trans-retinoic acid. FEBS Lett 371: 204–208, 1995
8. Takasawa S, Akiyama T, Nata K, Kuroki M, Tohgo A, Noguchi N, Kobayashi S, Kato I, Katada T, Okamoto H: Cyclic ADP-ribose and inositol 1,4,5-trisphosphate as alternate second messenger for intracellular Ca^{2+} mobilization in normal and diabetic β-cells. J Biol Chem 273: 2497–2500, 1998
9. Tohgo A, Munakata H, Takasawa S, Nata K, Akiyama T, Hayashi N, Okamoto H: Lysine 129 of CD38 (ADP-ribosyl cyclase/cyclic ADP-ribose hydrolase) participates in the binding of ATP to inhibit the cyclic ADP-ribose hydrolase. J Biol Chem 272: 3879–3882, 1997
10. Tohgo A, Takasawa S, Noguchi N, Koguma T, Nata K, Sugimoto T, Furuya Y, Yonekura H, Okamoto H: Essential cysteine residues for cyclic ADP-ribose synthesis and hydrolysis by CD38. J Biol Chem 269: 28555–28557, 1994
11. Nata K, Takamura T, Karasawa T, Kumagai T, Hashioka W, Tohgo A, Yonekura H, Takasawa S, Nakamura S, Okamoto H: Human gene encoding CD38 (ADP-riboseyl cyclase/cyclic ADP-ribose hydrolase): Organization, nucleotide sequence and alternative splicing. Gene 186: 285–292, 1997
12. Yagui K, Shimada F, Miura M, Hashimoto N, Suzuki Y, Tokoyama Y, Nata K, Tohgo A, Ikehata F, Takasawa S, Okamoto H, Makino H, Saito Y, Kanatsuka A: A missense mutation in the CD38 gene, a novel factor for insulin secretion: Association with Type II diabetes mellitus in Japanese subjects and evidence of abnormal function when expressed *in vitro*. Diabetologia 1998 (in press)
13. Hua S-Y, Tokimasa T, Takasawa T, Furuya Y, Nohmi M, Okamoto H, Kuba K: Cyclic ADP-ribose modulates Ca^{2+} release channels for activation by physiological Ca^{2+} entry in bullfrog sympathetic neurons. Neuron 12: 1073–1079, 1994
14. Ebihara S, Sasaki T, Hida W, Kikuchi Y, Oshiro T, Shimura S, Takasawa S, Okamoto H, Nishiyama A, Akaike N, Shirato K: Role of cyclic ADP-ribose in ATP-activated potassium currents in alveolar macrophages. J Biol Chem 272: 16023–16029, 1997
15. Noguchi N, Takasawa S, Nata K, Tohgo A, Kato I, Ikehata F, Yonekura H, Okamoto H: Cyclic ADP-ribose binds to FK506-binding protein 12.6 to release Ca^{2+} from islet microsomes. J Biol Chem 272: 3133–3136, 1997
16. Takasawa S, Ishida A, Nata K, Nakagawa K, Noguchi N, Tohgo A, Kato I, Yonekura H, Fujisawa H, Okamoto H: Requirement of calmodulin-dependent protein kinase II in cyclic ADP-ribose-mediated intracellular Ca^{2+} mobilization. J Biol Chem 270: 30257–30259, 1995
17. Berridge MJ, Irvine RF: Inositol phosphates and cell signalling. Nature 341: 197–205, 1989
18. Okamoto H, Takasawa S, Nata K: The CD38-cyclic ADP-ribose signalling system in insulin secretion: Molecular basis and clinical implications. Diabetologia 40: 1485–1491, 1997

Molecular and Cellular Biochemistry **193:** 119–125, 1999.

Newly discovered anti-inflammatory properties of the benzamides and nicotinamides

Ronald W. Pero,[1] Bengt Axelsson,[2] Dietmar Siemann,[3] David Chaplin[4] and Graeme Dougherty[5]

[1]*Section for Molecular Ecogenetics, Department of Cell and Molecular Biology, University of Lund, Wallenberg Laboratory, Lund;* [2]*Astra-Draco AB, Lund, Sweden;* [3]*Radiation Oncology, University of Florida; Shanda Cancer Center, Gainesville, Florida, USA;* [4]*Tumour Microcirculation Group, Gray Laboratory, Mount Vemon Hospital, Northwood, Middlesex, UK;* [5]*Terry Fox Laboratories, British Columbia Cancer Research Center, Vancouver, British Columbia, Canada*

Abstract

Our laboratory has concentrated on the possible regulation the benzamides and nicotinamides may have on the processes of DNA repair and apoptosis. Recent reports [14–16] have suggested that both apoptosis and inflammation are regulated by the transcription factor NF-kB. We have initiated studies regarding the hypothesis that the benzamides and nicotinamides could inhibit the production of tumor necrosis factor alpha (TNF_{alpha}) and the inflammatory response as well as induce apoptosis via inhibition of NF-kB. Our data have shown that nicotinamide and two N-substituted benzamides, metoclopramide (MCA) and 3-chloroprocainamide (3-CPA), gave dose dependent inhibition of lipopolysacharide induced TNF_{alpha} in the mouse within the dose range of 10–500 mg/kg. Moreover, lung edema was prevented in the rat by 3 × 50 mg/kg doses of 3-CPA or MCA, and 100–200 μM doses of MCA could also inhibit NF-kB in Hela cells. Taken together these data strongly support the notion that benzamides and nicotinamides have potent anti-inflammatory and antitumor properties, because their primary mechanism of action is regulated by inhibition at the gene transcription level of NF-kB, which in turn inhibits TNF_{alpha} and induces apoptosis. (Mol Cell Biochem **193**: 119–125, 1999)

Key words: benzamides, nicotinamides, apoptosis, inflammation, NF-kB, DNA repair

Abbreviations: PARP – poly (ADP-ribose) polymerase; NF-kB – nuclear fraction kappa B; D2 – dopamine D2; TNF_{alpha}, – tumor necrosis factor alpha; mu TNF_{alpha} – murine tumor necrosis factor alpha; LPS – lipopolysacharride; 5-HT3,5-hydroxytryptamine3; FCS – fetal calf serum; PBS – phosphate buffered saline; nMCA – neutral metoclopramide or Neu-Sensamide; 3-CPA – 3-chloroprocainamide; CAT – chloramphenicol acethyltransferase; VEGF – vascular endothelial growth factor.

Introduction

The benzamides and nicotinamides represent a class of drugs, which have experienced an unusually broad application for clinical development. So far these agents are being or have already been developed as clinically relevant antipsychotic (sulpiride), antiemetic (metoclopramide), antiarrhythmic (procainamide), diabetic (nicotinamide), local anesthetic (lidocaine), radio-/chemosensitizing (Neu-Sensamide), and antitumor (3-chloroprocainamide) drugs [1–4]. The diverse clinical applications of these drugs is paralleled by an equally diverse mode of action profile which involves effects on tumor microcirculation, inhibition of DNA repair and PARP (poly ADP ribose polymerase), binding to dopamine (D_2) and hydroxytryptamine ($5\text{-}HT_3$) receptors, effects on ADP-ribosyl cyclase and Ca^{2+} homeostasis, and selective induction of apoptosis [1–13].

Address for offprints: R.W. Pero, Section of Molecular Ecogenetics, Department of Cell and Molecular Biology, University of Lund, Wallenberg Laboratory, Box 7031, 220 07 Lund, Sweden

In an effort to better understand these broad based biochemical mechanisms of action, our laboratory has concentrated on the possible regulation these drugs may have on PARP activity and thereby the processes of DNA repair and apoptosis. In a series of scientific reports, it was established that (i) low dose radiation but not high dose radiation was sensitized by the model N-substituted benzamide, metoclopramide (MCA), (ii) repeated doses of radiation and MCA were more effective than single doses in combination as a radiosensitizing mixture, (iii) tumors with high constitutive levels of apoptosis were more sensitive to MCA treatment than tumors with lower spontaneous apoptosis, (iv) direct antitumor effects could be observed for, repeated but not single doses of MCA, and finally (v) MCA induced a saturating dose dependent cytotoxicity by apoptosis without affecting necrotic cell death of HL-60 cells *in vitro* [4, 10, 11]. These recent data are most consistent with the fact that at least the N-substituted benzamides and nicotinamides have, as an important component of their cytotoxic mechanism of action, a selective induction of apoptosis.

Further support for this contention can be found in the molecular biology of apoptosis [14–16]. DNA damage such as that induced by radiation and chemotherepeutic drugs activates the transcription factor, nuclear fraction kappa B (NF-kB), which in turn activates several other genes that protect cells from cytotoxic exposure to DNA damage. Thus, NF-kB activation protects cells against apoptotic induced cytotoxicity, and a drug that inhibits NF-kB could be a good inducer of apoptosis. Based on this knowledge of the molecular biology of apoptosis, it was reasoned that the benzamides and nicotinamides may express their cytotoxic properties through inhibition of NF-kB expression.

The control of the inflammatory cascade is also known to be mediated by NF-kB expression [14–17]. The major pro-inflammatory cytokine, TNF_{alpha}, is induced by NF-kB. Conversely if NF-kB is inhibited an anti-inflammatory response is elicited. In the present study the hypothesis that the benzamides and nicotinamides inhibit NF-kB expression was examined by evaluating their effect on (i) inhibition of TNF_{alpha} and the inflammatory response and (ii) the activation of apoptosis.

For more than 25 years, pyridinyl-N-substituted benzamides have been known to have antiulcerogenic, sedative and anti-inflammatory properties [18–21]. Because the N-pyridinyl substitution of the carboxamide is a rather bulky molecular addition, alternations in the benzamide portion of the structure have never been considered for drug development of anti-inflammatory drug properties. Contrary to the N-pyridinyl substitutions, the N-substituted analogs of benzamide are quite non-toxic having already been developed for a number of clinical indications. The newly discovered anti-inflammatory properties of the benzamides and nicotinamides reported on in this study offer an important opportunity to develop alternatives to steroid-directed anti-inflammatory therapies.

Materials and methods

In vivo *inhibition of* TNF_{alpha} *by benzamide and nicotinamide analogs*

Nicotinamide, 3-chloroprocainamide (3-CPA) and metoclopramide (neutral MCA, nMCA or Neu-Sensamide) at final doses of 10–500 mg/kg were injected intraperitoneally (i.p.) in CBA mice in volumes of 0.1 ml. saline per 10 grams body weight 15 min before a 1 mg/kg i.p. injection of lipopolysacharide (LPS, *E-coli* serotype 0111:B4). Blood samples were collected 2 h later into a heparinized syringe, centrifuged at 1614 × g for 5 min, the plasma removed from the supernatant fraction, and then stored at –20°C until analysis of TNF_{alpha}. The production of TNFalpha in plasma following a standardized inductive treatment *in vivo* by LPS was estimated using a murine TNF_{alpha} solid-phase ELISA and a multiple antibody sandwich principle. High binding flat-bottomed 96 well microtiter plates were pre-coated with 10 μg/ml (i.e. 50 μl per well) Hamster antimouse TNF_{alpha} (Genzyme 1221) in 0.1 M $NaHCO_3$ (pH 8.2) at 40°C which is used to capture any TNF_{alpha} present in the plasma samples or standards. The plates ware washed twice with 0.01% Tween 20 in phosphate buffered saline, and then blocked for 30 min at 37°C using 10% fetal calf serum (FCS) in phosphate buffered saline (PBS). Next the blocking buffer was removed by washing two times with the Tween 20 buffer. Standard murine TNF_{alpha} (i.e. 0–2000 picograms/ml, Genzyme) and plasma sample aliquots of 100 μl per well (diluted with 10% FCS in PBS) were left overnight at 4°C and then washed 4 times with Tween buffer. Followed by addition of 50 μl polyclonal rabbit antimouse TNF_{alpha} (1:500 dilution, Genzyme IP400) in 10% FCS in PBS, which binds to the captured TNF_{alpha}, was incubated for 2 h at room temperature and then washed 4 times with the Tween 20 buffer. Then 100 μl per microtiter-well of goat antirabbit IgG peroxidase conjugated polyclonal antibody (1:200, Sigma A6154) prepared in 10% FCS in PBS was added for 1 h at room temperature followed by 4 washes with the Tween 20 buffer. A substrate solution of 100 μl per microtiter-well containing 0.4 mg/ml paraphenyldiamine (Sigma P7288) in 0.05 M citrate-phosphate buffer, pH 5.0 and hydrogen peroxide (20 μl of 30% hydrogen peroxide per 50 ml of paraphenyldiamine solution) was added to initiate a peroxidase catalyzed color change that was subsequently stopped by addition of 25 μl of 3 M H_2SO_4. The $A_{492\ nm}$ was proportional to the concentration of TNF_{alpha} and the unknown levels in the plasma samples were determined by comparison to a standard curve of known concentrations of plasma-spiked, purified TNF_{alpha} samples.

Anti-inflammatory activity of the benzamides and nicotinamides

Lung edema was evaluated in rats weighing about 250 gm after Sephadex G-200 Superfine (Pharmacia-Upjohn, Sweden) was suspended in 0.9% NaCl (5 mg/ml) and instilled intratracheally (i.t., 5 mg/kg). The drugs, nMCA (Neu-Sensamide, Oxigene) and 3-CPA (supplied by Oxigene) were given three times orally (gavage, 3×50 mg/kg)) or intramuscularly (i.m., 3×50 mg/kg) at the same time, 8 and 18 h later. The next day the rats where killed, the lungs excised, and their weights recorded in mg. Means and standard error of the means were calculated. GLM two-tailed significance probability test was used for significance calculations.

Inhibition of the pro-inflammatory cytokine production by 3-CPA

THP-1 cells, a human monocytic cell line (ATCC), were maintained in RPMI 1640 medium supplemented with 10% fetal calf serum, glutamine (2 mM), $1 \times$ nonessential amino acids, Hepes (10 mM), sodium pyruvate (1 mM), penicillin (50 U/ml), streptomycin (50 µg/ml) and β-mercaptoethanol (0.5 µM). The THP-1 cells were plated at a density of 200,000 cells/well in 96 well plates. 3-CPA was added to the cells to final concentrations of 0.2, 1 or 5 mM, 1 h before stimulation with 100 ng LPS/ml. After 5 h of stimulation the plates were centrifuged, cell-free supernatants collected and stored at –80°C. In the supernatants, cytokines IL-1β, IL-6, IL-8 and TNF-α were analyzed using specific ELISAs (R&D Systems). Cytokine levels are expressed as percentage of values obtained in LPS stimulated control cultures.

Inhibition of NF-kB mediated transcription by MCA

The activation of NF-kB probably functions to transcriptionally up-regulate a gene or genes encoding for proteins involved in providing protection against apoptosis [14–16]. In an effort to explain the mode of action of MCA as an inducer of apoptosis and an inhibitor of the inflammatory cascade [17], we have studied whether the drug could inhibit NF-kB. Hela cells were transfected with a bacterial plasmid comprised of a promoter containing three NF-kB sites and a CAT (chloramphenicol acethyltransferase) reporter gene. The cells were treated with ± 100 or 200 µM nMCA 30 min prior to administering 5 Gy radiation to activate NF-kB, and 4 h later the levels of NF-kB expression were estimated by calculation in % CAT reporter gene activity relative to constitutive, non-irradiated control levels.

Results and discussions

Structural analysis of benzamides and nicotinamides that induce apoptosis

The benzamides and nicotinamides have been known for many years to inhibit DNA repair and PARP activity [2, 24]. However, the mode of action on DNA repair was believed to be via a competitive-type of inhibition whereby these agents had affinity for the NAD binding co-substrate site on the PARP enzyme. Later on, it was discovered that benzamides and nicotinamides could radiosensitize *in vivo* by increasing blood flow and overcoming tumor hypoxia whereas effects on DNA repair inhibition were not observed [2]. More recently our laboratory [8] published data supporting that the inhibition of DNA repair by the N-substituted benzamide, MCA, is non-competitive. In addition, MCA was shown to induce apoptosis *in vitro* in HL-60 cells at doses that are without any effects on necrosis [10]. In an effort to resolve these differing research experiences reported in the scientific literature, our laboratory has studied the mode of action of the antitumor effects of the benzamides and nicotinamides based on structure-function considerations. A summary of the results of a structure-function analysis is presented in Table 1. It was established that the mode of antitumor action of the N-substituted benzamides and nicotinamides (e.g. MCA, 3-CPA) could quite easily be distinguished from the non-N-substituted analogs (e.g. nicotinamide, 3-aminobenzamide). Not all N-substituted analogs (e.g. procainamide, remoxi-

Table 1. Mechanism of action differences between N-substituted and non-N-substituted benzamide and nicotinamide analogs

N-Substituted Analogs	Non-N-Substituted Analogs	References
1. Non-competitive inhibition of DNA repair	Competitive inhibition of DNA repair	8,9
2. Induce apoptosis but not necrosis	Do not induce apoptosis or necrosis	4, 10
3. Directly inhibit cell proliferation	Do not inhibit cell proliferation	4
4. Clonogenic inhibition	No clonogenic inhibition	unpublished data
5. Direct antitumor effects	No direct antitumor effects	2,11
6. Low dose radiosensitizer	High dose radiosensitizer	2,11
7. Generally react with radiation	Generally do not react with radiation	4,6
8. D_2 and 5-HT_3 binding	No D_2 and 5-HT_3 binding	1,3,5,10
9. Directly activates PARP	Directly inhibits PARP	4,24
10. No effect on tumor blood flow	Profound effect on tumor blood flow	2,4,25

pride, 4-N-succinyl-amino MCA) or non N-substituted analogs (e.g. isonicotinamide, picolinamide or nicotinic acid) had antitumor properties [4]. However, if they did, the N-substituted analogs: (i) non-competitively inhibited PARP, (ii) selectively induced apoptosis, (iii) inhibited cell proliferation and clonogenicity, (iv) had direct antitumor effects, (v) sensitized low but not high dose radiation, (vi) were sensitive to radiolysis, (vii) bound to D_2 and 5-HT_3 receptors, (viii) directly activated PARP *in vitro* by inducing cellular DNA damage, and (ix) did not increase blood flow. Typically the non N-substituted analogs had the opposite effect (Table 1).

The effect of benzamides and nicotinamides on the pro-inflammatory cytokine, TNF_{alpha}

Because at least the N-substituted benzamides and nicotinamides could selectively induce apoptosis (Table 1), it was hypothesized that the transcription factor, NF-kB, might become inhibited, thus preventing programmed cell death that resulted in an enhanced level of apoptosis. In order to examine this hypothesis we have studied if TNF_{alpha} production, which is under the same transcriptional control as the apoptotic prevention pathway [14–17], could be inhibited in vivo by exposure to benzamides and nicotinamides. For this purpose, we have induced TNF_{alpha} production in mice by giving i.p. injections of 1 mg/kg LPS where the mice had already been previously treated for 2 h with ± 10–500 mg/kg i.p. doses of nicotinamide, MCA or 3-CPA. As can be seen in Fig. 1, all 3 analogs reduced in a dose dependent manner, the production of Mu TNF_{alpha}. These data were taken as evidence that benzamides and nicotinamides could mediate the two pathways known to be involved under the transcriptional control of NF-kB.

However, the non-N-substituted analog, nicotinamide, could not induce apoptosis (Table 1, ref. [4]), although it was very effective at inhibiting mu TNF_{alpha} production (Fig. 1). In fact nicotinamide has been shown previously to inhibit nitric oxide production, but this was believed to be due to its PARP inhibition and not because of any consequential effect on inflammatory cytokines [26]. Hence, we have hypothesized that the TNF_{alpha} inhibition pathway may be yet another previously unknown mechanism shared by most of the benzamides and nicotinamides, which could convey antitumor as well as anti-inflammatory properties. For example, (i) TNF_{alpha} has been shown to promote NF-kB binding to the p53 promoter [27], (ii) p53 is also upregulated by hypoxia [28], and (iii) p53 in turn provides a growth advantage to turnors via stimulation of VEGF (vascular endothelial growth factor) production [29]. According to this hypothesis, inhibition of NF-kB by both the N-substituted and non-N-substituted benzamides and nicotinamides would also inhibit hypoxia-driven tumor angiogenesis due to suppressed p53 expression or direct modulation of the VEGF promoter by NF-kB. The prediction would be that the benzamides and nicotinamides in general could in turn have toxic effects on metastatic tumor growth, because new capillaries are necessary to grow tumors beyond the micrometastatic stage. However, it is clear that further research is needed into the molecular biology of these agents in order to substantiate our findings and speculations before any additional drug discovery is implemented.

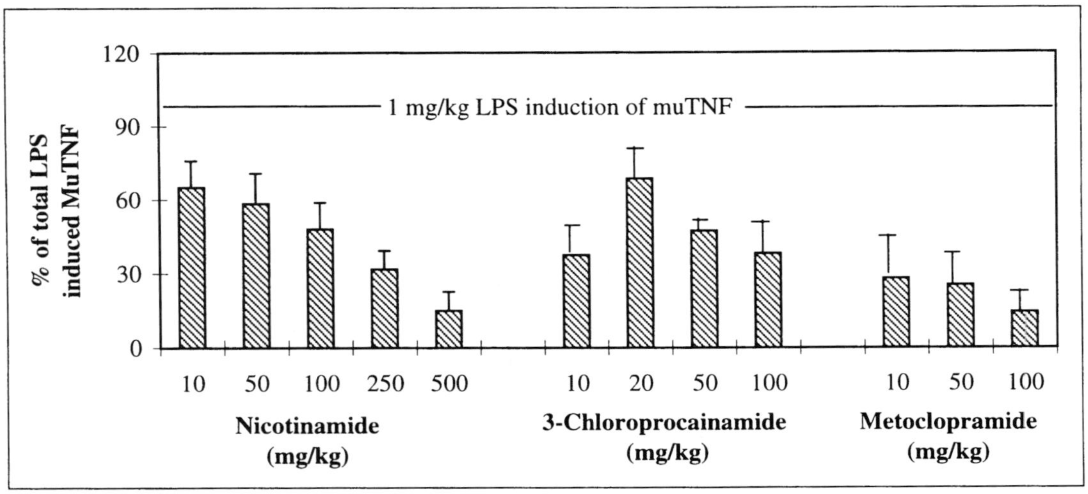

Fig. 1. Dose response inhibition of mouse TNF_{alpha}, production in plasma induced from a standardized i.p. injection of 1 mg/ml lipopolysacharride (LPS) administered 2 h before i.p. injections of 10–500 mg/kg of nicotinamide, metoclopramide (nMCA, Neu-Sensamide) and 3-chloro procainamide. Each bar represents the results from 9 mice. * all treatments are significantly different from LPS controls, $p < 0.05$, *t*-test.

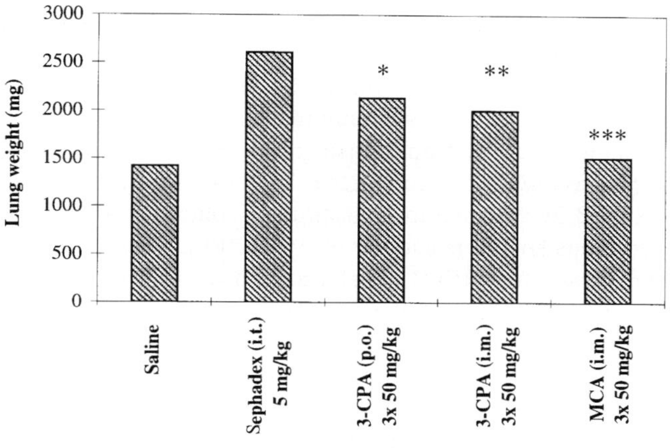

Fig. 2. Inhibition of Sephadex-induced lung ederna in the rat by metoclopramide and 3-chloro procainamide. $p < 0.001$ for Sephadex treated group compared to saline control group; *, **, *** = $p < 0.05$, 0.01, 0.001 for Sepadex drug-treated groups compared with the Sephadex treated group, respectively.

The effect of N-substituted benzamides on the inflammatory response

Although TNF_{alpha} is a major mediating cytokine of the inflammatory response cascade several other cytokines (e.g. interleukins), enzymes, adhesion molecules and receptors are also involved and needed to perpetuate the inflammatory response [17]. Therefore, it was important to demonstrate that the inhibition of TNF_{alpha}, by benzamides and nicotinamides also resulted in inhibition of the total inflammatory response. For this purpose inflammation was evaluated *in vivo* in a rat model where lung edema was induced by intratracheal instillation of Sephadex followed by simultaneous, 8 and 18 h oral or i.m. administration of the N-substituted benzamides, MCA or 3-CPA. The data displayed in Fig. 2. show that the Sephadex-induced lung edema estimated by lung weight was significantly inhibited by MCA and 3-CPA. It was concluded that these benzamide analogs, which could also be shown to inhibit TNF_{alpha}, were effective as anti-inflammatory agents.

Inhibition of pro-inflammatory cytokines IL-1B, IL-6, IL-8 and TNF_{alpha} by 3-CPA

Another way to indirectly establish that benzamide analogs have potent anti-inflammatory properties is to determine if other pro-inflammatory cytokines in addition to TNF_{alpha} could be inhibited by exposure to the model benzamide, 3-CPA. For this purpose, human monocytic cells (THP-1) were exposed to 3-CPA just before 5 h stimulation with LPS to induce cytokine production. Figure 3 clearly shows that a dose dependant inhibition of TNF_{alpha} was paralleled by an equivalent inhibition of IL-1B, IL-6 and IL-8. These data strongly support the anti-inflammatory properties of the benzamides at the biochemical level.

NF-kB inhibition by MCA

Generally speaking NF-kB is a transcription factor that is activated by DNA damaging agents such as radio- and

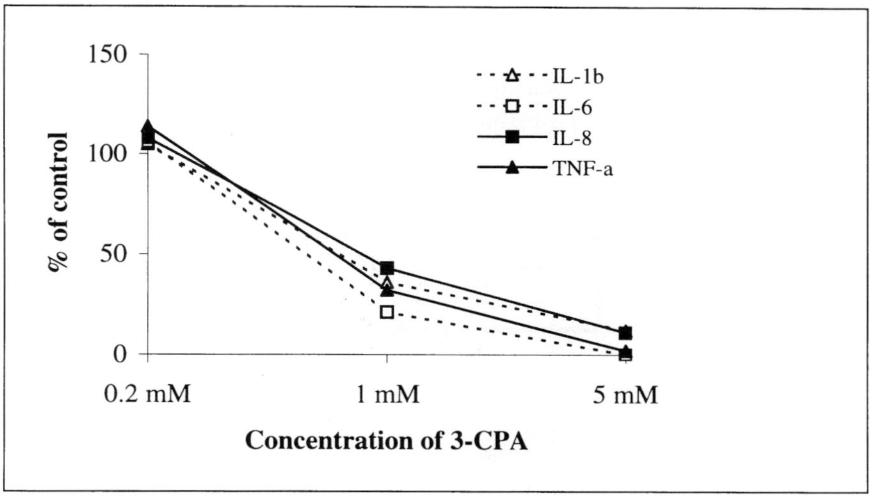

Fig. 3. Inhibition of lipopolysaccharide (LPS) induced pro-inflammatory cytokine production in TPH-1 human monocytic cells. Data points are the average of duplicate determinations.

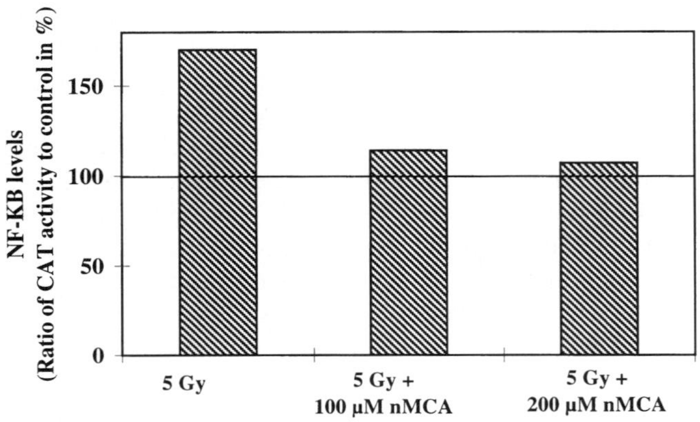

Fig. 4. Hela cells transfected with a bacterial plasmid comprised of a promoter containing three NF-kB sites and a CAT (chloramphenicol acethyltransferase) reporter gene were treated with +/- 100–200 μM metoclopramide (nMCA) 30 min prior to administering +/- 5 Gy radiation to activate NF-kB, and 4 h later the levels of NF-kB expression were estimated by calculation in % CAT reporter gene activity relative to constitutive, non-irradiated control levels.

chemotherapies, but the activation unfortunately results in protection against apoptotic tumor cell cytotoxicity as well as mediating the inflammatory response [14–17]. The data reported in this paper strongly support the hypothesis that benzamides and nicotinamides can modulate both the pathways known to be under the transcriptional control of NF-kB, namely apoptosis and inflammation. Therefore, we have directly examined this possibility by estimating the ability of the model N-substituted benzamide, nMCA, to inhibit the radiation-induced activation of NF-kB in Hela cells transfected with a bacterial plasmid comprised of a promoter containing three NF-kB binding sites and a CAT (chloramphenicol acethyltransferase) reporter gene. The results recorded in Fig. 4 showed that 5 Gy of radiation activated NF-kB 70% above the constitutive control (100%) level, whereas in the presence of 100–200 μM nMCA the 5 Gy-activated NFkB level was reduced to 14 and 7%, respectively. Based on these results and others in the scientific literature, a proposed mode of action scheme of the benzamides and nicotinamides is outlined in Fig. 5. This scheme accommodates what is known about the molecular toxicology of apoptosis and inflammation, and proposes that inhibition of NF-kB by the benzamides and nicotinamides represents a novel target for multifunctional drug development in the clinic.

There has been no explanation offered in the literature for why relatively simple molecules such as the benzamides and nicotinamides could be useful in treating so many diverse clinical disorders including cancer, emesis, psychosis, inflammation and arrhythmia. However, we would like to point out that these molecules share a very important naturally occurring analog, nicotinamide. Nicotinamide is not only the immediate precursor to the cellular energy source NAD, but it is the essential co-substrate to the nuclear located PARP enzyme [30]. PARP activity has been shown to be critical to DNA repair, proliferation, transcription and

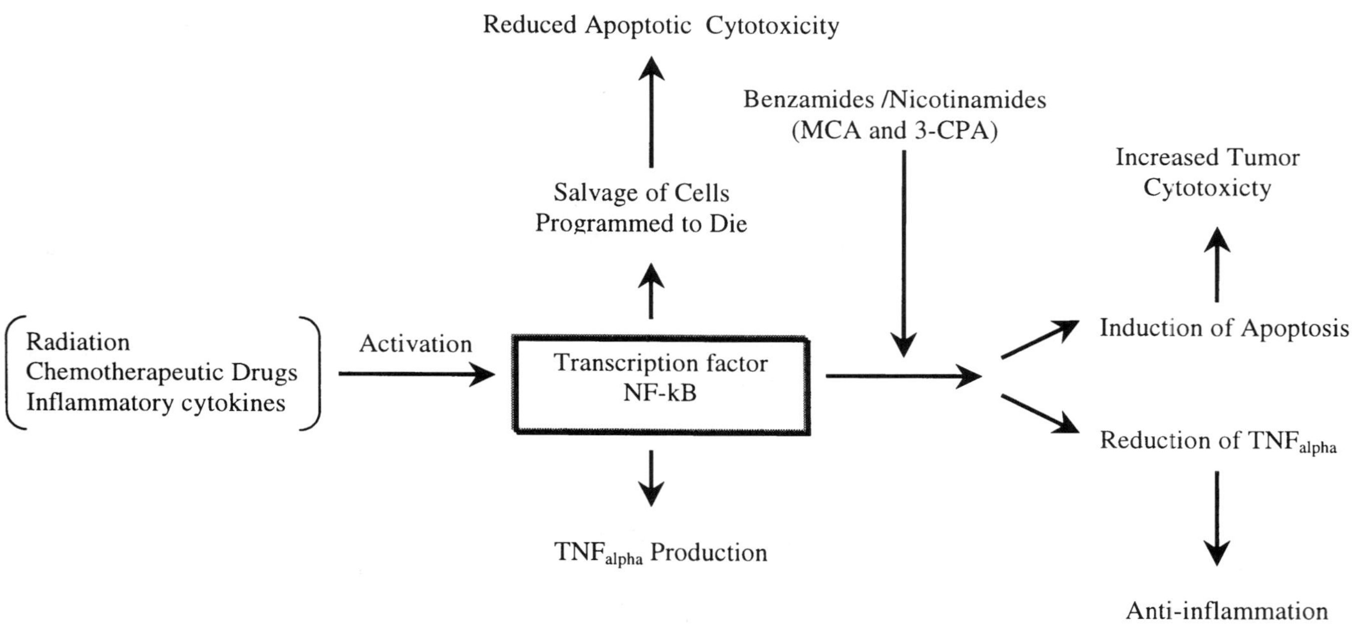

Fig. 5. Proposed scheme for the mode of action of N-Substituted benzamides and nicotinamides having both antitumor and anti inflammatory properties.

differentiation [30]. We believe one reasonable explanation why benzamides and nicotinamides have such broad based activity in the clinic is because PARP activity in itself is an important component to many human disorders via its important role in maintaining the integrity of nuclear function. For example, NF-kB activates p53 [27] but p53 activation in PARP-knockout mouse cells is diminished (F. Althaus *et al.*, 12th International ADP-Ribosylation Conference, May 12–14, 1997, Cancun, Mexico). Hence, it is possible that NF-kB expression in turn is regulated by PARP.

Acknowledgements

We wish to express our gratitude to Oxigene Europe AB, Astra-Draco AB, NIH grant CA55300 and the Swedish Cancer Society (1728-B95-07XAA) for supporting these studies.

References

1. Stanley, Rotrosen (eds). The Benzamides: Pharmacology, Neurobiology and Clinical Aspects. Raven Press, New York, 1982
2. Horsman MR: Nicotinamide and other benzamide analogs as agents for overcoming hypoxic cell radiation resistance in tumours. Acta Oncol 34: 571–587, 1995
3. Harrington RA, Hamilton CW, Brogden RN, Linkewich JA, Romankiewicz JA, Heel RC: Metoclopramide. An updated review of its pharmacological properties and clinical use. Drugs 25: 451–494, 1983
4. Pero RW, Olsson A, Amiri A, Chaplin D: Multiple modes of action of the benzamides and nicotinamides as sensitizers of radio- and chemotherapies. Opportunities for drug design. Cancer Prev Det 22(3): 225–236, 1998
5. King FD, Sanger GJ: 5-HT3 receptor antagonists. Drugs Future 14: 875–889, 1989
6. Pero RW, Olsson A, Sheng Y, Hua J, Mšller C, Kjellén E, Killander D, Marmor M: Progress in identifying clinical relevance of inhibition, stimulation and measurement of poly ADP-ribosylation. Biochimic 77: 385–393, 1995
7. Olsson A, Pero RW, Olofsson T: Specific binding and active transport of nicotinamide in human leukernic K562 cells. Biochem Pharmacol 45: 1191–1200, 1993
8. Olsson A, Sheng Y, Kjellén E, Pero RW: *In vivo* measurement of DNA damage, DNA repair and NAD pools as indicators of radiosensitization by metoclopramide. Carcinogenesis 16 (5): 1029–1035, 1995
9. Olsson A, Sheng Y, Pero RW, Chaplin D, Horsman M: DNA damage and repair in tumour and nontumour tissues of mice induced by nicotinamide. Brit J Cancer 74: 368–373, 1996
10. Pero RW, Olsson A, Simanaitis M, Amiri A, Andersen I: Pharmacokinetics, toxicity, side effects, receptor affinities and *in vitro* radiosensitizing effects of the novel metoclopramide formulations, Sensamiode and Neu-Sensamide. Pharmacol Toxicol 80: 231–239, 1997
11. Olsson AK, Hua J, Sheng Y, Pero RW: Neutral metoclopramide induces tumor cytotoxicity and sensitizes ionizing radiation of a human lung adenocarcinoma and virus induced sarcoma in mice. Acta Oncol 36 (3): 321–328, 1997
12. Jacobson MK, Cervantes-Lauran D, Strohm MS, Coyle DL, Bummer PM, Jacobson EL: NAD glycohydrolases and the metabolism of cyclic ADP-ribose. Biochimie 77: 341–344, 1995
13. Lee HC, Graef R, Walseth TF: Cyclic ADP-ribose and its metabolic enzymes. Biochimie 77: 345–355, 1995
14. Beg AA, Baltimore D: An essential role for NF-kB in preventing TNF_{alpha} induced cell death. Science 274: 782–784, 1996
15. Wang C-Y, Mayo MW, Baldwin AS: TNF- and cancer therapy-induced apoptosis: Potentiation by inhibition of NF-kB. Science 274: 784–787, 1996
16. Van Antwerp DJ, Martin SJ, Kafri T, Green DR, Verma IM: Suppression of TNF_{alpha}-induced apotosis by NF-kB. Science 274: 787–789, 1996
17. Barnes PJ, Adcock IM: NF-kB: A pivotal role in asthma and a new target for therapy. TIPS 18: 46–50, 1997
18. Robert-Piessard S, LeBaut G, Courant J, *et al.*: Non-acidic anti-inflammatory compounds: Activity of N-(4-6-dimethyl-2-pyridinyl) benzamides and derivatives. Eur J Med Chem 25: 9–19, 1990
19. Moffett RB, Robert A, Skaletzky LL: Antiulcer agents. p-aminobenzamido aromatic compounds. J Med Chem 14 (10): 963–968, 1971
20. Piriou R, Petit JY, Welin I: Benzamide potentiation of behavioral apomorphine-induced effects: Mechanism involved. Experientia 41: 1409–1410, 1985
21. Bouhayat S, Piessard S, LeBaut G, Sparfel L, Petit J-Y, Piriou F, Welin I: Synthesis and central dopaminergic effects of N-(4,6-dimethyl-2-pyrindyl) benzamides. J Med Chem 28: 555–559, 1985
22. Sampaio ER, Sarno EN, Galilly Z, Kaplan G: Thalidomide selectively inhibits tumor necrosis factor alpha production by stimulated human monocytes. J Med Chem 173: 699–703, 1991
23. Nishimura K, Hashimoto Y, Iwasaki S: Enhancement of phorbol ester-induced production of tumor necrosis factor alpha by thalidomide. Biochem Biophys Res Commun 199 (2): 455–460, 1994
24. Rankin PW, Jacobson EI, Benjamin RC, Moss J, Jacobson M: Quantitative studies of inhibitors of ADP-ribosylation *in vitro* and *in vivo*. J Biol Chem 264 (8): 4312–4317, 1989
25. Hirst DG, Joiner B, Hirst VK: Blood flow modification by nicotinamide and metoclopramide in mouse tumors growing in different sites. Br J Can 67: 1–6, 1993
26. Pellat-Deceurtynick C, Wietzerbin J, Drapier J-C: Nicotinamide inhibits nitric oxide synthase mRNA induction in acitvated macrophages. Biochem J 297: 53–58, 1994
27. Wu H, Lozano G: NF-kB activation of p53. J Biol Chem 269: 20067–20074, 1994
28. Graeber TG, Osamanian C, Jacks T, Houseman DE, Giaccia AJ: Hypoxia mediated selection of cells with diminished apoptoic potential in solid tumours. Nature 379: 88–91, 1996
29. Keiser A, Weich HA, Brandner G, Marme D, Kolch W: Mt p53 potentiates protein kinase C induction of vascular endothelial growth factor expression. Oncogene 9: 963–969, 1994
30. Poirier GG, Moreau P (eds): ADP-ribosylation reactions. Springer-Verlag, New York, pp. 1–410, 1992

Molecular and Cellular Biochemistry **193:** 127–135, 1999.

Poly(ADP-ribose) turnover in quail myoblast cells: Relation between the polymer level and its catabolism by glycohydrolase

E.B. Affar, R.G. Shah and G.G. Poirier

Health and Environment Unit, CHUL Research Center, CHUQ, Pavillon CHUL, Ste-Foy, Québec, Canada

Abstract

The concerted action of poly(ADP-ribose) polymerase (PARP) which synthesizes the poly(ADP-ribose) (pADPr) in response to DNA strand breaks and the catabolic enzyme poly(ADP-ribose) glycohydrolase (PARG) determine the level of polymer and the rate of its turnover. In the present study, we have shown that the quail myoblast cells have high levels of basal polymer as compared to the murine C3H10T1/2 fibroblasts. We have conducted this study to investigate how such differences influence polymer synthesis and its catabolism in the cells in response to DNA damage by alkylating agent. In quail myoblast cells, the presence of high MNNG concentration such as 200 μM for 30 min induced a marginal decrease of 15% in the NAD content. For C3H10T1/2 cell line, 64 μM MNNG provoked a depletion of NAD content by approximately 50%. The induction of the polymer synthesis in response to MNNG treatment was 6-fold higher in C3H10T1/2 cells than in quail myoblast cells notwithstanding the fact that 3-fold higher MNNG concentration was used for quail cells. The polymer synthesis thus induced in quail myoblast cells had a 4–5 fold longer half life than those induced in C3H10T1/2 cells. To account for the slow turnover of the polymer in the quail myoblast cells, we compared the activities of the polymer catabolizing enzyme (PARG) in the two cell types. The quail myoblast cells had about 25% less activity of PARG than the murine cells. This difference in activity is not sufficient to explain the large difference of the rate of catabolism between the two cell types implicating other cellular mechanisms in the regulation of pADPr turnover. (Mol Cell Biochem **193**: 127–135, 1999)

Key words: poly(ADP-ribose), polymerase, glycohydrolase, NAD, quail myoblast cells, C3H10T1/2 cells, alkylating agent

Introduction

Poly(ADP-ribose) polymerase (PARP) (EC 2.4.2.30) is a zinc finger nuclear enzyme present in most eukaryotes. Its activity is dependent on the presence of DNA stand breaks and has been implicated in many important genomic processes such as DNA replication, DNA repair, gene expression and apoptosis [1–8].

Under normal conditions, tissue culture cells display low basal level of polymer which can dramatically increase in cells exposed to DNA damaging agents (irradiation, alkylation, etc.). This increase of polymer synthesis is transient and is usually followed by a rapid degradation phase as indicated by the short half life values which can be less than 1 min [9, 10]. The low endogenous level of polymer in unstimulated cells and its rapid catabolism during DNA damage has been ascribed to high activity of polymer catabolizing enzyme poly(ADP-ribose) glycohydrolase (PARG). During heat shock of C3H10T1/2 cells, very high levels of polymer accumulation were observed as a result of glycohydrolase inhibition [11, 12].

However, the rapid cellular poly(ADP-ribose) (pADPr) turnover does not seem to be characteristic of all cell types. Previous work on murine lymphoma cells showed that LY-R and LY-S cell lines exhibit high basal levels of polymer, accompanied by a slow turnover rate and the absence of pADPr synthesis in the presence of UV_{254nm} [13]. Another study with chicken embryo tissues showed that liver and brain

Address for offprints: G. G. Poirier, Poly(ADP-ribose) metabolism group, Unité santé et environnement, Centre de recherche du CHUL, Centre Hospitalier Universitaire de Québec (CHUQ), 2705, Boul. Laurier, Ste-Foy, Québec Canada G1V 4G2

cells exhibit 3–4 fold higher constitutive PARP activity than mammalian thymic cells [14]. In addition, UV-light and mono functionally alkylating agents such as methyl methanesulfonate (MMS) did not significantly affect the PARP activity in chicken embryo cells. The present study was conducted in order to determine the relation between the basal level of polymer, the induction of its synthesis and its turnover. We have investigated and compared the characteristics of pADPr synthesis and its catabolism in an avian cell line, the quail myoblast cells with C3H10T1/2 fibroblasts. C3H10T1/2 fibroblasts were selected for comparison with quail cells for two reasons: One as a positive control where the induction of polymer by the alkylating agent MNNG and its catabolism thereafter has been well characterized [11, 12, 15, 16]. Secondly, the basal polymer level of C3H10T1/2 cells is in the lower range as compared to other cells. The quail cells were expected to have higher levels of basal polymer and thus these two cell lines could represent two different operative mechanisms with respect to polymer metabolism. In this study, basal polymer levels, MNNG induced polymer levels and the half lives of the induced polymer have been analyzed for both these cell lines. Additionally, the levels of PARG were determined in the two cell lines.

Materials and methods

Materials

[Adenylate-^{32}P]NAD (^{32}P-NAD; 30 Ci/mmol) was purchased from NEN Life Science Products (Guelph, ON, Canada), and [2,8-^{3}H]adenine (^{3}H-Ade; 31 Ci/mmol) was from ICN Pharmaceuticals Ltd (Montréal, QC, Canada). N-[2-hydroxyethyllpiperazine-N′-[2-ethanesulfonic acid] (HEPES), 3-aminophenylboronic acid hemisulfate, 1-ethyl-3-(3-dimethylamino-propyl) carbodiimide, activated calf thymus DNA, 1,5-isoquinolinediol (DHQ), 1-methyl-3-nitro-l-nitrosoguanidine (MNNG), phenylmethylsulfonylfluoride, 3-[4,5-dimethylthiazol-2-yl]-2,5 diphenyltetrazolium bromide, phenazine ethosulfate and alcohol dehydrogenase were all purchased from Sigma-Aldrich Canada Ltd (Oakville, ON, Canada). β′-NAD and dithiothreitol were from Boehringer Mannheim (Montreal, QC, Canada). Poly-Prep chromatography columns and a cation exchange Bio-Rex 70 Resin (200–400 mesh) were from Bio-Rad Laboratories (Richmond, CA, USA). Tryptose phosphate broth was from Difco Laboratory (Detroit, MI, USA). The culture media and the antimicrobial agents penicillin, streptomycin were from Life Technologies (Burlington, ON, Canada). PEI-300 cellulose TLC plates were from Machery-Nagel (Mandel Scientific Company Ltd., ON, Canada). The high performance liquid chromatography (HPLC) system, i.e., columns, pump, detectors and System Gold software for data collection and analysis were from Beckman (Berkeley, CA, USA).

Enzyme purifications

Calf thymus PARP was purified according to Zahradka and Ebisuzaki [17] up to the DNA- cellulose chromatography step. The active fractions were pooled and concentrated five time on sucrose granules. PARG was purified according to Thomassin *et al.* [18] up to single strand DNA chromatography. The enzymes were stored at –80°C until use.

Synthesis of cold and ^{32}P-labeled pADPr

Cold pADPr was synthesized essentially as described previously [19–21], except that the histones were omitted. The reaction mixture contained 100 mM Tris-HCl (pH 8.0), 10 mM $MgCl_2$, 8 mM DTT, 10% (v/v) glycerol, 10% (v/v) ethanol, 22.5 μg activated DNA, 1 mM NAD in a total volume of 900 μl. The reaction was started by addition of 20 U of PARP for 30 min at 30°C and was stopped by 300 mM sodium acetate (pH 5.2). The automodified PARP was precipitated with 41.2% isopropanol on ice for 20 min. The pellet was collected by centrifugation (13000 g/10 min) and dissolved in 250 mM ammonium acetate, 6 M guanidine-HCl, 10 mM EDTA, pH6.0.

^{32}P-pADPr preparation was carried according Ménard *et al.* [21] with minor modifications. The reaction mixture contained the same components at similar concentrations except that cold NAD and ^{32}P-NAD were added at 200 μM and 75 μCi respectively and the histones were omitted. The reaction was started with 20U of PARP and after 30 min of incubation at 30°C, the reaction was stopped by 33% of ice-cold trichloroacetic acid (TCA). The TCA-insoluble fraction was collected by centrifugation and washed with ethanol and ether, respectively.

Cold and ^{32}P-labeled pADPr were detached from PARP by alkali treatment and purified on dihydroxyboronyl Bio-Rex (DHBB) as described by Shah *et al.* [22] except that elution of polymer was done with water instead of 10 mM HCl.

Cell culture and treatments

Mouse embryo fibroblast cells C3H10T1/2 (clone 8) and Japanese quail myoblast cells (QM 7) were obtained from American Type Culture Collection (Rockville, MD, USA). C3H10T1/2 were cultured in Eagle's basal medium supplemented with 8% heat-inactivated fetal bovine serum (HyClone Laboratories, Logan, UT, USA), penicillin, streptomycin. The quail myoblast cells were grown in 80% medium 199 supplemented with 10% heat-inactivated fetal bovine serum and 10%

tryptose phosphate broth solution (29.5 grams/liter of deionized water and sterilized at 120°C for 15 min), penicillin, streptomycin. Both cells were grown at 37°C with 5% CO_2 in humidified atmosphere. The C3H10T1/2 cells were used between passages 9 and 14. Confluent monolayers of quail myoblasts cr C3H10T1/2 were incubated in 5 ml of fresh complete medium containing 20 µCi/ml of ^{3}H-adenine for radiolabeling of NAD pool. After 16 h of incubation, the medium was replaced with unlabeled complete medium containing MNNG. For polymer half life determinations, the MNNG medium was replaced with complete medium containing 1 mM DHQ. The cells were harvested at different time intervals. The solvent DMSO, used for preparing MNNG and DHQ solutions, was kept below 0.1% for the *in vivo* assays. After treatments, the monolayers were washed with cold HEPES saline and the cells were kept on ice for 15 min with 1 ml of 20% TCA. They were harvested by scraping with a rubber policeman. The plate was rinsed with another ml of 20% TCA and pooled with previous fraction. The TCA pellet was collected at 1800 × g for 10 min at 4°C and processed for pAPDr analysis. The TCA supernatant was used for NAD pool analyses.

Analysis of pADPr and NAD levels in cells

Confluent quail myoblast cells or C3H10T1/2 cells were labeled by incorporation of ^{3}H-adenine in the NAD pool. Untreated and MNNG-induced ^{3}H-pADPr synthesized in these cells was isolated and analyzed as ^{3}H-ribosyladenosine (RAdo).

The protein-bound ^{3}H-pADPr in the TCA pellet was dissolved in formic acid, and precipitated with 20% TCA (4°C, 15 min) after the addition of 1 mg bovine serum albumin and a spike of 1 nmol of cold pADPr. After centrifugation for 10 min at 1800 × g and 4°C, the TCA-insoluble pellet was processed further to isolate and analyze pADPr essentially as described by Shah *et al.* [22]. In brief, ^{3}H- pADPr was separated from the proteins by digestion at 37°C for 2 h with 1 M KOH, 50 mM EDTA, and purified by DHBB chromatography. The ^{3}H-pADPr preparation was digested at 37°C for 2 h with 25 U of RNase A to remove any contaminating RNA. It was then digested at 37°C overnight with 0.25 U of snake venom phosphodiesterase and 2.5 U of bacterial alkaline phosphatase to form ^{3}H-RAdo. ^{3}H-RAdo was resolved from other nucleosides by HPLC [16] on a C18 reverse-phase column (5 µm, 4.6×250 mm), eluted isocratically at room temperature using 7 mM ammonium formate, pH 5.8, 4% methanol at 1 ml/min. After UV detection at 254 nm, radioactivity in the effluent was manually monitored by mixing with PCS liquid scintillation cocktail (Amersham, Oakville, ON, Canada). The endogenous RAdo concentrations was estimated from radioactivity present in the RAdo peak and from the specific activity of the ^{3}H-NAD pool in the cells. Our improvements [22] over the original technique [16] enabled us to detect small quantities of pADPr formed in the control samples. The TCA supernatant was used for purification of total NAD after passing it through DHBB chromatography followed by a cycling microassay for NAD as we have recently described [22].

PARG assay

The cell extracts and the assay were carried out as described by Jonsson *et al.* [11], with slight modifications. The monolayers were washed with cold HEPES saline and 1 ml ice-cold extraction buffer (10 mM Tris-HCl, pH 7.5; 1 mM EDTA, 300 mM sucrose, 1 mM β-mercaptoethanol, 0.1% nonidet-P40) was added for every 0.75–2.5 × 10^6 cells. The flasks were kept on ice for 10 min and the cells were scraped with a rubber policeman, and homogenized using 10 strokes in a Dounce homogenizer with a tight fitting piston. The cell extract was aliquoted and stored at –80°C. The extraction buffer did not contain sodium fluoride, the inhibitor of ADP-ribose pyrophosphatase (ADPRibase) to detect its activity, if any [23, 24]. The sequential action of PARG and ADP-ribase would give AMP which can be detected as a separate band on the TLC. The reaction mixture for the assay contained: 50 mM potassium phosphate buffer (pH 7.5), 10 mM β-mercaptoethanol, 50 mM KCl, 100 µg/ml BSA, 0.1 mM PMSF, 10 µM ^{32}P- pADPr. The samples were preincubated for 5 min at 37°C and the reaction was started by the purified enzyme or the cell extracts in the final volume of 60 µl. The reaction was stopped by the addition of 0.1% SDS (final concentration). An aliquot of about 10,000 cpm was applied on PEI-cellulose TLC plate (10×20 cm). The plate was first developed at room temperature in methanol, dried and then developed in 0.3 M LiCl, 0.9 N acetic acid in the same direction. The TLC plate was dried and electronically autoradiographed on the Instant Image Analyzer (Packard Instrument Company, Meriden, CT, USA). The radioactivity in the ADP-ribose spot and the origin was used to calculate the activity. One unit of the enzyme is the amount which liberates 1 nmole of ADP-ribose per minute at 37°C.

Protein determinations

Protein determination were performed as per Vincent and Nadeau [25], a modified method of Bradford, which was adapted for 96-well microplate using bovine serum albumin (BSA, fraction V) as standard.

Results

Previous studies have shown that untreated tissue and cells in culture display variable constitutive levels of pADPr (Table 1).

Table 1. Poly (ADP-ribose) constitutive level of different cell type and tissue

Cell type and tissue	Poly(ADP-ribose) constitutive level	Method used	Reference
C3H10T1/2 cells	7 pmol[a]	^{3}H-RAdo/HPLC[c]	Shah *et al.*, 1996 [26]
	9.2 pmol[b]	∈ -RAdo/HPLC[d]	Singh *et al.*, 1985 [15]
Ehrlich ascites tumor cells	6.5–8.3 pmol[b]	PR-AMP/RIA[e]	Wielckens *et al.*, 1983 [27]
	6–25 pmol[a]	PR-AMP/RIA	Wielckens *et al.*, 1982 [10]
Human fibroblasts	12 pmol[a]	∈ -RAdo/HPLC	Jacobson *et al.*, 1983 [28]
Human fibroblasts (3229)	2–10 pmol[b]	∈ -RAdo/HPLC	Singh *et al.*, 1985 [15]
Human keratinocytes	63.0 pmol[b]	^{3}H-RAdo/HPLC	Malanga and Althaus, 1994 [7]
Human monocytes	25–50 pmol[a]	∈ -RAdo/HPLC	Singh *et al.*, 1985 [15]
Murine lymphoma cells, LY-R	60 pmol[b]	∈ -RAdo/HPLC	Kleczkowska *et al.*, 1990 [13]
Murine lymphoma cells, LY-S	160 pmol[b]	∈ -RAdo/HPLC	Kleczkowska *et al.*, 1990 [13]
Rat hepatocytes (primary culture)	23.3 pmol[b]	∈ -RAdo/HPLC	Alvarez-Gonzalez and Althaus, 1989 [9]
SV40 -3T3 fibroblasts (SVT2)	9 pmol[a]	∈ -RAdo/HPLC	Juarez-Salinas *et al.*, 1984 [29]
3T3 fibroblasts (proliferating)	5 pmol[b]	∈ -RAdo/HPLC	Hilz *et al.*, 1982 [30]
3T3 fibroblasts (resting)	55 pmol[b]	∈ -RAdo/HPLC	Hilz *et al.*, 1982 [30]
Yoshida hepatoma cells	7.3 pmol[a]	PR-AMP/RIA	Kreimeyer *et al.*, 1984 [31]
AH 130 cells (proliferating)	61 pmol[b]	PR-AMP/RIA	Wielckens *et al.*, 1981 [32]
AH 130 cells (resting)	31 pmol[b]	PR-AMP/RIA	Wielckens *et al.*, 1981 [32]
Mouse lymphoma cells (S49.1)	6.7 pmol[a]	PR-AMP/RIA	Wielckens and Delfs, 1986 [33]
Rat liver	39 pmol[b]	PR-AMP/RIA	Wielckens *et al.*, 1981 [32]
Calf thymus	60 pmol[b]	PR-AMP/RIA	Hilz *et al.*, 1982 [30]
Mouse heart	8 pmol[b]	PR-AMP/RIA	Hilz *et al.*, 1982 [30]
Mouse kidney	21 pmol[b]	PR-AMP/RIA	Hilz *et al.*, 1982 [30]
Mouse liver	28 pmol[b]	PR-AMP/RIA	Hilz *et al.*, 1982 [30]
Quail myoblast cells (QM7)	98 pmol[a,f]	^{3}H-RAdo/HPLC	Present study

[a]Results expressed per 10^8 cells. [b]Results expressed per mg of DNA. [c]Cellular radiolabeling with ^{3}H-adenine and purification of ^{3}H-pADPr using dihydroxyboronyl Bio-Rex. ^{3}H-RAdo, the product of enzymatic digestion of the polymer was determined by reversed-phase HPLC (Aboul-Ela *et al.*, 1988 [34]). [d]Polymer was purified using dihydroxyboronyl Sepharose and digested with phosphodiesterase and alkaline phosphatase. Etheno-RAdo, the fluorescent derivative of the RAdo was determined by fluorescence coupled HPLC (Juarez-Salinas *et al.*, 1979 [35]). [e]Purification of pADPr using DHBB and radioimmunological determination of PR-AMP, the enzymatic digestion product (Wielckens *et al.*, 1981 [32]). [f]average for three determinations.

To enable fair comparison of the polymer levels in different intact cells, results from highly specific and sensitive methods used for pADPr quantification have been compiled [7, 9, 10, 13, 15, 26–33] in Table 1. Studies which have used permeabilized cells for labeling of the polymer pool by adding radiolabeled NAD have not been included since high polymer levels have been observed in cells due to artifactual DNA strand breaks caused by permeabilization. This comparison provides a guideline for levels of polymer in different intact cells. Some cells such as human fibroblasts, C3H10T1/2 and Ehrlich ascites tumor cells display low level of polymer as compared to human keratinocytes, LY-R cells and LY-S cells. More than half the cell lines cited in the table have their polymer levels in the range of 2–30 pmoles/mg DNA or 10^8 cells. Few cell lines are in the range of 50–65 pmoles/mg DNA or 10^8 cells.

In the present study, confluent quail myoblast cells were labeled with ^{3}H-adenine, which was incorporated in the NAD pool, and ^{3}H-pADPr synthesized in these cells was isolated and analyzed for ^{3}H-ribosyladenosine (^{3}H-RAdo). Our results show quail myoblast cells to have approximately 100 pmol of pADPr per 10^8 cells and therefore can be considered to be a cell type having higher levels of basal pADPr. Under these conditions, pADPr metabolism was investigated in quail myoblast cells to determine if these cells exhibit unusual pADPr synthesis and/or catabolism as compared to C3H10T1/2 cells.

To examine if alkylating agents induce NAD depletion by stimulating pADPr synthesis, confluent quail myoblast cells and C3H10T1/2 cells were incubated with and without the common alkylating compound MNNG. The TCA soluble fraction was used for the determination of cellular NAD level by a highly specific cycling assay. The results are presented in Fig. 1. For the untreated culture, the cellular NAD level in quail myoblast cells was approximately 67 nmol per 10^8 cells and 82 nmol per 10^8 cells for C3H10T1/2 cells. In quail myoblast cells, the presence of 100 μM MNNG in culture medium did not significantly affect the NAD level after 15 and 30 min of treatment. The addition of 200 μM MNNG induced a discrete decrease in NAD content of about 10 and 15% of the control value in 15 min and 30 min respectively. This percentage of decrease remained approximately similar up to 60 min but was restored to control value in 120 min. On the other hand, in the C3H10T1/2 cell line, 64 μM MNNG provoked a depletion of NAD content by approximately 50

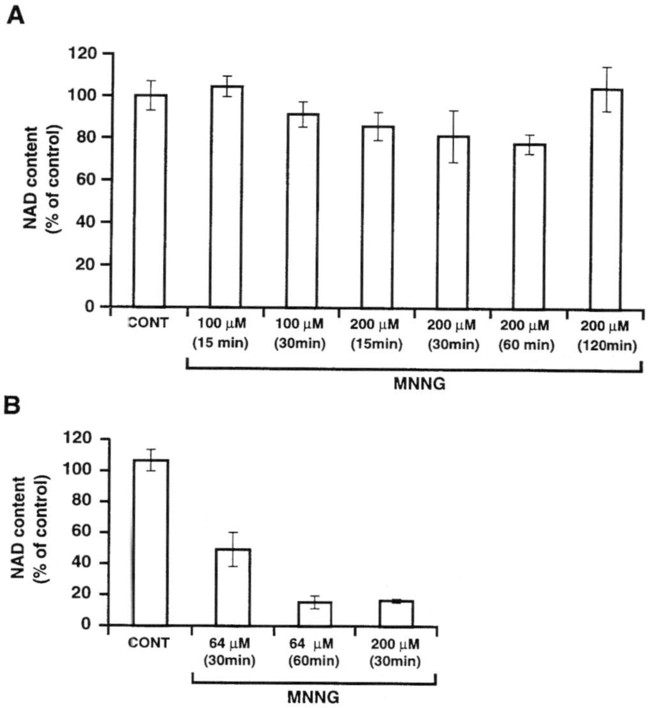

Fig. 1. NAD content in quail myoblast cells (A) and C3H10T1/2 fibroblast (B) in the presence and absence of MNNG. At the time indicated, the monolayers were treated with TCA and the acid soluble fractions were used for NAD determinations as mentioned in Materials and methods. Values are mean ± S.D. of three determinations. The control values for the quail and C3H10T1/2 cells were 67 ± 12 and 82 ± 10 nmoles/10^8 cells respectively.

and 80% of the control value in 30 and 60 min, respectively. When the cells were treated with 200 μM MNNG for 30 min, the effect on NAD was similar to that with 64 μM MNNG for 60 min. In order to determine if the decrease of NAD content was due to alkylation induced synthesis of the polymer by PARP, the two cell lines were treated with MNNG and NAD content determined before and after the addition of PARP inhibitor. As indicated in Fig. 2, DHQ arrested the NAD decrease in the two cell lines permitting its stabilization at about 80 and 70% of the control values for quail myoblast cells and C3H10T1/2 cells respectively. In addition, that the reduction in the NAD values was not due to leakage was verified by trypan blue exclusion assay. No cytotoxicity was detected under all experimental conditions.

The rate of pADPr synthesis in the presence of MNNG was determined for cells in the TCA insoluble fraction which was processed for pADPr analyses. Quail cells were treated with 200 μM MNNG for different time intervals up to 90 min (Fig. 3, solid line). pADPr synthesis was observed to be stimulated 2-fold reaching a peak at 20–30 min followed by decrease to control value at 60 min. In C3H10T1/2 cells, 64 μM MNNG produced a 12-fold stimulation in polymer synthesis at 30 min. Longer incubation with MNNG resulted in decrease in this stimulation (data not shown). It is interesting to note that the induction of polymer synthesis by MNNG was 6-fold higher in C3H10T1/2 cells than in the quail myoblast cells notwithstanding the fact that the MNNG concentration used for quail myoblast cells was about 3-fold higher. The NAD level does not seem to be a limiting factor for pADPr synthesis in quail cells because the MNNG did not provoke a substantial depletion of NAD (Fig. 1). Half life of MNNG induced polymer was determined by inhibiting its synthesis by addition of 1 mM DHQ and measuring the rate of polymer catabolism. Maximum polymer induction of confluent quail myoblast cells and C3H10T1/2 was achieved by MNNG treatment. At this stage, the cells were treated with the potent PARP inhibitor DHQ and processed as function of time for pADPr measurements (Fig. 3). As indicated in Fig. 3a, the pADPr catabolism in quail myoblast cells was biphasic. The first phase corresponding to a polymer half life value of ~4 min and a second phase corresponding to a polymer half life value of ~34 min. The C3H10T1/2 fibroblast cells had similar pattern (Fig. 3b) consisting of a biphasic polymer catabolism. By contrast, the velocity of the polymer degradation was different between the two cell types. The half lives of polymer were 4–5 fold higher in quail myoblast cells than in C3H10T1/2.

PARG is responsible for pADPr catabolism *in vivo*. To determine if the slow polymer catabolism was due to reduced activity of the enzyme, quail myoblast cell and C3H10T1/2 cell extracts were analyzed for PARG. The results in Table 2 show PARG activity in quail myoblast and C3H10T1/2 cells.

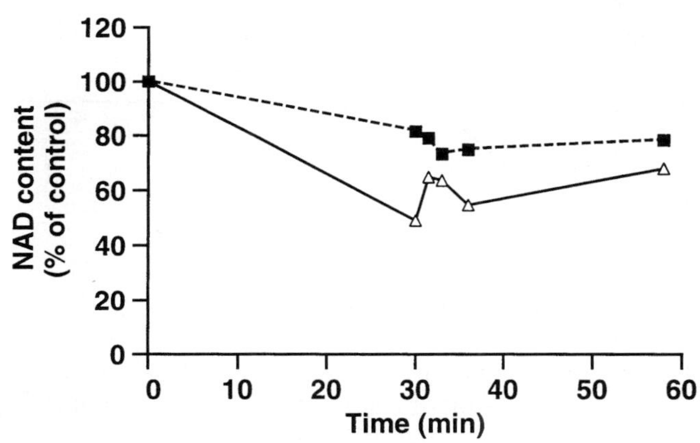

Fig. 2. NAD content in quail myoblast (-■-) and C3H10T1/2 (-Δ-) cells as a function of time subsequent to MNNG treatment followed by addition of DHQ. The monolayers were treated with 200 μM MNNG for quail myoblast cells and 64 μM for C3H10T1/2 cells (at 0 time in the curve) for 30 min and 1 mM DHQ was added at this time point as mentioned in Materials and methods. Values represent mean of at least two different readings.

Table 2. Poly(ADP-ribose) glycohydrolase activity in cultured cell lines

Cell type	% of ^{32}P-pADPr hydrolysis[a,b]	Activity U/10^6 cells[a,c]	Specific activity U/mg of protein[a,c]
C3H10T1/2 cells	11.4 ± 1.6	0.097± 0.014	0.48 ± 0.06
Quail myoblast cells	15.3 ± 2.2	0.078 ± 0.011	0.36 ± 0.09

[a]The values are average ± S.D. for three independent determinations. [b](CPM in ADP-ribose spot/CPM of the polymer remaining at origin) × 100. [c]One unit is the amount of enzyme which liberates 1 nmol of ADP-ribose per minute at 37°C under the assay conditions.

The assay conditions were such that the percentage of polymer hydrolysis in the two cell extracts was not more than 20% to maintain linearity of the reaction. It can be seen that PARG activity in quail myoblast cells was about 25% less than C3H10T1/2 cells. The difference in the activity between the two cell lines was irrespective of whether it was expressed as units/mg of protein or units/10^6 cells. No significant amount of AMP was detected in the two cell extracts indicating a paucity of ADP-ribose pyrophosphatase activity. However, ADPRibase activity was detected in chicken SL-29 cells when the PARG assay was similarly conducted using their cell extract (data not shown).

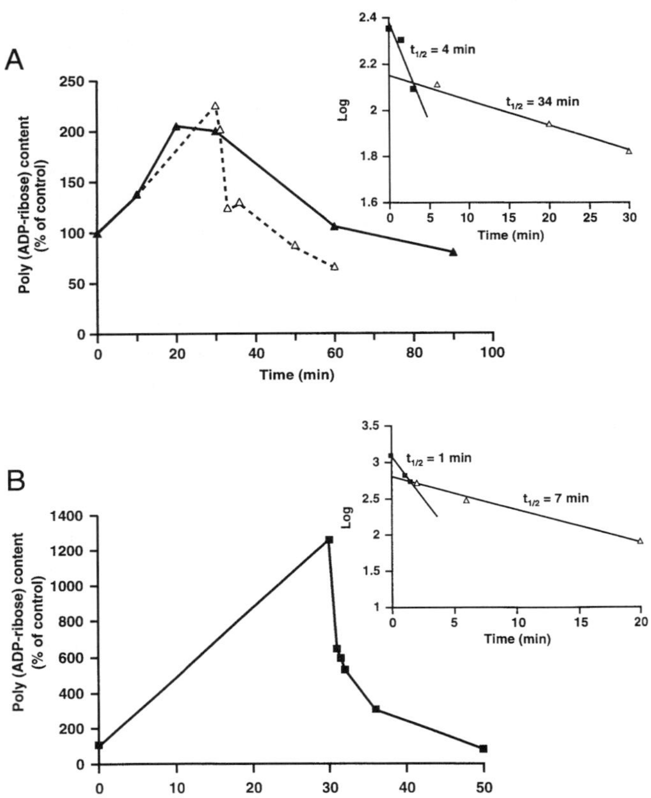

Fig. 3. Turnover of the poly(ADP-ribose) in the MNNG treated quail myoblast (A) and C3H10T1/2 cells (B). To determine optimal treatment time, quail cells were treated with 200 µM MNNG for different time intervals upto 90 min (Fig. 3A, -▲-). To determine half life of the MNNG induced polymer, 200 µM MNNG was added to quail cells (Fig. 3A, -Δ-) and 64 µM MNNG to C3H10T1/2 cells for 30 min (Fig. 3B), cells were treated with 1 mM DHQ$^-$ and the polymer levels were measured as a function of time. The insets show the half lives of the polymer determined according to Wielckens *et al.* [27]. The values represent the percentage of control values obtained following the DHBB purification performed as described in Materials and methods.

Discussion

Determination of the rate of pADPr synthesis and its catabolism are of particular interest with regard to the implication of the two antagonist enzyme PARP and PARG in important cellular event such as DNA repair, DNA replication and gene expression. Previous study in our laboratory using an *in vitro* pADPr turnover system [36–38] have shown that the combined and coordinated action of PARP and PARG determine the rate and velocity of polymer accumulation and its steady state concentrations. However pADPr metabolism in an *in vivo* system might be more complex due to the implication of other cellular parameters which could regulate pADPr level and its turnover.

The present study was conducted to examine the turnover of pADPr in presence of an alkylating agent in intact quail myoblast cells which have high level of basal pADPr. The high endogenous pADPr level of the unstimulated quail myoblast cells was determined by ^{3}H-adenine radiolabeling procedure coupled to DHBB purification of pADPr and the determination of its enzymatic digestion products with reverse phase HPLC. The basal level of pADPr was also confirmed by ELISA consisting of the fixation of pADPr on polylysine coated well and its revelation by anti-pADPr antibody (data not shown).

To determine the relationship between this high level of polymer in quail myoblast cells and its metabolism, the synthesis and catabolism of pADPr were investigated in quail myoblast cells and in C3H10T1/2 fibroblast which have low pADPr content. MNNG induced a 12-fold increase in pADPr synthesis in C3H10T1/2, in contrast the increase in quail myoblast cells was only 2-fold. Lower MNNG concentrations such as 30 and 60 µM and other alkylating agents like the MMS at 100–400 µM had no effect on pADPr synthesis and NAD level (data not shown) suggesting some 'resistance' of quail myoblast cells to alkylating agents.

The results presented show that the pADPr synthesis is weakly stimulated in quail myoblast cells which have high basal polymer level. Such inverse proportionality between the basal level of polymer and the induction rate of its synthesis by DNA damaging agents has been suggested in chicken embryo cells [14]. The permeabilized liver and brain cells of chicken embryo exhibited high constitutive PARP activity and showed very little polymer stimulation if at all in presence of DNA damaging agents such as H_2O_2. UV-light and alkylating agents. On the other hand, the rat thymic cells exhibit 3–4 fold lower PARP activity which could be stimulated 10 fold with H_2O_2. Interestingly, the study showed that chicken embryo cells had lower nucleoid sedimentation and viscosity of alkaline cell lysates suggesting lower chromatin condensation than rat thymic cells [14]. In addition, Kleczkowska *et al.* [13] showed that two murine lymphoma cells LY-R and LY-S, had high level of pADPr. The UV254 (32 j/m^2) had no effect on pADPr synthesis in the two cell lines. Generally, low pADPr basal levels were associated with substantial induced polymer synthesis on DNA damage. For example, high stimulation of the polymer (17-fold) was observed with UV irradiation at 5 j/m^2 in human fibroblast [28]. Alkylation with 200 μM dimethyl sulfate induced a 10-fold accumulation of pADPr in Ehrlich ascites tumor cells and 50 μM MNNG caused a 21-fold increase in hepatocytes [9]. In Yoshida hepatoma cells, 400 μM dimethyl sulfate induced about 33-fold increase in polymer synthesis [31]. These results indicate a correlation between the constitutive level of the polymer and the rate of its synthesis in response to DNA damaging agents. A primary culture of human keratinocytes was an exception since 30 μM MNNG stimulated a 15–27 fold increase in polymer synthesis in spite of the relatively high polymer content in these cells [39].

Low stimulation of PARP by alkylating agents in quail myoblast cells could indicate a high preexisting level of physiological DNA strand breaks. Such DNA breaks could induce continuous stimulation of PARP resulting in a steady state of automodified PARP reaching significant proportions. The considerable level of automodified PARP decreases its total activity, which in turn decreases the rate of stimulation by DNA damaging agents. Kawaichi *et al.* [40] showed a downregulation of PARP activity by its automodification. Results indicating a coexistence of two populations of PARP in quail myoblast cells has been observed using Western blot analysis with monoclonal anti-polymer 10H antibody (data not shown, personal communication of Dr. G.M. Shah). One population of PARP molecules is in native state while the other is automodified. It is likely that the automodified PARP molecules help in maintaining an open chromatin structure and that the native PARP molecules act as DNA damage sensors, permitting response to external stimulus.

The pADPr catabolism curve in C3H10T1/2 cells is biphasic as determined by a successive treatment with MNNG and DHQ with the half life values of 1 and 7 min for the first and the last phase respectively. Similar results were reported for Ehrlich ascites tumor cells [10] and human keratinocytes [7] with 1 min for the first phase and 6 min for the residual fraction of the polymer. The quail myoblast cells also displayed biphasic catabolic pattern but the two half lives ascertained for these cells were 4–5 fold higher in comparison with C3H10T1/2 cells. The possibility that a slower catabolism of polymer is due to partial inhibition of PARP which in turn could over estimate the half life of polymer degradation, can be ruled out. The inhibition of PARP is probably complete since DHQ is a potent inhibitor of PARP, both *in vitro* and *in vivo*. This agent can completely inhibit oxidant induced pADPr synthesis in C3H10T1/2 fibroblast cells and is more effective than other inhibitors such as benzamide derivatives which do not obliterate PARP activity completely [26].

In an attempt to determine the relation between the biphasic behavior of pADPr catabolism and the chain length and complexity, Malanga *et al.* [7] showed that in human keratinocytes, the first quick phase of catabolism, corresponds to longer polymer. The second slower phase corresponds to the small polymer with a half life which is approximately 6-fold higher than the first phase. The biphasic catabolic pattern in quail cells probably corresponds to the successive degradation of long and small polymer. The difference in catabolism rates between the two phases is probably due to the difference in the affinity of PARG to small vs. long polymer. These results are in agreement with the result reported by Hatekeyama *et al.* [41] who used purified protein free polymer as the substrate. Purified PARG catabolized the polymer utilizing two different mechanisms, one rapid and processive during which the long polymers are preferentially degraded, and the other distributive, where small polymers are degraded. The difference between PARG affinity constant for long and small polymers could represent a mechanism that differentially regulates the size of polymer in the cell.

Determinations of PARG activity showed about 25% lower activity in quail myoblast cells as compared to C3H10T1/2. This difference of activity did not seem to be sufficient to explain the high level of pADPr and the slow catabolism rate in quail myoblast cells. A previous study in our laboratory using an *in vitro* pADPr turnover system [37] have shown that the polymer half life was inversely proportional to the enzyme amount. An increase in the amount of PARG from 1.25–5 U induced a decrease in half life from 55–26 sec. Similar results were obtained for *in vivo* system [11, 12]. Treatment of C3H10T1/2 with heat shock provoked a 10-fold decrease of PARG activity and 5-fold increase of polymer content. In addition, when the cells were returned for recovery, 40% of control level of PARG was sufficient to restore pADPr basal level and its normal half life. As there is only 25% difference of PARG activity between C3H10T1/2 and quail myoblast cells, it can be concluded that the PARG content is not the limiting factor in pADPr catabolism in quail cells but other factors are involved in the regulation of polymer catabolism. The activity of PARG could be influenced by different cellular factors such as the presence of high concentrations of ADP-ribose which is known to inhibit the enzyme [41, 42].

Such a situation could be operational *in vivo* for the quail cells since no ADPRibase was detected in the PARG assay. Whether the high basal polymer in quail cells is a cause or the consequence of such a condition cannot be determined based on this study, since the C3H10T1/2 cells also lack the ADPRibase and yet express a high turnover of the polymer in spite of not so different levels of PARG. Thus, other factors may regulate the rate of catabolism e.g. polymer acceptor proteins. It was reported by Uchida *et al.* [43] that the kinetic parameters of PARG may change with the nature of the polymer acceptor. Further investigations at the chromatin level are necessary to understand the physiological significance of this high level of pADPr and its slow turnover in quail myoblast cells.

Acknowledgments

We are grateful to Dr. G. M. Shah for his critical comments, unpublished data and helpful discussions; to A. Gaudreau and J. Lagueux for their help in the performance of HPLC experiments, and to D. Poirier for her excellent secretarial assistance. This work was supported by Medical Research Council of Canada.

References

1. Boulikas T: Relation between carcinogenesis, chromatin structure and poly(ADP-ribosylation) (review). Anticancer Res 11: 489–527, 1991
2. de Murcia G, Schreiber V, Molinete M, Saulier B, Poch O, Masson M, Niedergang C, Menissier de Murcia J: Structure and function of poly(ADP-ribose) polymerase. Mol Cell Biochem 138: 15–24, 1994
3. de Murcia G, Menissier de Murcia J: Poly(ADP-ribose) polymerase: A molecular nick-sensor. Trends Biochem Sci 19: 172–176, 1994
4. Takahashi A, Alnemri ES, Lazebnik YA, Fernandes-Alnemri T, Litwack G, Moir RD, Goldman RD, Poirier GG, Kaufmann SH, Earnshaw WC: Cleavage of lamin A by Mch2∝ but not CPP32: Multiple ICE-related proteases with distinct substrate recognition properties are active in apoptosis. Pro Natl Acad Sci USA 93: 8395–8400,1996
5. Lazebnik YA, Kaufmann SH, Desnoyers S, Poirier GG, Earnshaw WC: Cleavage of poly(ADP-ribose) polymerase by a proteinase with properties like ICE. Nature 371: 346–347, 1994
6. Lautier D, Lagueux J, Thibodeau J, Ménard L, Poirier GG: Molecular and biochemical features of poly (ADP-ribose) metabolism. Mol Cell Biochem 122: 171–193, 1993
7. Malanga M, Althaus FR: Poly(ADP-ribose) molecules formed during DNA repair *in vivo*. J Biol Chem 269: 17691–17696, 1994
8. Poirier GG, Moreau P: ADP-Ribosylation Reactions. Springer-Verlag, New York, 1992
9. Alvarez-Gonzalez R, Althaus FR: Poly (ADP-ribose) catabolism in mammalian cells exposed to DNA-damaging agents. Mutat Res 218: 67–74, 1989
10. Wielckens K, Schmidt A, George E, Bredehorst R, Hilz H: DNA fragmentation and NAD depletion: Their relation to the turnover of endogenous to mono(ADP-ribosyl) and poly(ADP-ribosyl) proteins. J Biol Chem 257: 12872–12877, 1982
11. Jonsson GG, Jacobson EL, Jacobson MK: Mechanism of alteration of poly(adenosine diphosphate-ribose) metabolism by hyperthermia. Cancer Res 48: 4233–4239, 1988
12. Jonsson GG, Ménard L, Jacobson EL, Poirier GG, Jacobson MK: Effect of hyperthermia on poly(adenosine diphosphate-ribose) glycohydrolase. Cancer Res 48: 4240–4243, 1988
13. Kleczkowska HE, Szumiel I, Althaus F: Differential poly(ADP-ribose) metabolism in repair-proficient and repair-deficient murine lymphoma cells. Mutat Res 235: 93–99, 1990
14. Ignatius A, Hund M, Tempel K: Poly(ADP-ribose) polymerase-activity of chicken embryo cells exposed to nucleotoxic agents. Toxicology 76: 187–196, 1992
15. Singh NS, Poirier GG, Cerutti PA: Tumor promoter phorbol-12-myristate-13 acetate induces poly(ADP-ribosyl)ation in fibroblasts. EMBO J 4: 1491–1494, 1985
16. Jacobson EL, Smith JY, Mingmuang M, Meadows R, Sims JL, Jacobson MK: Effect of nicotinamide analogues on recovery from DNA damage in C3H10T1/2 cells. Cancer Res 44: 2485–2492, 1984
17. Zahradka P, Ebisuzaki K: Poly(ADP-ribose) polymerase is a zinc metalloenzyme. Eur J Biochem 142: 503–509, 1984
18. Thomassin H, Jacobson MK, Guay J, Verreault A, Aboul-Ela N, Ménard L, Poirier GG: An affinity matrix for the purification of poly (ADP-ribose) glycohydrolase. Nucleic Acid Res 18: 4691–4694, 1990
19. Brochu G, Duchaine C, Thibeault L, Lagueux J, Shah GM, Poirier GG: Mode of action of poly(ADP-ribose) glycohydrolase. Biochim Biophys Acta 1219: 342–350, 1994
20. Brochu G, Shah GM, Poirier GG: Purification of poly(ADP-ribose) glycohydrolase and detection of its isoforms by a zymogram following one- or two-dimensional electrophoresis. Anal Biochem 218: 265–272, 1994
21. Ménard L, Poirier GG: Rapid assay of poly (ADP-ribose) glycohydrolase. Biochem Cell Biol 65: 668–673, 1987
22. Shah GM, Poirier D, Duchaine C, Brochu G, Desnoyers S, Lagueux J, Verreault A, Hoflack JC, Kirkland JB, Poirier GG: Methods for biochemical study of poly(ADP-ribose) metabolism *in vitro* and *in vivo*. Anal Biochem 227: 1–13, 1995
23. Canales J, Pinto RM, Costas MJ, Hernandez MT, Miro A, Bernet D, Fernandez A, Cameselle JC: Rat liver nucleoside diphosphosugar or diphosphoalcohol pyrophosphatases different from nucleotide pyrophosphatase or phosphodiesterase I: substrate specificities of Mg(2+)- and/or Mn(2+) -dependent hydrolases acting on ADP-ribose. Biochim Biophys Acta 1246: 167–177, 1995
24. Miwa M, Nakatsugawa K, Hara K, Taijiro M, Sugimura T: Degradation of poly(adenosine diphosphate ribose) by homogenates of various normal tissues and tumors of rats. Arch Biochem Biophys 167: 54–60, 1975
25. Vincent R, Nadeau D: A micromethod for the quantification of cellular proteins in Percoll with the Coomassie brilliant blue dye-binding assay. Anal Biochem 135: 355–362, 1983
26. Shah GM, Poirier D, Desnoyers S, Saint-Martin S, Hoflack JC, Rong P, ApSimon M, Kirkland JB, Poirier GG: Complete inhibition of poly(ADP-ribose) polymerase activity prevents the recovery of C3H10T1/2 cells from oxidative stress. Biochim Biophys Acta 1312: 1–7, 1996
27. Wielckens K, George E, Pless T, Hiltz H: Stimulation of poly(ADP-ribosyl)ation during Ehrlich ascites tumor cell 'starvation' and suppression of concomitant DNA fragmentation by benzamide. J Biol Chem 258: 4098–4104, 1983
28. Jacobson EL, Antol KM, Juarez-Salinas H, Jacobson MK: Poly(ADP-ribose) metabolism in ultraviolet irradiated human fibroblasts. J Biol Chem 258: 103–107, 1983
29. Juarez-Salinas H, Duran-Torres G, Jacobson MK: Alteration of poly(ADP-ribose) metabolism by hyperthermia. Biochem Biophys Res Commun 122: 1381–1388, 1984

30. Hilz H, Wielckens K, Bredehorst R: Quantitation of mono(ADP-ribosyl) and poly(ADP-ribosyl) proteins. In: O. Hayaishi, K. Ueda (eds) ADP-ribosylation reactions: Biology and Medicine. Academic Press, New York 1982, pp 305–321
31. Kreimeyer A, Wielckens K, Adamietz P, Hilz H: DNA repair-associated ADP-ribosylation *in vivo*. J Biol Chem 259: 890–896, 1984
32. Wielckens K, Bredehorst R, Adamietz P, Hilz H: Protein-bound polymeric and monomeric ADP-ribose residues in hepatic tissues. Comparative analyses using a new procedure for the quantification of poly(ADP-ribose). Eur J Biochem 117: 69–74, 1981
33. Wielckens K, Delfs T: Glucocorticoid-induced cell death and poly[adenosine diphosphate(ADP)-ribosyl]ation: Increased toxicity of dexamethasone on mouse S49.1 lymphoma cells with the poly(ADP-ribosyl)ation inhibitor benzamide. Endocrinology 119: 2383–2392, 1986
34. Aboul-Ela N, Jacobson EL, Jacobson MK: Labeling methods for the study of poly- and mono(ADP-ribose) metabolism in cultured cells. Anal Biochem 174: 239–250, 1988
35. Juarez-Salinas H, Sims JL, Jacobson MK: Poly(ADP-ribose) levels in carcinogen treated cells. Nature 282: 740–741, 1979
36. Lagueux J, Menard L, Candas B, Brochu G, Potvin F, Verreault A, Cook PF, Poirier GG: Equilibrium model in an *in vitro* poly(ADP-ribose) turnover system. Biochim Biophys Acta 1264: 201–208, 1995
37. Ménard L, Thibeault L, Poirier GG: Reconstitution of an *in vitro* poly (ADP-ribose) turnover system. Biochim Biophys Acta 1049: 45–58, 1990
38. Thomassin H, Menard L, Hengartner C, Kirkland JB, Poirier GG: Poly(ADP-ribosyl)ation of chromatin in an *in-vitro* poly(ADP-ribose)-turnover system. Biochim Biophys Acta 1137: 171–181, 1992
39. Malanga M, Bachmann S, Panzeter PL, Zweifel B, Althaus FR: Poly(ADP-ribose) quantification at the femtornole level in mammalian cells. Anal Biochem 228: 245–251, 1995
40. Kawaichi M, Ueda K, Hayaishi O: Multiple autopoly(ADP-ribosyl)ation of rat liver poly(ADP-ribose) synthetase. J Biol Chem 256: 9483–9489, 1981
41. Hatakeyama K, Nemoto Y, Ueda K, Hayaishi O: Purification and characterization of poly(ADP-ribose) glycohydrolase. Different modes of action on large and small poly(ADP-ribose). J Biol Chem 261: 14902–14911, 1986
42. Tanuma S, Kawashima K, Endo H: Purification and properties of an (ADP-ribose)n glycohydrolase from guinea pig liver nuclei. J Biol Chem 261: 965–969, 1986
43. Uchida K, Suzuki H, Maruta H, Abe H, Aoki K, Miwa M, Tanuma S: Preferential degradation of protein-bound (ADP-ribose)n by nuclear poly(ADPribose) glycohydrolase from human placenta. J Biol Chem 268: 3194–3200, 1993

Molecular and Cellular Biochemistry **193**: 137–148, 1999.

Involvement of PARP and poly(ADP-ribosyl)ation in the early stages of apoptosis and DNA replication

Cynthia Marie Simbulan-Rosenthal, Dean S. Rosenthal, Sudha Iyer, Hamid Boulares and Mark E. Smulson

Department of Biochemistry and Molecular Biology, Georgetown University School of Medicine, Washington, DC, USA

Abstract

We have focused on the roles of PARP and poly(ADP-ribosyl)ation early in apoptosis, as well as during the early stages of differentiation-linked DNA replication. In both nuclear processes, a transient burst of PAR synthesis and PARP expression occurs early, prior to internucleosomal DNA cleavage before commitment to apoptosis as well as at the round of DNA replication prior to the onset of terminal differentiation. In intact human osteosarcoma cells undergoing spontaneous apoptosis, both PARP and PAR decreased after this early peak, concomitant with the inactivation and cleavage of PARP by caspase-3 and the onset of substantial DNA and nuclear fragmentation. Whereas 3T3-L1, osteosarcoma cells, and immortalized PARP +/+ fibroblasts exhibited this early burst of PAR synthesis during Fas-mediated apoptosis, neither PARP-depleted 3T3-L1 PARP-antisense cells nor PARP -/- fibroblasts showed this response. Consequently, whereas control cells progressed into apoptosis, as indicated by induction of caspase-3-like PARP-cleavage activity, PARP-antisense cells and PARP -/- fibroblasts did not, indicating a requirement for PARP and poly(ADP-ribosyl)ation of nuclear proteins at an early reversible stage of apoptosis. In parallel experiments, a transient increase in PARP expression and activity were also noted in 3T3-L1 preadipocytes 24 h after induction of differentiation, a stage at which ~95% of the cells were in S-phase, but not in PARP-depleted antisense cells, which were consequently unable to complete the round of DNA replication required for differentiation. PARP, a component of the multiprotein DNA replication complex (MRC) that catalyzes viral DNA replication *in vitro*, poly(ADP-ribosyl)ates 15 of ~40 MRC proteins, including DNA pol α, DNA topo I, and PCNA. Depletion of endogenous PARP by antisense RNA expression in 3T3-L1 cells results in MRCs devoid of any DNA pol α and DNA pol δ activities. Surprisingly, there was no new expression of PCNA and DNA pol α, as well as the transcription factor E2F-1 in PARP-antisense cells during entry into S-phase, suggesting that PARP may play a role in the expression of these proteins, perhaps by interacting with a site in the promoters for these genes. (Mol Cell Biochem **193**: 137–148, 1999)

Key words: PARP, poly(ADP-ribosyl)ation, apoptosis, DNA replication

Introduction

Poly(ADP-ribose) polymerase (PARP), which is catalytically activated by DNA strand breaks, plays an auxiliary role in nuclear processes, such as DNA repair [1–5] , DNA replication [6–11], cellular differentiation [10–15], and more recently in apoptosis. PARP catalyzes the poly(ADP-ribosyl)ation of various nuclear proteins, with NAD as substrate, and undergoes proteolytic cleavage into 89- and 24-kDa fragments that contain the active site and the DNA-binding domain of the enzyme, respectively, during drug-induced [16] and spontaneous apoptosis [17, 18]. More recently, PARP has been implicated in the induction of both p53 expression and apoptosis [19] , with the specific proteolysis of the enzyme thought to be a key apoptotic event [17, 20, 21]. Caspase-3, a member of the family of aspartate-specific cysteine pro-

Address for offprints: M.E. Smulson, Department of Biochemistry and Molecular Biology, Georgetown University School of Medicine, Basic Science Building, Rm. 351, 3900 Reservoir Road NW, Washington, DC 20007, USA

teases, is composed of two subunits of 17 and 12-kDa derived from a common proenzyme (CPP32), plays a central role in the execution of the apoptotic program [22], and is responsible for the cleavage of PARP during cell death [17, 21, 23]. Human osteosarcoma cells that undergo confluence-associated apoptosis over a 10-day period, exhibit a peak of caspase-3 activity, measured with a specific [^{35}S] PARP-cleavage assay *in vitro*, 6–7 days after initiation of apoptosis, concomitant with the onset of internucleosornal DNA fragmentation [17].

We recently examined the time course of PARP activation and cleavage during apoptosis in intact osteosarcoma cells by immunofluorescence microscopy [18] as well as by immunoblot analysis with antibodies to PARP and to poly(ADP-ribose) (PAR). The results showed that poly(ADP-ribosyl)ation of relevant nuclear proteins may be required at an early reversible stage of apoptosis but that degradation of PARP may be necessary at a later stage, suggesting a negative regulatory role for PARP in apoptosis. We have now investigated this hypothesis utilizing well-characterized cell lines stably transfected with inducible PARP- antisense constructs [11, 24, 25] as well as with immortalized fibroblasts derived from PARP knockout mice [26]. Induction of PARP antisense RNA in these cells have consistently shown significant depletion of endogenous PARP protein and activity. Using these PARP-depleted antisense cells, we have shown that PARP and poly(ADP-ribosyl)ation facilitates the initial rate of DNA repair in HeLa cells [1] and keratinocytes [24]. Although the PARP knock out mice from which the PARP -/- fibroblasts used in our studies were derived from were viable, fibroblasts and thymocytes from these mice exhibited proliferation deficiencies following DNA damage [26]; primary PARP -/- splenocytes derived from other knockout mice also showed abnormal apoptosis implicating a role for PARP in the regulation of cell death [27].

PARP and poly(ADP-ribosyl)ation has also been implicated in cellular differentiation since PARP inhibitors (i.e. nicotinamide and benzamide) markedly inhibit differentiation of 3T3-L1 preadipocytes into adipocytes by preventing a transient increase in PARP activity that appears essential for entering the differentiation program [28]. Consistently, we have shown that inducible stably transfected 3T3-L1 cells expressing PARP antisense RNA do not exhibit the increase in PARP protein and activity normally apparent 24 h after exposure to differentiation inducers and they fail to differentiate into adipocytes [11] . The inability of the antisense cells to synthesize PARP and PAR during the early stages of differentiation correlates with their inability to undergo the round of DNA replication required for onset of terminal differentiation. We have now extended these studies to better clarify the role of PARP in differentiation-linked DNA replication and its putative roles as a component of multiprotein DNA replication complexes (known as MRC or the DNA synthesome) in cells [10].

Results

PARP and PAR are synthesized early during apoptosis of human osteosarcoma cells

Human osteosarcoma cells were induced to undergo confluence-associated apoptosis for 10 days, and examined by immunofluorescence staining for PAR synthesis with antibodies to PAR (left panel), and PARP (middle panel); the levels of DNA strand breaks in cells were also assessed by staining cells with biotinylated PARP DBD and Texas red-conjugated streptavidin (right panel; [29]. Representative samples from immediate (day 1), early (day 3), mid- (day 6), and late (day 10) stages of apoptosis are shown (Fig. 1). The synthesis of PAR from NAD increased early and peaked 3 days after initiation of apoptosis, prior to the appearance of internucleosomal DNA cleavage and before the cells became irreversibly committed to apoptosis. During the same early period, new expression of full-length PARP was detected. At 6 days, however, the amounts of both PAR and PARP decreased markedly, and PAR was not observed during days 8–10, despite the presence of abundant DNA strand breaks (right panel), potential activators of intact PARP, during this time. Significant intemucleosomal DNA fragmentation was also shown to occur at days 6–9 of apoptosis in these cells [17, 18]. Thus, a transient burst of PARP expression and poly(ADP-ribosyl)ation of nuclear proteins occurs early, prior to commitment to death, in osteosarcoma cells and is

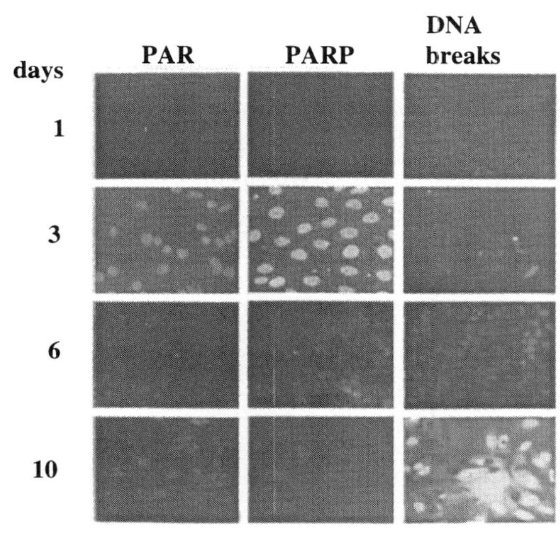

Fig. 1. Time courses of PAR synthesis, PARP expression, and the presence of DNA strand breaks during spontaneous apoptosis of human osteosarcoma cells. Cells were cultured without any changes in medium for 1–10 days, fixed, and subjected to immunofluorescence analysis with antibodies to PAR (left panel) and PARP (middle panel), as well as biotinylated recombinant PARP DBD which was used to detect DNA strand breaks (right panel).

followed by caspase-3 -mediated cleavage and inactivation of PARP.

Different cells undergo a transient burst of poly(ADP-ribosyl)ation of nuclear proteins early during apoptosis followed by a decline concomitant with the onset of PARP cleavage

In agreement with the immunofluorescence results shown in Fig. 1, immunoblot analysis with antibody to PAR of osteosarcoma cells at various stages of apoptosis also showed an early burst of poly(ADP-ribosyl)ation of nuclear proteins at day 4 of apoptosis, followed by a marked decline in poly(ADP-ribosyl)ation at later time points, coincident with the onset of PARP cleavage activity and intemucleosomal DNA cleavage.

Consistently, other cell lines induced into Fas-mediated apoptosis (Figs. 2C and D) or by the topoisomerase inhibitor, camptothecin (Fig. 2B), also exhibited the same early burst of poly(ADP-ribosyl)ation of nuclear proteins during an early reversible stage of apoptosis, when the cells were still viable (as assessed by exclusion of trypan blue) and could still be replated. HL-60 cell extracts derived at early time points during camptothecin-induced apoptosis (Fig. 2B) as well as 3T3-L1 and PARP +/+ fibroblasts induced into apoptosis by a combination of antibody to Fas and cycloheximide (Fig. 2, C and D, respectively) all showed a similar transient peak of poly(ADP-ribosyl)ation of nuclear proteins 1–4 h after induction; PAR moieties that were bound to these proteins were degraded and no further modification was apparent thereafter, concomitant with the onset of caspase-3 catalyzed cleavage of PARP.

PARP-depleted 3T3-L1 antisense cells and immortalized PARP -/- fibroblasts do not exhibit the transient burst of PAR synthesis at the early stages of Fas-mediated apoptosis

To investigate the role(s) of PARP during the early stages of apoptosis, we examined PAR synthesis in PARP-depleted 3T3-L1 antisense and PARP -/- cells during this period. 3T3-L1 antisense cells that had been preincubated with dexamethasone (Dex) for 72 h to deplete endogenous PARP as well as fibroblasts derived from PARP knockout mice were exposed to anti-Fas and cycloheximide for various times and then subjected to immunofluorescence staining (data not shown) and immunoblot analysis with antibody to PAR. In control 3T3-L1 and PARP +/+ fibroblasts, PAR levels in the nucleus were significantly elevated at 4 and 1 h, respectively, after induction of apoptosis, a stage wherein all the cells were still viable and could be replated. This was followed by a

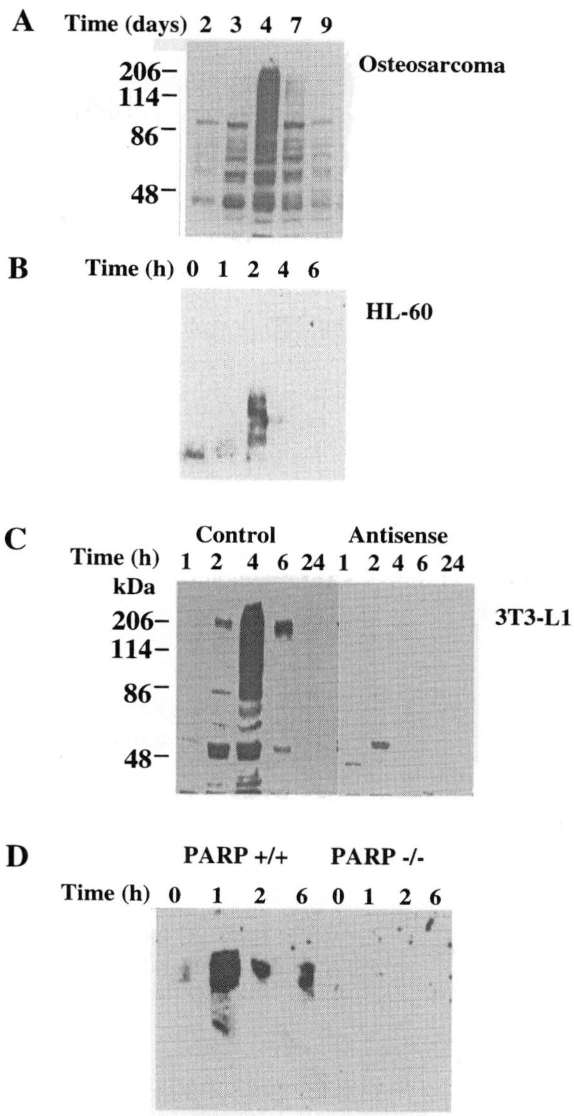

Fig. 2. Time course of poly(ADP-ribosyl)ation of nuclear proteins during apoptosis induced by various apoptotic stimuli. (A) Osteosarcoma cells were allowed to undergo confluence associated spontaneous apoptosis for 9 days; (B) HL-60 cells were induced into apoptosis with the topoisomerase inhibitor, camptothecin (10 μM); and (C) 3T3-L1 control and PARP-antisense fibroblasts as well as (D) immortalized PARP +/+ and PARP -/- fibroblasts were induced into apoptosis with a combination of anti-Fas (100 ng/ml) and cyclohexamide (10 μg/ml). 3T3-L1 control and antisense cells were preincubated with Dex for 72 h prior to apoptosis induction. At the indicated times, equal amounts of total cellular protein (30 μg) were subjected to immunoblot analysis with monoclonal antibody to PAR (1:250). The positions of the molecular size standards (in kilodaltons) are indicated.

marked decline in poly(ADP-ribosyl)ation at later time points, concomitant with the induction of PARP cleavage activity. In contrast, anti-Fas and cycloheximide did not induce this burst of poly(ADP-ribosyl)ation of nuclear proteins during early apoptosis in PARP-depleted 3T3-L1 antisense cells (Fig. 2C) nor in PARP -/- cells (Fig. 2D), suggesting that the PARP and poly(ADP-ribosyl)ation may play a role at this early stage in Fas-mediated apoptosis in these cells.

Depletion of PARP by antisense RNA expression blocks progression of apoptosis in 3T3-L1 and Jurkat cells

PARP depletion induced by antisense RNA expression in 3T3-L1 cells under the present conditions were confirmed by immunoblot analysis with antibody to PARP (Fig. 3A). Whereas PARP levels in mock-transfected control 3T3-L1 cells were not affected by incubation with Dex for 72 h (Fig. 3A), Dex induced significant depletion of endogenous PARP

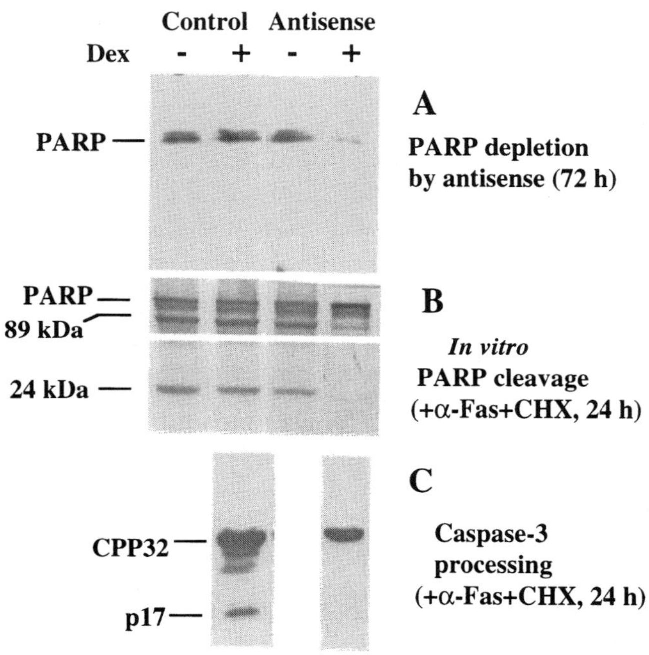

Fig. 3. PARP depletion by antisense RNA expression inhibits induction of anti-Fas mediated apoptosis in 3T3-L1 cells as monitored by induction of *in vitro* PARP cleavage activity and proteolytic processing of CPP32 to caspase-3. (A) Mock-transfected (control) and PARP-antisense 3T3-L1 cells were incubated in the absence or presence of 1 μM Dex for 72 h, after which equal amounts of total cellular protein (30 μg) were subjected to immunoblot analysis with antibody to full-length PARP. After preincubation in the absence or presence of 1 μM Dex, control and antisense cells were incubated with anti-Fas (100 ng/ml) and cycloheximide (10 μg/ml) for 24 h; equal amounts of cytosolic extracts were then assayed for *in vitro* PARP-cleavage activity with [^{35}S]PARP as substrate (B) or subjected to immunoblot analysis with a monoclonal antibody to the p17 subunit of caspase-3 (C). The positions of full length PARP and its 89- and 24-kDa cleavage products as well as those of CPP32 (32-kDa) and its active form (caspase-3, p17) are indicated.

in antisense cells, with only ~5% of the protein remaining after 72 h. Essentially equal protein loading and transfer among lanes was confirmed by Ponceau S staining for total protein on the same immunoblot (data not shown).

3T3-L1 control and antisense cells were then preincubated in the presence or absence of Dex for 72 h, and subsequently induced into apoptosis with a combination of antibody to Fas and cycloheximide for various times. Using an *in vitro* caspase-3 mediated PARP-cleavage assay [17], incubation of cells with anti-Fas and cycloheximide for 24 h were shown to result in a marked induction of caspase-3 activity (Fig. 3B), as indicated by generation of the 89- and 24-kDa cleavage fragments of PARP. There was no effect of anti-Fas by itself or cycloheximide alone on the induction of apoptosis in these cells [71]. Cycloheximide has been used to overcome resistance to Fas-mediated apoptosis in various cell lines [30–32], an effect which does not appear to be mediated by translational inhibition because resistant cells are sensitized to anti-Fas by subinhibitory concentrations.

Anti-Fas and cycloheximide induced a marked time-dependent increase in *in vitro* PARP-cleavage activity in control 3T3-L1 cells that had been preincubated either in the absence or presence of Dex (Fig. 3B), an effect which was maximal 24 h after induction of apoptosis. On the other hand, whereas PARP-antisense 3T3-L1 cells that were not exposed to Dex showed a similar increase in caspase-3 activity in response to anti-Fas and cycloheximide, no such increase in caspase-3 activity was apparent in PARP-antisense cells that had been depleted of PARP by preincubation with Dex and then induced for apoptosis (Fig. 3B).

Similar to other members of the caspase family, caspase 3 is expressed in cells as an inactive 32-kDa proenzyme (CPP32), which is activated during apoptosis by cleavage at specific Asp residues, with the mature active enzyme (caspase-3) consisting of a large 17-kDa subunit (p17), containing the catalytic domain, and a 12-kDa subunit (p12) [17]. Proteolytic processing of CPP32 to p17 during the course of apoptosis was then confirmed in 3T3-L1 cells by immunoblot analysis with a monoclonal antibody to the p17 subunit of caspase-3 (Fig. 3D). Whereas, in control cells, CPP32 was proteolytically processed to p17 (caspase-3) by 24 h, coinciding with the peak of *in vitro* PARP cleavage activity, proteolytic cleavage, processing, and activation of CPP32 to p17 was not apparent in the PARP-depleted antisense cells, again suggesting that PARP as well as presumably poly(ADP-ribosyl)ation plays a role in some early event in apoptosis, upstream of the proteolytic cascade mechanism for processing the caspase-3 precursor to its active form. Consistent with these results, whereas control 3T3-L1 cells as well as antisense cells preincubated in the absence of dexamethasone showed changes in nuclear morphology typical of apoptosis after exposure to anti-Fas and cycloheximide i.e. chromatin condensation, nuclear and DNA fragmentation, antisense cells depleted of PARP by preincubation with Dex did not [71].

Essentially the same results were obtained with another cell line, Jurkat T cells, that were stably transfected with either the PARP antisense RNA construct or the empty vector. Immunoblot analysis confirmed that preincubation of Jurkat antisense cells for 72 h with Dex resulted in depletion of endogenous PARP by ~99% (Fig. 4A). Anti-Fas alone was used to induce apoptosis in these cells since Jurkat cells are Fas-sensitive and express high levels of the Fas antigen [30, 33]. After 4 h of exposure to anti-Fas, mock-transfected cells preincubated in the absence or presence of Dex as well as antisense cells in the absence of Dex showed a significant increase in caspase-3 activity, as indicated by the *in vitro* cleavage of exogenous PARP into 89- and 24-kDa fragments, however, caspase-3 activity remained negligible in PARP-depleted antisense cells treated with anti-Fas for up to 24 h (Fig. 4B). Furthermore, while anti-Fas induced changes in nuclear morphology consistent with apoptosis in control Jurkat cells preincubated with or without Dex as well as in antisense cells preincubated without Dex, no such changes were evident in PARP-depleted antisense cells treated with anti-Fas [71]. Depletion of PARP by antisense RNA expression

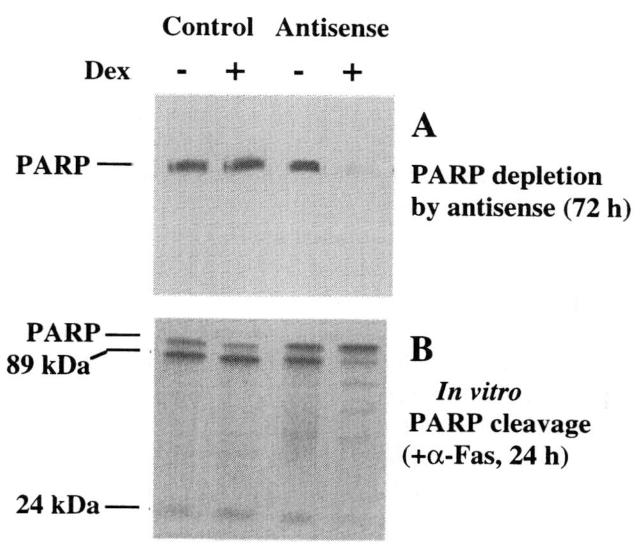

Fig. 4. PARP-depleted Jurkat antisense cells are unable to undergo Fas-mediated apoptosis as measured by caspase-3 activity assays. (A) Jurkat control and antisense cells were incubated in the absence or presence of 1 μM Dex for 72 h, after which equal amounts of total cellular protein (30 μg) were subjected to immunoblot analysis with anti-PARP. (B) Control and PARP antisense Jurkat cells were preincubated in the absence or presence of 1 μM Dex and then exposed to anti-Fas (50 ng/ml) for 24 h. Cytosolic extracts were prepared and equal amounts were assayed for PARP-cleavage activity with [^{35}S]PARP as substrate. The positions of full length PARP and its 89- and 24-kDa cleavage products are indicated.

apparently blocks progression of Fas-mediated apoptosis in both 3T3-L1 and Jurkat cells.

Fibroblasts derived from PARP-knockout mice are resistant to Fas-mediated apoptosis

PARP knockout mice generated by disruption of the PARP gene by homologous recombination in embryonic stem cells [26] were used to derive immortalized fibroblasts by the 3T3 protocol, which were confirmed to be devoid of PARP (Fig. 5A) as evidenced by immunoblot analysis with antibody to PARP.

When immortalized PARP +/+ and PARP -/- fibroblasts were induced into apoptosis with a combination of anti-Fas

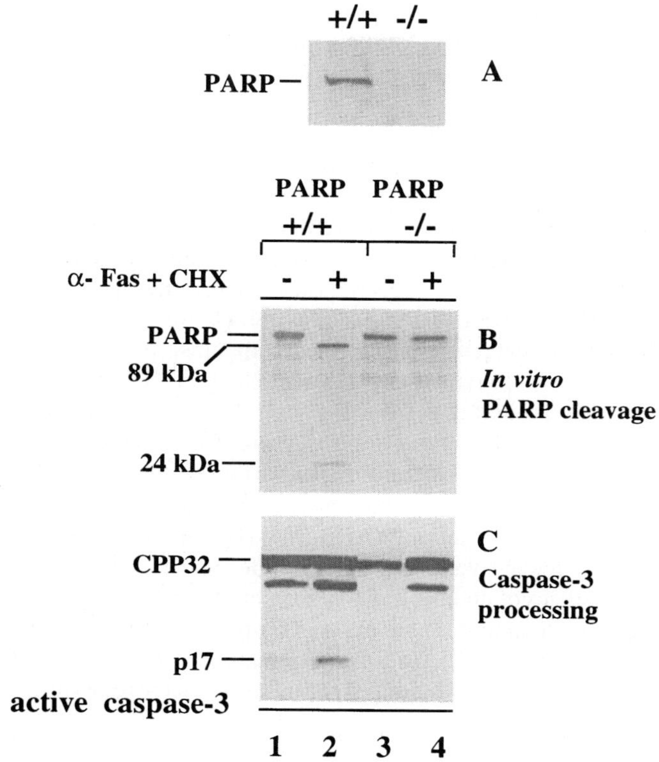

Fig. 5. Fibroblasts derived from PARP knockout mice do not exhibit any increase in caspase-3 activity and proteolytic processing of CPP32 to caspase-3. (A) Immortalized fibroblasts (PARP -/-) derived from knockout mice as well as fibroblasts from wild type mice (PARP +/+) were subjected to immunoblot analysis with anti-PARP. (B) PARP +/+ and PARP -/- fibroblasts were incubated with anti-Fas (100 ng/ml) and cycloheximide (10 μg/ml) for 24 h, after which cytosolic extracts were assayed for *in vitro* PARP-cleavage activity with [^{35}S]PARP as substrate. (C) Cytosolic extracts in (B) were also subjected to immunoblot analysis with a monoclonal antibody to the p17 subunit of caspase-3. The positions of full length PARP and its 89- and 24-kDa cleavage products as well as those of CPP32 (32-kDa) and its active form (p17) are indicated.

and cycloheximide, PARP +/+ cells showed a rapid apoptotic response, as indicated by a marked increase in *in vitro* PARP-cleavage activity; the activity was maximal 24 h after induction as indicated by the complete cleavage of PARP in to 89 kDa- and 24 kDa cleavage fragments (Fig. 5B). In contrast, PARP -/- cells showed no such increase in caspase-3 mediated PARP cleavage after 24 h exposure to anti-Fas and cycloheximide. Immunoblot analysis with the antibody to the p17 subunit of caspase-3 further showed that CPP32 was proteolytically processed to p 17 in the PARP +/+ fibroblasts (Fig. 5C), but not in the PARP -/- cells exposed to anti-Fas. Furthermore, whereas PARP +/+ fibroblasts were ~99% apoptotic 24 h after induction as shown by development of nuclear apoptotic morphology, PARP -/- cells exhibited normal, non-apoptotic morphology even after 24 h of exposure to anti-Fas [71]. Altogether, these results confirm the earlier data obtained with the 3T3-L1 and Jurkat control and PARP antisense cells and provide further evidence for a role for PARP at an early event in apoptosis.

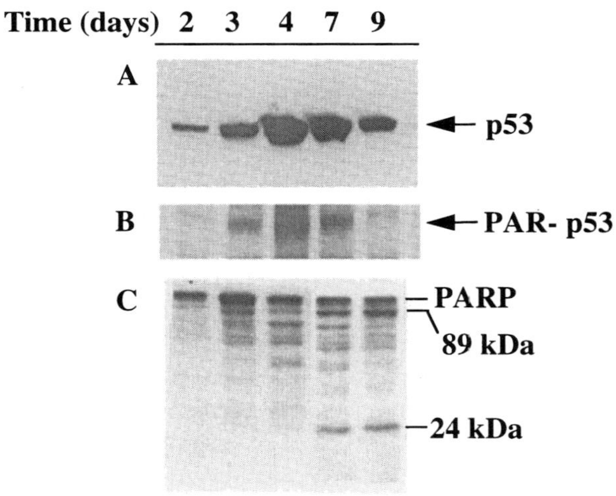

Fig. 6. Time courses of accumulation (A) and poly(ADP-ribosyl)ation (B) of p53 as well as caspase-3 mediated PARP-cleavage (C) during spontaneous apoptosis in osteosarcoma cells. Human osteosarcoma cells were induced to undergo spontaneous apoptosis for 9 days, and, at the indicated times, cytosolic extracts were derived and equal amounts of protein (30 μg) were subjected to immunoblot analysis with monoclonal antibody to p53 (A) or assayed for *in vitro* PARP-cleavage activity, using [^{35}S]PARP as a substrate (C). The positions of full length PARP and its 89- and 24-kDa cleavage products are indicated. The immunoblot in (A) was stripped with a buffer containing 100 mM 2-mercaptoethanol, 2% SDS, and 62.5 mM Tris-HCl pH 6.7 for 30 min at 50°C, blocked, and reprobed with monoclonal antibody to PAR (B).

Poly(ADP-ribosyl)ation of p53 during the early stages of apoptosis in osteosarcoma cells

The apparent role for early PAR synthesis during apoptosis was further clarified by investigating a well-characterized apoptotic response to DNA strand breaks, the rapid accumulation of p53. Osteosarcoma cells were plated under conditions that result in spontaneous apoptosis over a 10-day period, with maximal caspase-3 mediated PARP-cleavage apparent around days 7–9 (Fig. 6C). Immunoblot analysis showed that the amount of p53 protein was markedly increased at days 4–7 and declined thereafter (Fig. 6A). Reprobing of the same immunoblot with antibodies to PAR revealed significant poly(ADP-ribosyl)ation of p53 at days 3–4 coincident with the burst of PAR synthesis during the early stages of apoptosis, and the extent of p53 modification declined thereafter at the onset of caspase-3 activity at days 7–9 (Fig. 6B and C). These experiments were not performed with Jurkat and HL-60 cells since these cells have been shown to be p53-null [34, 35].

A transient peak of PARP expression and poly(ADP-ribosyl)ation is also observed during the early stages of differentiation-linked DNA replication in control 3T3-L1 cells, but not in PARP depleted antisense cells

A transient peak of PARP expression occurs in postconfluent cultures of 3T3-L1 cells induced to differentiate to adipocytes for 7–8 days by exposure to insulin, Dex, and methylisobutylxanthine [11]. The time courses of PARP activity and % of cells in S-phase after exposure to inducers of differentiation were then compared in 3T3-L1 control cells and antisense cells (Fig. 7). FACS analysis was performed on control and antisense cells at various times after induction and the data quantitated. Prior to induction of differentiation (zero time), 70–80% of postconfluent control and antisense cells were in growth-arrest induced by contact inhibition. More than 95% of the control cells synchronously entered S phase by 24 h, coincident with the peak of poly(ADP-ribosyl)ation observed 24 h after induction of differentiation (Fig. 7). While the control cells progressed through one round of replication prior to the onset of terminal differentiation, the antisense cells did not show the burst of PAR synthesis apparent in the control cells and did not enter S phase. When the transient peak of PARP expression and poly(ADP-ribosl)ation is prevented in the antisense cells, these cells are unable to enter S-phase and undergo the round of DNA replication at the early stages of differentiation, and consequently, these cells do not differentiate, as confirmed by their inability to synthesize and accumulate cytoplasmic triglyceride [10, 11]. PARP is apparently required for a round of DNA replication that occurs within the first 24 h of differentiation and is necessary for differentiation in these cells.

Confocal microscopy further revealed that, during this early stage of differentiation, when essentially all control cells

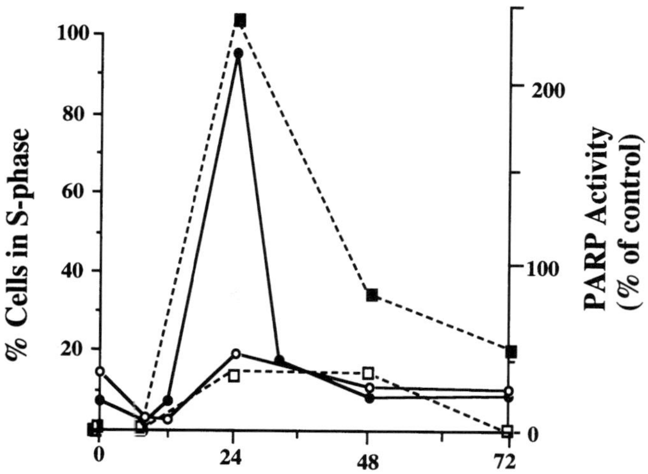

Fig. 7. PARP depleted- 3T3-L1 antisense cells do not undergo a peak of PAR synthesis and are unable to enter S-phase during the round of DNA replication at the early stages of differentiation. Control (solid symbols) and antisense (open symbols) cells were exposed to inducers of differentiation and harvested at the indicated times. Duplicate cell pellets were assayed for either PARP activity (dotted lines) or % cells in S-phase (solid lines). Cells were assayed for PARP activity by measuring [^{32}P]NAD incorporation into acid-insoluble acceptors at 25°C for 1 min, with 20 μg of protein per determination and triplicate determinations per time point. [^{32}P]NAD, incorporation of uninduced control 3T3-L1 cells was taken as baseline, and PARP activity values were calculated and plotted based on this control value. % of control and PARP antisense cells in S phase phase of the cell cycle was quantitated from flow cytometric (FACS) data of nuclei stained with propidium iodide (0.42 mg/ml).

have entered the S phase of the cell cycle, PARP is localized within distinct intranuclear granular foci that are associated with replication centers [10]. Consistently, we recently showed that PARP is a component of the multiprotein DNA replication complexes (MRC) or DNA synthesomes that catalyze replication of viral DNA *in vitro,* and contain replicative enzymes such as DNA polymerases α and δ (pol α and pol δ), DNA primase, DNA helicase, DNA ligase, and topoisomerase I and II, as well as accessory proteins such as proliferating-cell nuclear antigen (PCNA), RFC, and RPA. PARP catalyzes the poly(ADP-ribosyl)ation of about 15 of the ~40 MRC proteins, including pol α, DNA topoisomerase I, and PCNA [10]. To further clarify the involvement of PARP within the DNA synthesome during the round of DNA replication in the early stages of differentiation in 3T3-L1 cells, we have now purified and characterized replicative complexes from control cells that had entered S phase after induction of differentiation and from cells depleted of PARP by expression of PARP antisense RNA [72].

PARP regulates the expression of components of the DNA synthesome during entry into S-phase at the early stages of differentiation-linked DNA replication

The roles of PARP in modulating the composition and enzyme activities of the DNA synthesome were further investigated by characterizing the complex purified from 3T3-L1 cells before and 24 h after induction of a round of DNA replication required for differentiation of these cells; at the latter time point, ~95% of the cells are in S phase and exhibit a transient peak of PARP activity (Fig. 7). The MRC fraction was also purified from similarly treated 3T3-L1 cells depleted of PARP by expression of antisense RNA; these cells do not undergo DNA replication or terminal differentiation. Both PARP protein and activity and essentially all of the pol α and pol δ activities exclusively cosedimented with the MRC fractions from S phase control cells, and were not detected in the MRC fractions from PARP-antisense or uninduced control cells [72].

SDS-polyacrylamide gel electrophoresis and silver staining of MRC fractions prepared from control and PARP-antisense cells, before or 24 h after induction of differentiation, revealed that the most prominent difference was the presence of a 116-kDa protein, corresponding to the size of intact PARP, in the fraction from control cells in S phase but not in those from antisense cells or uninduced control cells. Consistent with these observations and with the PARP activity data, immunoblot analysis with polyclonal antibodies to PARP showed that PARP protein was present exclusively in the MRC fraction from control 3T3-L1 cells in S phase, and not in the MRC fractions nor total cell extracts from PARP-antisense cells or non-replicating control cells ([72], Fig. 8).

We next investigated the effects of PARP depletion on the composition of the MRC by immunoblot analysis with antibodies to specific MRC components. Immunoblot analysis of cell extracts revealed that while the amount of PCNA was markedly increased in control cells upon exposure to inducers of differentiation, there was no new expression in the PARP antisense cells. Although PCNA was also present in the antisense cells and uninduced control cells (Fig. 8), PCNA as well as topoisomerase I appeared to be present in antisense cells and uninduced control cells only in the free, uncomplexed form, because they were detected in the MRC fraction only from replicating control cells, indicating that PARP may play a role in their recruitment into the DNA synthesome [72].

Immunoblot analysis of total cell extracts and MRC fractions with antibody to pol α further revealed induction of expression of this replicative enzyme during early S-phase in control cells and its absence from PARP-antisense cells (Fig. 8). These results indicate that PARP may play a role in the expression of pol α and PCNA during early S phase. We thus investigated the effect of PARP depletion on the abun-

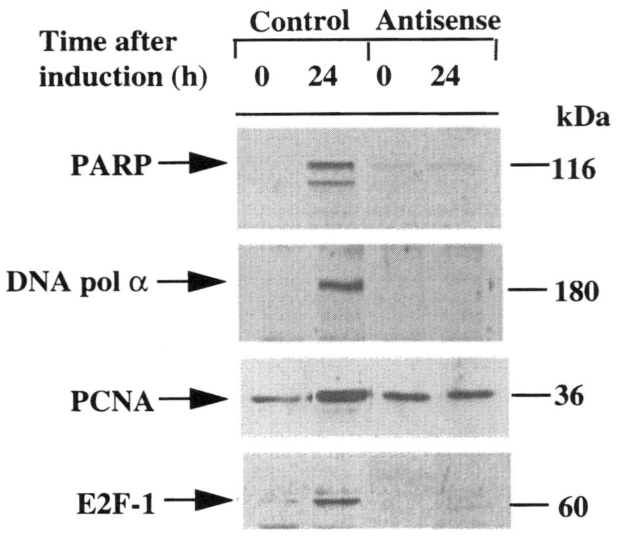

Fig. 8. PARP depleted-3T3-L1 antisense cells do not exhibit new expression of PARP, pol α, PCNA, and E2F-1 during early S-phase. Total cell extracts from 3T3-L1 control and PARP antisense cells were prepare before and 24 h after induction of differentiation and equal amounts of total protein (30 μg) were subjected to immunoblot analysis with antibodies to PARP (1:2000 dilution), pol α (1:200), PCNA (1:250), and E2F-1 (1:1000). Arrows indicate the positions of the various proteins, and their molecular sizes (in kilodaltons) are indicated on the right.

dance of E2F-1, a transcription factor that positively regulates the expression of the pol α and PCNA genes during entry of cells into S phase. Immunoblot analysis of total cell extracts revealed that, whereas control cells exhibited a marked increase in the expression of E2F-1, PARP-depleted antisense cells contained negligible amounts of E2F-1 during the 24 h exposure to inducers of differentiation (Fig. 8). PARP may therefore play a role in the expression of pol α and PCNA by indirectly regulating the expression of E2F-1.

Discussion

PARP and poly(ADP-ribosyl)ation in the early stages of apoptosis

We have previously established and characterized several mammalian cell lines, including HeLa cells [25], keratinocytes [24], and 3T3-L1 fibroblasts [11] that are stably transfected with PARP antisense cDNA under the control of a Dex-inducible promoter in order to assess, without the use of possibly nonspecific chemical inhibitors [36–38], the potential roles of PARP and poly(ADP-ribosyl)ation in a variety of nuclear processes. Depletion of endogenous PARP protein and activity at specific times by PARP antisense RNA expression has enabled us to investigate the roles of PARP in DNA repair, recovery of cells from exposure to mutagenic agents [24, 25], gene amplification [39], and differentiation-linked DNA replication [10, 11, 72].

Recently, PARP cleavage by caspase-3 has been shown to be necessary for apoptosis [17, 21] since the cleavage and inactivation of PARP as well as apoptotic events are inhibited by a peptide inhibitor of this protease, suggesting a negative regulatory role for PARP in apoptosis. Aside from Fas activation, multiple apoptotic signals can activate caspase-3 by proteolytic cleavage of CPP32, including serum withdrawal, ionizing radiation, and various pharmacological agents [40–44]. The peptide inhibitor of caspase-3 (AcDEVD-CHO) blocks the apoptotic program in response to various stimuli [17, 45], and addition of active caspase-3 to normal cytosol activates the apoptotic program [46].

We now show that poly(ADP-ribosyl)ation of nuclear proteins occurs early in the course of apoptosis induced by different apoptotic signals, including Fas activation (3T3-L1, PARP +/+, and Jurkat cells), pharmacological agents such as camptothecin (HL-60 cells), as well as serum depletion (osteosarcoma cells). This burst of PAR synthesis occurs prior to commitment to death, and is followed by cleavage and inactivation of PARP; only small amounts of PAR remain during the later stages of apoptosis, despite the presence of a large number of DNA strand breaks. By depleting the normally abundant PARP from 3T3-L1 and Jurkat cells by antisense RNA expression prior to the induction of apoptosis, or with the use of immortalized fibroblasts derived from PARP knockout mice, we further demonstrate that, whereas PARP cleavage is required in the later stages of apoptosis, the transient burst of poly(ADP-ribosyl)ation of relevant nuclear proteins appears to also play a role during the early stages of the death program. PARP-depleted 3T3-L1 and Jurkat PARP antisense cells as well as in PARP -/- fibroblasts did not undergo the burst of PAR synthesis early in apoptosis and progression into apoptosis was prevented in these cells, as assessed by the lack of *in vitro* PARP cleavage activity, proteolytic processing and activation of CPP32 to caspase-3, development of nuclear apoptotic morphology and internucleosomal DNA fragmentation [71].

A potential role for PAR synthesis at an early reversible stage in apoptosis is suggested by the substantial extent of nuclear poly(ADP-ribosyl)ation apparent at this time in 3T3-L1, HL60, and osteosarcoma cells and is consistent with the presence of large (1 Mb) chromatin fragments at this reversible stage [20] since the activity of PARP is dependent on DNA strand breaks. This stage of apoptosis may correspond to a stage described previously during which the initial events of nuclear breakdown occur and in which poly(ADP-ribosyl)ation may play an accessory role [20]; exposure of cells to chemical inhibitors of PARP at this time, but not later, prevents cell death [47]. A drop in NAD concentration, indicative of increased PAR synthesis, followed by a recovery in NAD levels has also been previously noted to occur prior to the onset of internucleosomal DNA cleavage [47]. Furthermore, previous studies have suggested a correlation between poly(ADP-ribosyl)ation of nuclear proteins and internucleosomal DNA fragmentation during apoptosis, since PARP inhibitors have a suppressive effect on DNA fragmentation [48, 49]. However, the relevant target proteins for PARP during apoptosis have yet to be identified. During the early stages of apoptosis, poly(ADP-ribosyl)ation of histone H1, for example, was shown to facilitate internucleosomal DNA fragmentation by enhancing chromatin susceptibility to cellular endonucleases [50]. Although the concomitant loss of poly(ADP-ribosyl)ation of target proteins resulting from PARP cleavage and inactivation are consistent with the later stages of apoptosis during which cells become irreversibly committed to death, the biological significance and the biochemical consequences of PARP cleavage and its consequent inactivation remain unclear.

We showed that extensive poly(ADP-ribosyl)ation of p53 is apparent early during spontaneous apoptosis in osteosarcoma cells, and that degradation of PAR attached to these apoptosis-related proteins occurred concomitant with caspase-3 catalyzed PARP cleavage (Fig. 6). These results are consistent with recent studies showing substantial poly(ADP-ribosyl)ation of p53, with polymer chain lengths from 4–30 residues, in cells undergoing apoptosis in response to DNA damage [51, 52]. Electrophoretic mobility-shift analysis further showed that ADP-ribose polymers attached to p53 blocked its sequence specific binding to a 26-bp oligonucleotide containing the palindromic p53 consensus binding sequence, suggesting that poly(ADP-ribosyl)ation of p53 may negatively regulate p53-mediated transcriptional activation of genes important in the cell cycle and apoptosis [53]. Induction of spontaneous apoptosis in osteosarcoma cells is associated with an increase in the intracellular levels of p53 (Fig. 6). It remains to be determined whether this accumulation of p53 is due to induced expression of the protein or stabilization by reduced degradation as a result of posttranslational modification. Nevertheless, the extensive poly(ADP-ribosyl)ation of p53 that occurs early in apoptosis and the subsequent degradation of PAR attached to p53 concomitant with caspase-3 activity suggest that this post-translational modification may play a role in the regulation of p53 function or, alternatively, its degradation. The stabilization and accumulation of p53 has recently been shown to play a key role in apoptosis induced by proteasome inhibitors [54]. Poly(ADP-ribosyl)ation as well as caspase-3-catalyzed PARP cleavage and inactivation may contribute to the regulation of proteasome-mediated degradation of p53, by promoting p53 stabilization and accumulation at a stage when it is required in the apoptotic program.

PARP expression and poly(ADP-ribosyl)ation in the early stages of differentiation-linked DNA replication

We had previously shown that PARP depletion by antisense RNA expression inhibits the differentiation of 3T3-L1 preadipocytes, including the differentiation-linked round of DNA replication [11]. The differentiation of 3T3-LI preadipocytes as well as Friend erythroleukemia cells [55] is prevented by blocking this round of DNA replication at the early stages of differentiation. The failure of PARP-depleted 3T3-L1 cells to undergo terminal differentiation into adipocytes is thus correlated to their inability to undergo replication in the early stages of this process, indicating that PARP plays a role in this replication. We have also shown that PARP is tightly associated with the core proteins of MRC purified from HeLa and FM3A cells [10] that have been shown to support viral DNA replication *in vitro* and migrate as discrete, high molecular weight complexes on native polyacrylamide gel electrophoresis [56]. PARP has been thought to play a regulatory role within these complexes because it is capable of modulating the catalytic activity of some of the replicative enzymes or factors by physical association [pol α [9] or by catalyzing their poly(ADP-ribosyl)ation [pol α [57], DNA topoisomerase I and II [58–60], and RPA [61]. Consistently, we showed that 15 of the ~ 40 polypeptides of the MRC including pol α, topoisomerase I, and PCNA, are poly(ADP-ribosyl)ated by immunoprecipitation and immunoblot analysis with antibodies to PAR and these replication proteins [10].

To further clarify the role of PARP and the significance of the early burst of PAR synthesis during differentiation-linked DNA replication, we purified the MRC from 3T3 -L1 cells harvested prior to and 24 h after exposure to inducers of differentiation-linked DNA replication; at the latter time point, ~95% of control cells are in S phase and exhibit a transient peak of PARP activity (Fig. 7). DNA pol α and pol δ catalyze the synthesis of lagging and leading strands of DNA, and PCNA is required for pol δ-mediated synthesis of the leading strand [62]. Interestingly, only the MRC fraction from S phase control cells, not those from uninduced control cells nor from PARP-antisense cells, contained PARP protein and activity as well as pol α and pol δ activities. Thus, PARP depletion results in a replicative complex deficient in these two replicative enzyme activities. These results are consistent with our earlier data showing that *in vivo* DNA replication, as assessed by incorporation of bromodeoxyuridine or [^{3}H]-thymidine into newly synthesized DNA, occurred only in replicating 3T3-L1 control cells, but not in the PARP-depleted antisense cells. We next tried to determine whether this effect was attributed to present, but inactive enzymes or due to absence of these enzymes in the MRC or in the cells.

Immunoblot analysis revealed that PCNA and DNA topoisomerase I were present in the MRC fraction only from S phase control cells, although they were both detected in total cell extracts from control and PARP-antisense cells, indicating that PARP may play a role in the recruitment of these replicative proteins into the DNA synthesome during entry into S phase [72]. The absence of PCNA in the antisense MRC may account in part for the lack of any significant DNA pol δ activity in these MRCs. Immunoblot analysis with anti-PARP showed that, consistent with the PARP activity assays (Fig. 7), PARP protein was present exclusively in the MRC fractions from control 3T3-L 1 cells at S phase, but was essentially depleted in MRC from nonreplicating control cells or from the PARP-antisense cells [72]. Furthermore, pol α was absent from total cell extracts from PARP-antisense cells, suggesting that PARP may be also play a role in the expression of the pol α gene during entry of cells into S phase. Surprisingly, depletion of PARP by antisense RNA expression also down-regulated the expression of E2F-1 during the early stage of differentiation in 3T3-L1 cells. In mammalian cells, the transcription factor E2F-1 binds to a specific recognition site (5′-TTTCGCGC) and thereby activates the promoters of several genes that regulate S phase entry, such as c-myc [63], cyclin D, and cyclin E [64], and genes required for DNA synthesis, such as dihydrofolate reductase [65], thymidine kinase [66], pol α, and PCNA [64, 67]. Transcription of the E2F-1 gene, in turn, is also regulated during the cell cycle by E2F-DNA binding sites within its own promoter [68]. PARP has recently been shown to enhance activator-dependent transcription probably by interacting with RNA polymerase II-associated factors [69] and a basal eukaryotic transcription factor, THIIF, was reported to be a highly specific substrate for poly(ADP-ribosyl)ation [70]. Experiments are now underway to determine whether PARP plays a role in the transcription of the pol α, PCNA, and E2F-1 genes during entry into S-phase.

Acknowledgements

We thank Dr. H. Hilz for the polyclonal antibody to murine PARP, Drs. M. Miwa and T. Sugimura for the antibody to PAR, and Dr. Z.Q Wang for the immortalized PARP +/+ and PARP -/- fibroblasts. This work was supported in part by grants CA25344 and POI CA74175 from the National Cancer Institute, the United States Air Force Office of Scientific Research (grant AFOSR-89-0053), and the United States Army Medical Research and Development Command (contract DAMD17-90-C-0053) (to M.E.S.) and DAMD17-96-C-6065 (to D.S.R.).

References

1. Ding R, Stevnsner T, Bohr V, Smulson M: Preferential gene repair and HN2 damage: Depletion of PADPRP by antisense RNA expression. Proceedings of the 1993 Medical Defense Bioscience Review, 1: 217–226, 1993
2. Molinete M, Vermeulen W, Burkle A, de Murcia JM, Kupper J, Hoeijmakers J, de Murcia G: Overproduction of the poly(ADP-ribose) polymerase DNA-binding domain blocks alkylation-induced DNA repair synthesis in mammalian cells. EMBO J 12: 2109–2117, 1993
3. Satoh M, Lindahl T: Role of poly(ADP-ribose) formation in DNA repair. Nature 356: 356–358, 1992
4. Satoh M, Poirier G, Lindahl T: NAD$^+$-dependent repair of damaged DNA by human cell extracts. J Biol Chem 268: 5480–5487, 1993
5. Smulson M, Istock N, Ding R, Cherney B: Deletion mutants of poly(ADP-ribose) polymerase support a model of cyclic association and dissociation of enzyme from DNA ends during DNA repair. Biochemistry 33: 6186–6191, 1994
6. Burzio L, Koide S: Activation of the template activity of isolated rat liver nuclei for DNA synthesis and its inhibition by NAD. Biochem Biophys Res Commun 53: 572–579, 1973
7. Anachkova B, Russev G, Poirier G: DNA replication and poly(ADP-ribosyl)ation of chromatin. Cytobios 58: 19–28, 1989
8. Cesarone C, Scarabelli L, Giannini P, Orunesu M: Differential assay and biological significance of poly (ADP-ribose) polymerase activity in isolated liver nuclei. Mutat Res 245: 157–63, 1990
9. Simbulan C, Suzuki M, Izuta S, Sakurai T, Savoysky E, Kojima K, Miyahara K, Shizuta Y, Yoshida S: Poly (ADP-ribose) polymerase stimulates DNA polymerase α. J Biol Chem 268: 93–99, 1993
10. Simbulan-Rosenthal CMG, Rosenthal D, Hilz H, Hickey R, Malkas L, Applegren N, Wu Y, Bers G, Smulson M: The expression of poly(ADP-ribose) polymerase during differentiation-linked DNA replication reveals that this enzyme is a component of the multiprotein DNA replication complex. Biochemistry 35: 11622–11633, 1996
11. Smulson M, Kang V, Ntambi J, Rosenthal D, Ding R, Simbulan CMG: Requirement for the expression of poly(ADP-ribose) polymerase during the early stages of differentiation of 3T3-L1 preadipocytes, as studied by antisense RNA induction. J Biol Chem 270: 119–127, 1995
12. Farzaneh F, Meldrum R, Shall S: Transient formation of DNA strand breaks during the induced differentiation of a human promyelocytic leukemic cell line, HL-60. Nucleic Acids Res 15: 3493–3502, 1987
13. Bhatia K, Pommier Y, Giri C, Fornace A, Imaizumi M, Breitman T, Chemey B, Smulson M: Expression of the poly(ADP-ribose) polymerase gene following natural and induced DNA strand breakage and effect of hyperexpression on DNA repair. Carcinogenesis 11: 123–128, 1990
14. Suzuki H, Uchida K, Shima H, Sato T, Okamoto T, Kimura T, Sugimura T, Miwa M: Molecular cloning of cDNA for human placental poly(ADP-ribose) polymerase and decreased expression of its gene during retinoic acid-induced granulocytic differentiation of HL-60 cells. In: M.K. Jacobson, E.L. Jacobson, (eds), ADP-Ribose Tranfer Reactions: Mechanisms and Biological Significance, New York: SpringerVerlag, 1989, pp. 511–514
15. Caplan A, Rosenberg M: Interrelationship between poly(ADP-rib) synthesis, intracellular NAD levels, and muscle or cartilage differentiation from mesodermal cells of embryonic chick limb. Proc Nacl Acad Sci 72: 1852–1857, 1975
16. Kaufmann S, Desnoyers S, Ottaviano Y, Davidson N, Poirier G: Specific proteolytic cleavage of poly(ADP-ribose)polymerase: An early marker of chemotherapyinduced apoptosis. Cancer Res 53: 3976–3985, 1993
17. Nicholson D, Ali A, Thomberry N, Vaillancourt J, Ding C, Gallant M, Gareau Y, Griffin P, Labelle M, Lazebnik Y, Munday N, Raju S, Smulson M, Yamin T, Yu V, Miller D: Identification and inhibition of the ICE/CED-3 protease necessary for mammalian apoptosis. Nature 376: 37–43, 1995
18. Rosenthal D, Ding R, Simbulan-Rosenthal CM, Vaillancourt J, Nicholson D, Smulson M: Intact cell evidence to the early snthesis, and subsequent late apopainmediated suppression, of poly(ADP-ribose) during apoptosis. Exp Cell Res 232: 313–321, 1997

19. Whitacre C, Hashimoto H, Tsai ML, Chatterjee S, Berger S, Berger N: Involvement of NAD-poly(ADP-ribose) metabolism in p53 regulation and its consequences. Cancer Res 55: 3697–3701, 1995
20. Neamati N, Fernandez A, Wright S, Kiefer J, McConkey D: Degradation of lamin B1 precedes oligonucleosomal DNA fragmentation in apoptotic thymocytes and isolated thymocyte nuclei. J Immunology 154: 3788–3795, 1995
21. Tewari M, Quan L, O'Rourke K, Desnoyers S, Zeng Z, Beidler D, Poirier G, Salvesen G, Dixit V: Yarna/CPP32b, a mammalian homolog of CED-3, is a crmA-inhibitable protease that cleaves the death substrate poly(ADP-ribose) polymerase. Cell 81: 801–809, 1995
22. Alnenmi E, Livingston D, Nicholson D, Salvesen G, Thornberry N, Wong W, Ytian J: Human ICE/CED-3 protease nomenclature [letter]. Cell 87: 171, 1996
23. Ytian J, Shaham S, Ledoux S, Ellis H, Horvitz H: The C. elegans death gene ced-3 encodes a protein similar to mammalian interleukin-1-b-converting enzyme. Cell 75: 641–652, 1993
24. Rosenthal D, Shima T, Celli G, De Luca L, Smulson M: An engineered human skin model using poly(ADP-ribose) polymerase antisense expression shows a reduced response to DNA damage. J Invest Dermatol 105: 38–44, 1995
25. Ding R, Pommier Y, Kang V, Smulson M: Depletion of poly(ADP-ribose) polymerase by antisense RNA expression results in a delay in DNA strand break rejoining. J Biol Chem 267: 12804–12812, 1992
26. Wang ZQ, Auer B, Stingl L, Berghammer H, Haidacher D, Schweiger M, Wagner E: Mice lacking ADPRT and poly(ADP-ribosyl)ation develop normally but are susceptible to skin disease. Genes Dev 9: 509–520, 1995
27. de Murcia G, Masson M, Trucco C, Dantzer F, Oliver, Flatter E, Niedergang C, Ricoul M, Dutrillaux B, Dierich A, LeMeur M, Waltzinger C, Chambon P, de Murcia J: Poly(ADP-ribose) polymerase interacts with base excision repair enzymes and is required in recovery from DNA damage in mice and cells. The 12th International Symposium on ADP-ribosylation reactions, 1997
28. Lewis J, Shimizu Y, Shimizu N: Nicotinamide inhibits adipocyte differentiation of 3T3-L1 cells. FEBS Lett 146: 37–41, 1982
29. Rosenthal D, Ding R, Simbulan-Rosenthal CM, Cherney B, Vanek P, Smulson M: Detection of DNA breaks in apoptotic cells utilizing the DNA binding domain of poly(ADP-ribose) polymerase with fluorescence microscopy. Nucleic Acids Res 25: 1437–1441, 1997
30. Totpal K, Singh S, Lapushin R, Aggarwal B: Qualitative and quantitative differences in the cellular responses mediated through Fas antigen and TNFR. J Interferon Cytokine Res 16: 259–267, 1996
31. Ogasawara J, Suda T, Nagata S: Selective apoptosis of CD4+CD8+ thymocytes by the anti-Fas antibody. J Exp Med 181: 485–491, 1995
32. Natoli G, Ianni A, Costanzo A, De Petrillo G, Ilari I, Chirillo P, Balsano C, Levrero M: Resistance to Fas-mediated apoptosis in human hepatoma cells. Oncogene 11: 1157–1164, 1995
33. Chiba T, Takahashi S, Sato N, Ishii S, Kikuchi K: Fas-mediated apoptosis is modulated by intracellular glutathione in human T cells. Eur J Immunol 26: 1164–1169, 1996
34. Ronen D, Schwartz D, Teitz Y, Goldfinger N, Rotter V: Induction of HL-60 cells to undergo apoptosis is determined by high levels of wild-type p53 protein whereas differentiation of cells is mediated by lower p53 levels. Cell Growth Diff 7: 21–30, 1996
35. Yamato K, Yamamoto M, Hirano Y, Tsuchida N: A human temperature-sensitive p53 mutant p53Val-138: Modulation of the cell cycle, viability, and expression of p53-responsive genes. Oncogene 11: 1–6, 1995
36. Cleaver J, Borek C, Milam K, Morgan W: The role of poly(ADP-ribose) synthesis in toxicity and repair of DNA damage. Pharmacol Ther 31: 269–293, 1985
37. Hunting D, Gowans B, Henderson J: Effects of 6-aminonicotinamide on cell growth, poly(ADP-ribose) synthesis and nucleotide metabolism. Biochem Pharmacol 34: 3999–4003, 1985
38. Milam K, Cleaver J: Inhibitors of poly (adenosine diphosphate-ribose) synthesis: Effect on other metabolic processes. Science 223: 589–591, 1984
39. Ding R, Smulson M: Depletion of nuclear poly(ADP-ribose) polymerase by antisense RNA expression: Influences on genomic stability, chromatin organization and carcinogen cytotoxicity. Cancer Res 54: 4627–4634, 1994
40. Chinnaiyan A, Orth K, O'Rourke K, Duan H, Poirier G, Dixit V: Molecular ordering of the cell death pathway. Bcl-2 and Bcl-xL function upstream of the CED-3-like apoptotic proteases. J Biol Chem 271: 4573–4576, 1996
41. Darmon A, Ley T, Nicholson D, Bleackley R: Cleavage of CPP32 by granzyme B represents a critical role for granzyme B in the induction of target cell DNA fragmentation. J Biol Chem 271: 21709–21712, 1996
42. Erhardt P, Cooper G: Activation of CPP32 apoptostic protease byu distinct signalling pathways with differential sensitivity to Bcl-xL. J Biol Chem 271: 17601–17604, 1996
43. Martin S, Amarante-Mendes G, Shi L, Chuang T, Casiano C, O'Brien G, Fitzgerald P, Tan E, Bokoch G, Greenberg A, Green D: The cytotoxic cell protease granzyme B initiates apoptosis in a cell-free system by proteolytic processing and activation of the ICE/CED-3 family protease, CPP32, via a novel two-step mechanism. EMBO J 15: 2407–2416, 1996
44. Schlegel J, Peters I, Orrenius S, Miller D, Thornberry N, Yamin T, Nicholson D: CPP32/apopain is a key interleukin 1 beta converting enzyme-like protease involved in Fas-mediated apoptosis. J Biol Chem 271: 1841–1844, 1996
45. Dubrez L, Savoy I, Hamman A, Solary E: Pivotal role of a DEVD-sensitive step in etoposide-induced and Fas-mediated apoptosis. EMBO J 15: 5504–5512, 1996
46. Enari M, Talanian R, Wong W, Nagata S: Sequential activation of ICE-like and CPP32-like proteases during Fas-mediated apoptosis. Nature 380: 723–726, 1996
47. Nosseri C, Coppola S, Ghibelli L: Possible involvement of poly(ADP-ribosyl) polymerase in triggering stress-induced apoptosis. Exp Cell Res 212: 367–373, 1994
48. Jones D, McConkey D, Nicotera P, Orrenius S: Calcium-activated DNA fragmentation in rat liver nuclei. J Biol Chem 264: 6398–6403, 1989
49. Bertrand R, Solary E, Jenkins J, Pommier Y: Apoptosis and its modulation in human promyelocytic HL-60 cells treated with DNA topoisomerase I and II inhibitors. Exp Cell Res 207: 388–397, 1993
50. Yoon Y, Kim J, Kang K, Kim Y, Choi K, Joe C: Poly(ADP-ribosyl)ation of histone H1 correlates with intemucleosomal DNA fragmentation during apoptosis. J Biol Chem 271: 9129–9134, 1996
51. Kumari S, Mendoza-Alvarez H, Alvarez-Gortzalez R: Poly(ADP-ribosyl)ation of p53 in apoptotic cells following DNA damage. The 12th International Symposium on ADP-ribosylation reactions, 1997
52. Nozaki T, Masutani M, Nishiyarna E, Shimokawa T, Wkabayashi K, Sugimura T: Interactions between poly(ADP-ribose) polymerase and p53. The 12th International Symposium on ADP-ribosylation reactions, 1997
53. Malanga M, Pleschke J, Kleczkowska H, Althaus F: Poly(ADP-ribose) binds to specific domains of p53 and alters its DNA binding functions. J Biol Chem 273: 11839–11843, 1998
54. Lopes U, Erhardt P, Yao R, Cooper G: p53-dependent induction of apoptosis by proteosome inhibitors. J Biol Chem 272: 12893–12896, 1997
55. Spriggs L, Hill S, Jeter J: Proliferation is required for induction of terminal differentiation of Friend erythroleuken-iia cells. Biochem and Cell Biol 70: 1992
56. Tom T, Malkas L, Hickey R: Identification of multiprotein compexes containing DNA replication factors by native immunoblotting of HeLa cell protein preparations with T-antigen-dependent SV40 DNA replication activity. J Cell Biochem 63: 259, 1996

57. Yoshihara K, Itaya A, Tanaka Y, Ohashi Y, Ito K, Teraoka H, Tsukada K, Matsukage A, Kamiya T: Inhibition of DNA polymerase α, DNA polymerase β, terminal nucleotidyltransferase and DNA ligase II by poly (ADP-ribosyl)ation reaction *in vitro*. Biochem Biophys Res Commun 128: 61–67, 1985
58. Kasid U, Halligan B, Liu L, Dritschilo A, Smulson M: Poly(ADP-ribose)-mediated post-translational modification of chromatin-associated human topoisomerase I. Inhibitory effects on catalytic activity. J Biol Chem 264: 18687–18692, 1989
59. Ferro A, Olivera B: Poly (ADP-ribosylation) of DNA topoisomerase I from calf thymus. J Biol Chem 259: 547–554, 1984
60. Darby M, Schmitt B, Jongstra B, Vosberg H: Inhibition of calf thymus type II DNA topoisomerase by poly(ADP-ribosylation). EMBO J 4: 2129–2134, 1985
61. Eki T, Hurwitz J: Influence of poly(ADP-ribose) polymerase on the enzymatic synthesis of SV40 DNA. J Biol Chem 266: 3087–3100, 1991
62. Tsurimoto T, Stillman B: Replication factors required for SV40 DNA replication *in vitro*. I. DNA structure-specific recognition of a primer-template junction by eukaryotic DNA polymerase and their accessory proteins. J Biol Chem 266: 1950, 1991
63. Hiebert S, Lipp M, Nevins J: E1A-dependent transactivation of the human MYC promoter is mediated by the E2F factor. Proc Natl Acad Sci USA 86: 3594–3598, 1989
64. DeGregori J, Leoni G, Miron A, Jakoi L, Nevins J: Distinct roles for E2F proteins in cell growth control and apoptosis. Proc Nacl Acad Sci USA 94: 7245–7250, 1997
65. Blake M, Azizkhan J: Transcription factor E2F is required for efficient expression of hamster dihydrofolate-reductase gene *in vitro* and *in vivo*. Mol Cell Biol 9: 4994–5002, 1989
66. Ogris E, Rotheneder H, Mudrak I, Pichler A, Wintersberger E: A binding site for transcription factor E2F is a target for transactivation of murine thymidine kinase by polyoma large T antigen and plays an important role in growth regulation of the gene. J Virol 67: 1765–1771, 1993
67. Pearson B, Nasheuer H,Wang T: Human DNA polymerase α gene: Sequences controlling expression in cycling and serum-stimulated cells. Mol Cell Biol 11: 2081–2095, 1991
68. Neuman E, Flemington E, Sellers W, Kaelin W: Transcription of the E2F-1 gene is rendered cell cycle dependent by E2F DNA-binding sites within its promoter. Mol Cell Biol 15: 4660, 1995
69. Meisterernst M, Stelzer G, Roeder R: Poly(ADP-ribose) polymerase enhances activator-dependent transcription *in vitro*. Proc Natl Acad Sci USA 94: 2261–2265, 1997
70. Rawling J, Alvarez-Gonzalez R: TEIF, a basal eukaryotic transcription factor, is a substrate for poly(ADP-ribosyl)ation. Biochem J 324: 249–253, 1997
71. Simbulan-Rosenthal C, Rosenthal D, Iyer S, Boulares H, Smulson M: Transient poly(ADP-ribosylation) of nuclear proteins and role of poly(ADP-ribose) polymerase in the early stages of apoptosis. J Biol Chem 273: 13703–13712, 1998
72. Simbulan-Rosenthal C, Rosenthal D, Boulares H, Hickey R, Malkas L, Coll J, Smulson M.: Regulation of the expression or recruitment of components of the DNA synthesome by poly(ADP-ribose) polymerase. Biochemistry 37: 9363–9370, 1998

Molecular and Cellular Biochemistry **193:** 149–152, 1999.

Function of poly(ADP-ribose) polymerase in response to DNA damage: Gene-disruption study in mice

Mitsuko Masutani,[1] Tadashige Nozaki,[1] Eiko Nishiyama,[1] Takashi Shimokawa,[1] Yumiko Tachi,[1] Hiroshi Suzuki,[3] Hitoshi Nakagama,[1] Keiji Wakabayashi[2] and Takashi Sugimura[1]

[1]*Biochemistry Division;* [2]*Cancer Prevention Division, National Cancer Center Research Institute, Tokyo;* [3]*Chugai Pharmaceutical Co., Shizuoka, Japan*

Abstract

To elucidate the biological functions of poly(ADP-ribose) polymerase (PARP, [EC 2.4.2.30]) in DNA damage responses, genetic and biochemical approaches were undertaken. By disrupting exon 1 of the mouse PARP gene by a homologous recombination, PARP-deficient mouse embryonic stem (ES) cell lines and mice could be produced without demonstrating lethality. $PARP^{-/-}$ ES cells showed complete loss of PARP activity and increased sensitivity to γ-irradiation and an alkylating agents, indicating a physiological role for PARP in the response to DNA damage. p53, a key molecule in cellular DNA damage response, was found to stimulate PARP activity and became poly(ADP-ribosyl)ated in the presence of damaged DNA. However, $PARP^{-/-}$ ES cells showed *p21* and *Mdm-2* mRNA induction following γ–irradiation, indicating that PARP activity is not indispensable for *p21* and *Mdm-2* mRNA induction in the established p53-cascade. On the other hand, in a reconstituted reaction system, purified PARP from human placenta suppressed the pRB-phosphorylation activity in the presence of NAD and damaged DNA. Human PARP expressed in *E. coli* showed a similar effect on pRB-phosphorylation activity of cdk2. These findings suggest a direct involvement of PARP in the regulation of cdk activity for cell-cycle arrest. (Mol Cell Biochem **193**: 149–152, 1999)

Key words: poly(ADP-ribose) polymerase, gene-disruption, embryonic stem cell, DNA damage responses, p53, pRB, cdk2

Introduction

Our understanding of the DNA damage response in mammalian cells has been advanced by a variety of biochemical and genetic evidences. One system of interest is poly(ADP-ribose), which is produced immediately after DNA damage in nuclei using NAD as a substrate. Poly(ADP-ribose) polymerase (PARP) modifies PARP itself and various other nuclear proteins by poly(ADP-ribosyl)ation. Since the elucidation of poly(ADP-ribose) formation in the 1960s [1], it has been suggested that PARP is involved in initiating DNA damage response networks, including DNA repair [2–4], cell cycle arrest [5–7] and apoptosis [8, 9]. However, the precise functions of PARP *in vivo* have not been fully elucidated yet. To clarify this, the PARP gene was disrupted in mouse ES cells and intact mice. In parallel with these genetic approaches, biochemical approaches were used to ascertain the interaction of PARP with other molecules in DNA damage response. Specifically the influence of PARP on cyclin-dependent kinase (cdk) activity and functional interaction between PARP and p53 was studied using reconstituted systems.

Address for offprints: M. Masutani, Biochemistry Division, National Cancer Center Research Institute, 1-1, Tsukiji, 5-chome, Chuo-ku, Tokyo, 104-0045 Japan

PARP gene-disruption: Double knock-out mouse ES cells and PARP deficient mice

In order to inactivate the mouse PARP gene, a *neomycin-resistance gene (neo^r)* cassette was introduced into one allele of PARP exon 1 in mouse ES J1 cells. For this purpose, a gene-targeting vector containing a 6 kb region encompassing PARP exon 1 with the inserted *neo* cassette was constructed. Using the resulting $PARP^{+/-}$ ES clones, a similarly constructed gene-targeting vector featuring a *puromycin-resistant gene ($puro^r$)* cassette [10] was introduced into the remaining allele to produce $PARP^{-/-}$ ES clones. PARP double knock-out in ES clones are not lethal. Loss of PARP activity in $PARP^{-/-}$ ES clones were documented using two methods, namely activity-gel analysis and NAD-derived radioactivity incorporation into acid-insoluble fraction. Western blotting analysis also indicated a lack of PARP expression at the protein level in $PARP^{-/-}$ cells. In $PARP^{+/-}$ ES clones, half the PARP activity of the parental ES clones were detected, demonstrating an equal contribution of both alleles to PARP gene expression in mouse ES cells. The morphology was not altered in either the $PARP^{+/-}$ or $PARP^{-/-}$ ES cells. Using these ES clones, cytotoxicity of DNA damaging agents was assessed in a colony formation assay. $PARP^{-/-}$ clones demonstrated enhanced sensitivity to γ-irradiation and the alkylating agent, methyl methanesulfonate, compared to $PARP^{+/-}$ and parental ES cells (Fig. 1). These results indicate that PARP activity is an important factor determining the survival of cells from DNA damage. Analysis of DNA damage response in $PARP^{-/-}$ clones in detail, including extent of DNA damage and DNA repair ability, will lead to the clarification of this mechanism.

In parallel with this double knock-out study in ES cells, germline chimeras were obtained by the aggregation of heterozygous ES clones with cells from embryos of ICR mice. Subsequently $PARP^{+/-}$ mouse strains were established. Intercrossing $PARP^{+/-}$ mice of 129/SJ and ICR-mixed genetic backgrounds yielded 12, 18 and 9 of $PARP^{-/-}$, $PARP^{+/-}$ and wild-type mice, respectively, as determined by genotyping analysis (Fig. 2). Thus the *PARP* disruption in mice caused no lethality. There was no apparent difference in phenotypes, including size, indicating the *PARP* is not essential for development and life-maintenance. Further studies on the biological consequences of *PARP* gene-disruption, including changes in the cellular machinery responsible for repairs of DNA damage, are now in progress.

Suppression of cdk activity by purified PARP

Cdks play important regulatory roles in cell-cycle check-point control in nuclei through the phosphorylation of key molecules including pRB [11]. Therefore the possibility of direct regulation of cdk activities by PARP was investigated. For this study, PARP fraction purified from human placenta by hydroxylapatite, phosphocellulose, 3-aminobenzamide affinity column chromatography was used. This fraction contains a single 113 kDa PARP band when subjected to Coomassie-brilliant-blue staining after SDS/PAGE. When purified PARP fraction was subjected to pRB-phosphorylation reaction with γ-^{32}P-ATP, definite pRB-phosphorylation was observed due to the included cdk activity in the PARP fraction. As 3-aminobenzamide affinity column is highly specific for PARP, there is a possibility that some PARP molecules physically associate with cdk molecules. In the DNA damaging condition, namely by addition of activated DNA and NAD, this cdk activity in PARP fraction was suppressed. Neither NAD nor activated DNA alone caused such decrease of the pRB-phosphorylation activity, suggesting that poly(ADP-ribosyl)ation is necessary for this effect. However, it is noteworthy that the addition of a PARP inhibitor, 3-aminobenzamide, did not inhibit the decrease of pRB-phosphorylation activity. As purified recombinant human PARP expressed in *E. coli* possessed no pRB-phosphorylation activity, the effects on cdk2 activity were tested using cdk2 immunoprecipitated from mouse fibroblast cells. The

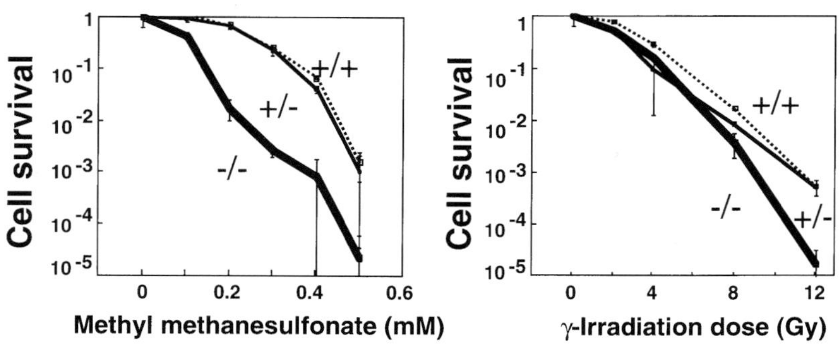

Fig. 1. Survival curve of ES cells following treatment with methyl methanesulfonate and γ-irradiation. Colony formation ability was tested on STO feeder cells.

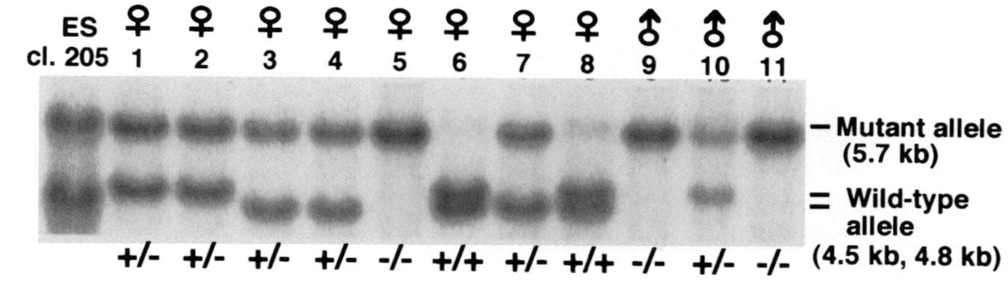

Fig. 2. PARP gene-disruption in mice. Genotyping analysis of offspring of a *PARP*$^{+/-}$ intercross. Southern blotting analysis of tail DNA was performed with a 3′ probe for the homologous recombination junction.

recombinant human PARP demonstrated similar suppression of cdk2 activity in the presence of NAD and activated DNA. Again, 3-aminobenzamide did not inhibit suppressive effect of PARP on cdk2 activity.

This evidence indicates that PARP can directly block the cell cycle under DNA damaging conditions by inhibition of cdk activity on pRB-phosphorylation (Fig. 3). The fact of no inhibition with 3-aminobenzamide suggests that PARP binding to DNA breakage with an NAD molecule into its catalytic site is sufficient for this function. Analysis of pRB-phosphorylation level after DNA damage in *PARP*$^{-/-}$ ES cells will be helpful to understand this result.

Interactions between p53 and PARP

As p53 is a key molecule for controlling DNA-damage dependent cell-cycle check-point control and apoptosis [12], any functional interactions with PARP in DNA damage signal-transduction would be of interest. When wild-type human p53 expressed and purified from baculovirus transfected Sf9 cells was added to a poly(ADP-ribose) synthesis reaction system at an equal molecular ratio to human PARP, p53 stimulated PARP activity 1.5–2 fold in the presence of either activated DNA or single stranded 16-mer DNA. There was no significant difference in the mean chain length of poly(ADP-ribose) and branching site frequency in the presence and in the absence of p53. Therefore p53 did not substantially change the poly(ADP-ribose) synthesis reaction kinetics. Instead, these results suggested the possibility of p53 acting as a substrate for poly(ADP-ribosyl)ation. Indeed, when p53 molecules were immunoprecipitated after the reaction, poly(ADP-ribosyl)ated p53 with 4 to over 20 residues of poly(ADP-ribose) attached to p53 molecules. When the effects of PARP on p53 binding to consensus binding DNA sequence was examined, no inhibition was observed. Therefore, these observations indicate that p53 and PARP do not interfere each other in DNA damage responses. Recently, Wesierska-Gadek *et al.* also reported that wild-type p53 is poly(ADP-ribosyl)ated and that binding of p53 to the consensus DNA sequences prevented this modification [13]. However, the biological relevance of the observed interaction between PARP and p53 remains to be clarified.

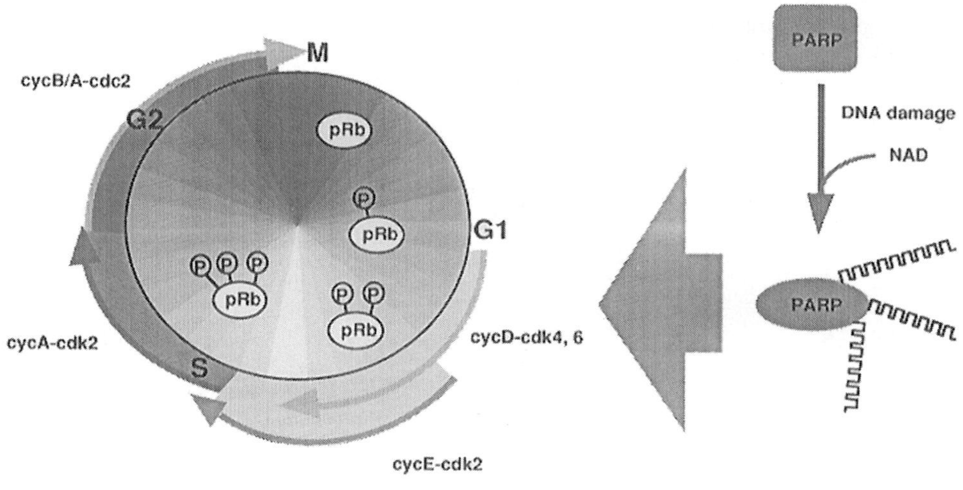

Fig. 3. A scheme for cell cycle arrest by suppression of pRB-phosphorylation activity by PARP under DNA damaging conditions.

Since the *PARP*$^{-/-}$ and *PARP*$^{+/-}$ cells are useful tools for studying functional relationship between PARP and p53, p53 responses to DNA damages were studied in *PARP*$^{-/-}$ ES cells. Notably, three out of four clones showed a decreased basal level of p53 protein. However, it could not be precluded that this decrease was due to the transfection or subcloning. In all *PARP*$^{-/-}$ clones, an increase in the p53 protein level was observed after γ-irradiation. Following an 8 Gy irradiation, induction of *p21* and *Mdm-2* mRNA was also observed in *PARP*$^{-/-}$ and *PARP*$^{+/-}$ cells as well as in parental wild-type cells. Thus the poly(ADP-ribosyl)ation reaction does not appear to be required for p53 stabilization and induction of *p21* and *Mdm-2* mRNA.

Future perspectives

The genetic approaches using PARP gene-disruption at exon 1 demonstrated that lethality or abnormal development is not caused in mice. This is in line with the results for exon 2 and exon 4 disruption published by Wang *et al.* [14] and de Murcia *et al.* [15], respectively. No difference in DNA repair of UV and alkylating agent damage was reported for exon 2 disrupted *PARP*$^{-/-}$ mice cells [14]. Heller *et al.* observed an elevated resistance of islet cells from exon 2 disrupted *PARP*$^{-/-}$ mice to nitric oxides and oxygen radicals [16]. In mouse *PARP*$^{-/-}$ ES clones, PARP clearly contributes to determine cellular sensitivity to DNA damage including γ-irradiation. This result is in accordance with recent observations in exon 4-disrupted mice [15]. Clarification of the underlying mechanisms requires further investigation.

In addition, elucidation of the molecular interactions of PARP with factors involved in DNA damage signal transduction will be important. Although wild-type p53 is efficiently poly(ADP-ribosyl)ated, PARP is not indispensable for downstream events in the DNA damage-dependent p53 response. Since PARP suppresses pRB phosphorylation activity under DNA damaged condition, it may be directly involved in regulation of cdk activity. Many studies have so far indicated the poly(ADP-ribose) synthesizing activity in animal cell nuclei, through mammalian to protozoa but not in prokaryotes [17]. This well conserved aspect of poly(ADP-ribose) metabolism suggests that this system plays a fundamental biochemical role in DNA damage response network.

Acknowledgements

This work was supported in parts by a Grant-in-Aid for Cancer Research from the Ministry of Health and Welfare, and Grants-in-Aid from Sankyo Foundation of Life-Science and from Haraguchi Memorial Fund for Cancer Research.

References

1. Sugimura T: Poly(adenosine diphosphate ribose). Prog Nucleic Acid Res Mol Biol 13: 127–1512, 1973
2. Durkacz, BW, Omidiji O, Gray DA, Shall S: (ADP-ribose)n participates in DNA excision repair. Nature 283: 593–596, 1980
3. Ding R, Pommier Y, Kang VH, Smulson M: Depletion of poly(ADP-ribose) polymerase by antisense RNA expression results in a delay in DNA strand break rejoining. J Biol Chem 267: 12804–12812, 1992
4. Satoh, MS, Lindahl T: Role of poly(ADP-ribose) formation in DNA repair. Nature 356: 356–358, 1992
5. Jacobson EL, Meadows R, Measel J: Cell cycle perturbations following DNA damage in the presence of ADP-ribosylation inhibitors. Carcinogenesis 6: 711–714, 1985
6. Nozaki T, Masutani M, Akagawa T, Sugimura T, Esumi H: Suppression of G1 arrest and enhancement of G2 arrest by inhibitors of poly(ADP-ribose) polymerase: Possible involvement of poly(ADP-ribosyl)ation in cell cycle arrest following γ-irradiation. Jpn J Cancer Res 85: 1094–1098, 1994
7. Masutani M, Nozaki T, Wakabayashi K, Sugimura T: Role of poly(ADP-ribose) polymerase in cell-cycle checkpoint mechanisms following γ-irradiation. Biochimie 77: 462–465, 1995
8. Lazebik YA, Kaufmann SH, Desnoyers S, Poirier GG, Earnshow WC: Cleavage of poly(ADP-ribose) polymerase by a protease with properties like ICE. Nature 371: 346–347, 1994
9. Schreiber V, Hunting D, Trucco C, Gowans B, Grunwald D, de Murcia G: A dominant-negative mutant of human poly(ADP-ribose) polymerase affects cell recovery, apoptosis, and sister chromatid exchange following DNA damage. Proc Natl Acad Sci USA 92: 4753–4757, 1995
10. Watanabe S, Kai N, Yasuda M, Kohmura N, Sanbo M, Mishina M, Yagi T: Stable production of mutant mice from double gene converted ES cells with puromycin and neomycin. Biochem Biophys Res Commun 213: 130–137, 1995
11. Hartwell LH: Defects in a cell cycle checkpoint may be responsible for the genomic instability of cancer cells. Cell 71: 543–546, 1992
12. Kastan MB, Onyekwere O, Sidransky D, Vogelstein B, Craig RW: Participation of p53 protein in the cellular response to DNA damage. Cancer Res 51: 6304–6311, 1991
13. Wesierska-Gadek J, Schmid G, Cerni C: ADP-ribosylation of wild-type p53 *in vitro*: Binding of p53 protein to a specific p53 consensus sequence prevents its modification. Biochem Biophys Res Commun 224: 96–102, 1996
14. Wang ZQ, Auer B, Sting L, Berghammer H, Haidacher D, Schweiger M, Wanger EF: Mice lacking ADPRT and poly(ADP-ribosyl)ation develop normally but are susceptible to skin disease. Genes Dev 9: 509–520, 1995
15. de Murcia JM, Niedergang C, Trucco C, Ricoul M, Dutrillaux B, Mark M, Oliver FJ, Masson M, Dierich A, LeMeur M, Walztinger C, Chambon P, de Murcia G: Requirement of poly(ADP-ribose) polymerase in recovery from DNA damage in mice and in cells. Proc Natl Acad Sci USA 94: 7303–7307, 1997
16. Heller B, Wang ZQ, Wagner EF, Randons J, Bürkle A, Fehsel K, Burkart V, Kolb H: Inactivation of the poly(ADP-ribose) polymerase gene affects oxygen radical and nitric oxide toxicity in islet cells. J Biol Chem 270: 11176–11180, 1995
17. Scovassi AI, Franchi E, Isernia P, Brusamolino E, Stefanini M, Bertazzoni U: Activity gels of poly(ADP-ribose) polymerase: Phylogenetic studies and variations in human blood. In: F.R. Althaus, H. Hilz, S. Shall (eds). ADP-ribosylation of Proteins. Springer, Berlin, 1985, pp 111–115

Molecular and Cellular Biochemistry **193:** 153–157, 1999.

Activation of toxin ADP-ribosyltransferases by eukaryotic ADP-ribosylation factors

Joel Moss and Martha Vaughan
Pulmonary-Critical Care Medicine Branch, National Heart, Lung, and Blood Institute, National Institutes of Health, Bethesda, Maryland, USA

Abstract

ADP-ribosylation factors (ARFs) are members of a multigene family of 20-kDa guanine nucleotide-binding proteins that are regulatory components in several pathways of intracellular vesicular trafficking. The relatively small (~180-amino acids) ARF proteins interact with a variety of molecules (in addition to GTP/GDP, of course). Cholera toxin was the first to be recognized, hence the name. Later it was shown that ARF also activates phospholipase D. Different parts of the molecule are responsible for activation of the two enzymes. In vesicular trafficking, ARF must interact with coatomer to recruit it to a membrane and thereby initiate vesicle budding. ARF function requires that it alternate between GTP- and GDP-bound forms, which involves interaction with regulatory proteins. Inactivation of ARF-GTP depends on a GTPase-activating protein or GAP. A guanine nucleotide-exchange protein or GEP accelerates release of bound GDP from inactive ARF-GDP to permit GTP binding. Inhibition of GEP by brefeldin A (BFA) blocks ARF activation and thereby vesicular transport. In cells, it causes apparent disintegration of Golgi structure. Both BFA-sensitive and insensitive GEPs are known. Sequences of peptides from a BFA-sensitive GEP purified in our laboratory revealed the presence of a Sec7 domain, a sequence of ~200 amino acids that resembles a region in the yeast Sec7 gene product, which is involved in Golgi vesicular transport. Other proteins of unknown function also contain Sec7 domains, among them a lymphocyte protein called cytohesin-1. To determine whether it had GEP activity, recombinant cytohesin-1 was synthesized in *E. coli*. It preferentially activated class I ARFs 1 and 3 and was not inhibited by BFA but failed to activate ARF5 (class II). There are now five Sec7 domain proteins known to have GEP activity toward class I ARFs. It remains to be determined whether there are other Sec7 domain proteins that are GEPs for ARFs 4, 5, or 6. (Mol Cell Biochem **193**: 153–157, 1999)

Key words: ADP-ribosylation factor, guanine nucleotide-exchange protein, brefeldin A, Sec7 domain, cytohesin-1

By the late 1970s, the ADP-ribosyltransferase activity of cholera toxin had become widely employed to activate and/or radiolabel Gs, the α subunit of the GTP-binding protein that activates adenylyl cyclase. After Gilman and his coworkers had purified Gαs, they found that in order to obtain toxin-catalyzed ADP-ribosylation of the pure protein it was necessary to add back a cellular component [1]. This component was termed 'ADP-ribosylation factor' or ARF and its purification was reported by Kahn and Gilman in 1984 [2]. They believed for some time that ARF interacted with Gαs to make it a better substrate for the toxin. In retrospect, it is easy to understand why that seemed quite a plausible conclusion because of the way that ARF was discovered. Later, probably chiefly because of the development of several other assays for the enzymatic activity of CT employing substrates other than Gαs, it was established that ARF is a GTP-dependent activator of CT, whether or not Gαs is the ADP-ribose acceptor [3].

We now know, as reviewed in ref. [4], that ARFs are members of a highly conserved multigene family of 20-kDa guanine nucleotide-binding proteins and have learned that they are regulatory components in intracellular vesicular trafficking systems. ARF proteins are ubiquitous in eukaryotic cells. Their structures are highly conserved from Giardia, believed to be the most ancient existing eukaryote, to primates. Six mammalian ARFs that have been identified by

Address for offprints: J. Moss, Building 10, Room 5N307, 9000 Rockville Pike, Bethesda, Maryland 20892-1434, USA

cDNA cloning fall into three classes based on size, deduced amino acid sequences, and phylogenetic analysis.

These relatively small, ~180-amino acid ARF proteins are known to interact with (in addition to GTP of course), a variety of other molecules, only some of which are listed in Table 1. The first recognized was CTA. Considerably later it was shown that ARF also activates PLD. Work by several people in our group has established that two different parts of the ARF molecule are responsible for activation of the two enzymes [5]. Wieland and coworkers [6] have demonstrated the direct interaction of ARF with βCOP, which is one of seven proteins in a complex termed coatomer that is recruited to the Golgi by ARF to initiate vesicle budding [7]. ARF action requires that it interact with proteins called GAP- for GTPase-activating protein - and GEP for guanine nucleotide-exchange protein - that are discussed below. A specific requirement for phosphatidylinositol 4,5-bisphosphate (PIP_2) in several ARF actions had been known for some time but only more recently has Randazzo [8] reported experiments with recombinant mutant ARF proteins that show specifically how it interacts with the ARF molecule. Based on the deduced amino acid sequence of ARF and knowledge of the crystal structure of the related Ras p21 protein, it was believed that specific amino acids from four different segments of the linear sequence come together in the folded protein to form the nucleotide-binding site. These amino acids are identical in all known ARF molecules and share a high degree of identity with the cognate sequences in other families of guanine nucleotide-binding regulatory proteins, such as α subunits of the heterotrimeric G proteins that transduce signals from the cell surface receptors with seven membrane-spanning regions. The similarity of the nucleotide-binding sites was confirmed by the crystal structure of ARF·GDP [9], which showed also the amino and carboxyl-terminal α helices positioned in parallel on the surface of the molecule. The sites of interaction with PIP_2, i.e., lysines 15, 16, and 181 and Arg 178, as demonstrated by Randazzo [8], are on these helices.

GTP-binding proteins, or GTPases, that are molecular switches for the selection, amplification, timing, and delivery of diverse signals share several generic properties. They function via differences in conformation that depend on whether GTP or GDP is bound. Vectorial signaling is the result of sequential GTP binding, hydrolysis of bound GTP, and release of bound GDP. Figure 1 depicts these processes in the regulation specifically of ARF activity by GEP and GAP. ARF in the cytosol is in an inactive state with GDP bound. Under physiological conditions, GDP dissociation is very slow and the activation of ARF depends on its interaction with a GEP or *g*uanine nucleotide-*e*xchange *p*rotein that accelerates GDP release. This is followed by GTP binding to produce active ARF-GTP, which then associates with the membrane. *In vitro*, ARF-GTP can associate with artificial phospholipid membranes or liposomes [10]. ARF itself has no demonstrable GTPase activity and its interaction with ARF-GAP or *G*TPase-*a*ctivating *p*rotein [11, 12] is required for hydrolysis of the bound GTP that results in formation of inactive ARF-GDP, which moves back to the cytosol.

ARF guanine nucleotide-exchange proteins (GEPs) that accelerate GTP binding and activation exist in soluble and particulate forms, which may or may not be inhibited by brefeldin A. BFA, as it is called, is a fungal metabolite that reversibly inhibits vesicular trafficking in many cells and causes apparent disintegration of Golgi structure. Tsai *et al.* had purified two cytosolic GEPs [13, 14] that were insensitive to BFA and more recently a soluble BFA-sensitive GEP has been purified from bovine brain [15]. Based on amino acid sequences of tryptic peptides, a cDNA was cloned. Its deduced amino acid sequence contains a so-called Sec7 domain.

Table 1. Molecules that interact with ARF

Molecules	References
CTA (cholera toxin A1 peptide)	3, 5
PLD (phospholipase D)	5, 23–25
βCOP (coatomer)	6
GAP (GTPase-activating protein)	11, 12, 26, 27
GEP (guanine nucleotide-exchange protein)	13–15, 19–21
PIP_2 (phosphatidylinositol4,5-bisphosphate)	8
Arfaptin	28
G protein βγ	29, 30
Gαs	29

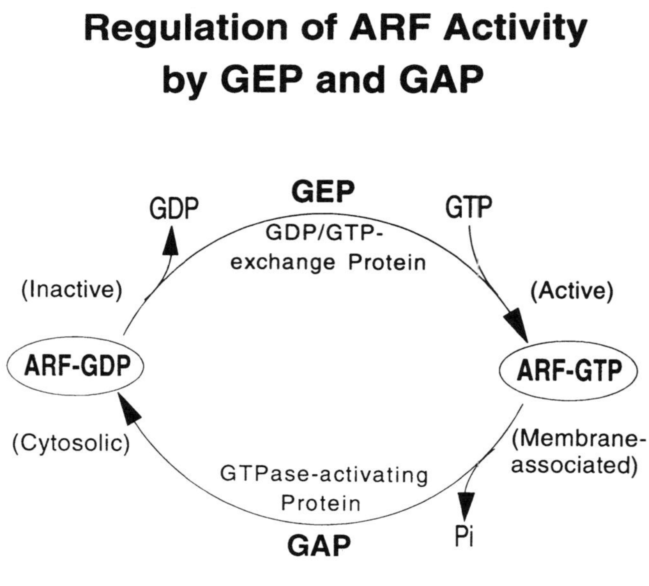

Fig. 1. Roles of GEP and GAP in the activation and inactivation of ARF.

Sec7 was identified in a collection of conditionally lethal yeast mutants as a gene involved in Golgi vesicular trafficking. The Sec7 gene product is a ~230-kDa phosphoprotein that shuttles between membrane and cytosolic fractions [16]. Two different Sec7 mutant alleles were rescued by human ARF4 or by overexpression of yeast ARF1 or yeast ARF2 [17]. Sec7 domains were noted in a few other proteins of uncertain function, among them a lymphocyte protein called cytohesin-1, which was believed to be involved in signaling initiated by integrin interaction with extracellular matrix proteins [18].

Knowing that an ARF GEP contains a Sec7 domain and that ARF rescued Sec7-defective *Saccharomyces cerevisiae,* we asked the question, 'Do Sec7 domains have GEP activity?' To answer it, recombinant histidine-tagged cytohesin-1 was synthesized in *E. coli* and its ARF GEP activity was assessed [19]. Figure 2 shows the effect of cytohesin-1 on binding of [^{35}S]GTPγS to native ARF3 as a function of time. GTPγS binding was clearly accelerated by the presence of increasing amounts of cytohesin. The effect of cytohesin-1 on the release of bound [^{35}S]GTPγS from native ARF3 is shown in Fig. 3. At the beginning of the release period (zero minutes), 15 pmol of ARF3 with - 3 pmol of GTPγS bound were transferred to medium containing 50 μM unlabeled nucleotide with or without 0.5 μg of cytohesin. Remaining bound nucleotide was determined after incubation at 36.5°C. Release in the absence of added nucleotide was very slow, as was release in the presence of nucleotide but without cytohesin. In the presence of cytohesin, release was more rapid with GTPγS or GDP than it was with GDPβS. Similarly, when assessed after a 40-min incubation, GTPγS and GDP, at equal concentrations, were equally effective. In other experiments in which BFA inhibition of a partially purified GEP [15] served as a positive control, the GEP activity of cytohesin was not inhibited by brefeldin A.

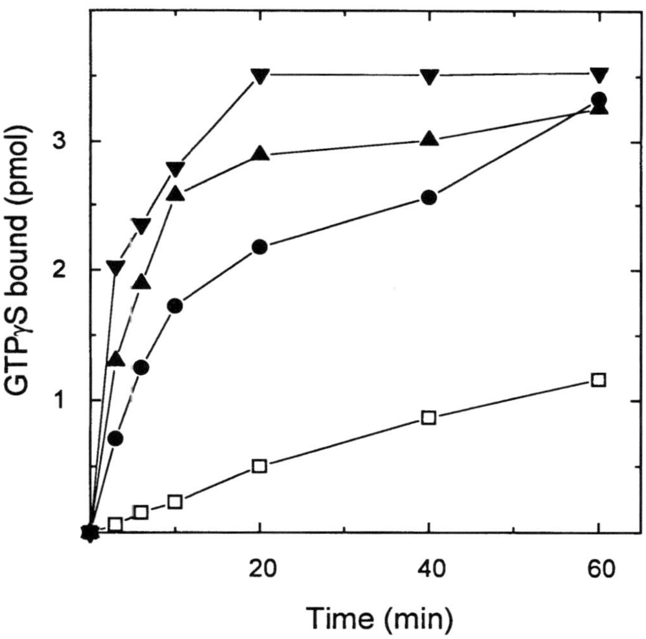

Fig. 2. Effect of cytohesin-1 on binding of [^{35}S]GTPγS to native ARF3 as a function of time. Native ARF3 (0.25 μg, =12.5 pmol) was incubated with zero (□), 0.06 (●), 0.12 (▲) or 0.24 μg (▼) of cytohesin-1 and 4 μM [^{35}S]GTPγS (2×10^6 cpm) in 100 μl containing 20 mM Tris, pH 8.0, 10 mM imidazole, 100 mM NaCl, 10 μg of phosphatidylserine, 40 μg of bovine serum albumin, 10 mM dithiothreitol and 3.1mM $MgCl_2$ at 36.5°C for the time indicated, then at 0°C for a total of 60 min. Proteins were collected on nitrocellulose filters for radioassay of bound GTPγS. GTPγS bound to samples without ARF3 or cytohesin-1 after 60 min at 0°C was 0.1 pmol, to ARF3 was 0.14 pmol, and to cytohesin-1 alone was 0.11 pmol. After 60 min at 0°C, 12.5 pmol of ARF3 plus 0.06, 0. 1 2, or 0.24 μg of cytohesin-1 bound 0.99, 1.2, and 1.53 pmol, respectively. Control values have been subtracted from the data presented.

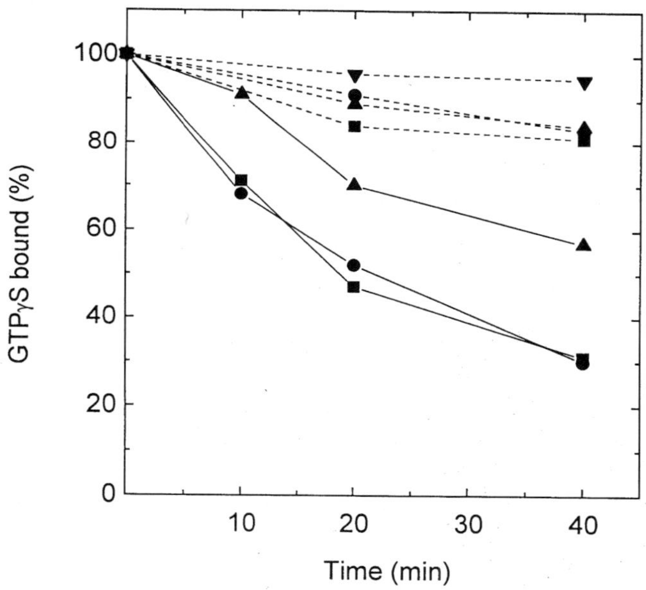

Fig. 3. Effect of cytohesin-1 on release of bound [^{35}S]GTPγS from native ARF3. Samples of native ARF3 (1.5 μg, 75 pmol) were incubated with 4 mM [^{35}S]GTPγS (2×10^6 cpm) as described [19] in 100 μl containing 20 mM Tris (pH 8.0), 50 mM NaCl, 20 μg of phosphatidylserine, 50 μg of bovine serum albumin, 12 mM dithiothreitol, 0.65 mM $MgCl_2$, and 1 mM EDTA for 40 min at 36.5°C. Tubes were placed in ice and 20-μl samples (15 pmol of ARF3) were transferred to tubes containing 50 μM unlabeled GTPγS (■), GDP (●) GDPβS (▲), or no nucleotide (▼), with (solid line) or without (interrupted line) cytohesin-1 (0.5 μg) and other additions as indicated in Fig. 2 in a final volume of 100 μl. After incubation for the indicated time, samples were filtered for radioassay of protein-bound [^{35}S]GTPγS 100% binding (3.1–3.2 pmol) was the mean of values from samples of ARF3 plus or minus cytohesin-1 and unlabeled nucleotide kept at 0°C for 60 min. Without ARF3, 0.03–0.05 pmol of [^{35}S]GTPγS was bound; these control values have been subtracted from the data presented.

In summary, cytohesin, a 47-kDa protein that is highly expressed in natural killer (NK) lymphocytes and T cells contains N-terminal Sec7 and C-terminal PH or pleckstrin homology domains. The protein, synthesized in *E. coli,* exhibited BFA-insensitive ARF GEP activity [19], similar to a ~55-kDa ARF GEP isolated earlier from spleen cytosol [14]. Cytohesin-1, like other known ARF GEPs, preferentially activated ARF1 and ARF3. It failed to activate recombinant myristoylated ARF5 (data not shown). After these studies were completed, ARF GEP activity of other Sec7 domain proteins was reported. Chardin *et al.* [20], described ARNO (or *AR*F nucleotide-binding site *o*pener), a ~47-kDa human protein that displayed BFA-insensitive GEP activity and was activated by PIP_2 via its pleckstrin homology (PH) domain. Gea1 (for *g*uanine nucleotide *e*xchange on *A*RF), is one of two larger yeast proteins described by Peyroche *et al.* [21]. Its GEP activity was BFA-sensitive. Most recently Klarlund *et al.* [22] reported a 46-kDa mouse protein that they called GRP1 (*g*eneral *r*eceptor of *p*hosphoinositides). It is very similar in structure to cytohesin-1 and specific binding of phosphatidylinositol 3,4,5,-trisphosphate to its PH domain was demonstrated.

There are now five Sec7 domain proteins known to have ARF GEP activity. All of them activate class I ARFs. The two that have been tested failed to activate ARF5, although human ARF4, (like ARF5, a class II ARF), rescued SEC7-defective yeast. It remains to be determined whether there are other Sec7 domain proteins that can act as GEPs for ARFs 4, 5, and 6. Some of these studies of Sec7 domain proteins have provided information also about binding of PIP_2 and other polyphosphoinositides to their pleckstrin homology domains that may help us to understand exactly how and why these specific phospholipids are so important in ARF action and in vesicular trafficking.

It has been intriguing to learn that within the relatively small ARF protein of ~180 amino acids there are so many domains for specific interactions (not all simultaneous), with phospholipid and with other proteins. In fact, self-association to form functional ARF dimers or tetramers has recently been suggested by Wieland *et al.* [6]. More precise definition of structure/function relationships and complex interactions of the ARF molecule may also help us to understand the mechanism of action of the cholera toxin ADP-ribosyltransferase, which is probably analogous to those of several other toxin transferases, and also perhaps the eukaryotic NAD:arginine ADP-ribosyltransferases.

Acknowledgments

We thank Ms. Carol Kosh for expert secretarial assistance. Figures 2 and 3 are reprinted by permission from reference [19].

References

1. Schleifer LS, Kahn RA, Hanski E, Northup JK, Sternweis PC, Gilman AG: Requirements for cholera toxin-dependent ADP-ribosylation of the purified regulatory component of adenylate cyclase. J Biol Chem 257: 20–23, 1982
2. Kahn RA, Gilman AG: Purification of a protein cofactor required for ADP-ribosylation of the stimulatory regulatory component of adenylate cyclase by cholera toxin. J Biol Chem 259: 6228–6234, 1984
3. Noda M, Tsai S-C, Adamik R, Moss J, Vaughan M: Mechanism of cholera toxin activation by a guanine nucleotide-dependent 19 kDa protein. Biochim Biophys Acta 1034: 195–199, 1990
4. Moss J, Vaughan M: Structure and function of ARF proteins: Activators of cholera toxin and critical components of intracellular vesicular transport processes. J Biol Chem 270: 12327–12330, 1995
5. Zhang G-F, Patton WA, Lee F-JS, Liyanage M, Han J-S, Rhee SG, Moss J, Vaughan M: Different ARF domains are required for the activation of cholera toxin and phospholipase D. J Biol Chem 270: 21–24, 1995
6. Zhao L, Helms JB, Brügger B, Harter C, Martoglio B, Graf R, Brunner J, Wieland FT: Direct and GTP-dependent interaction of ADP-ribosylation factor 1 with coatomer subunit β. Proc Natl Acad Sci USA 94: 4418–4423, 1997
7. Rothman JE: Mechanisms of intracellular protein transport. Nature 372: 55–63, 1994
8. Randazzo PA: Functional Interaction of ADP-ribosylation factor 1 with phosphatidylinositol 4,5-bisphosphate. J Biol Chem 272: 7688–7692, 1997
9. Amor JC, Harrison DH, Kahn RA, Ringe D: Structure of the human ADP-ribosylation factor 1 complexed with GDP. Nature 372: 704–708, 1994
10. Walker MW, Bobak DA, Tsai S-C, Moss J, Vaughan M: GTP but not GDP analogues promote association of ADP-ribosylation factors, 20-kDa protein activators of cholera toxin, with phospholipids and PC-12 cell membranes. J Biol Chem 267: 3230–3235, 1992
11. Cukierman E, Huber I, Rotman M, Cassel D: The ARF1 GTPase-activating protein: Zinc finger motif and Golgi complex localization. Science 270: 1999–2002, 1995
12. Ding M, Vitale N, Tsai S-C, Adamik R, Moss J, Vaughan M: Characterization of a GTPase-activating protein that stimulates GTP hydrolysis by both ADP-ribosylation factor (ARF) and ARF-like proteins. J Biol Chem 271: 24005–24009, 1996
13. Tsai S-C, Adamik R, Moss J, Vaughan M: Identification of a brefeldin A-insensitive guanine nucleotide-exchange protein for ADP-ribosylation factor in bovine brain. Proc Natl Acad Sci USA 91: 3063–3066, 1994
14. Tsai S-C, Adamik R, Moss J, Vaughan M: Purification and characterization of a guanine nucleotide-exchange protein for ADP-ribosylation factor from spleen cytosol. Proc Natl Acad Sci USA 93: 305–309, 1996
15. Morinaga N, Tsai S-C, Moss J, Vaughan M: Isolation of a brefeldin A-inhibited guanine nucleotide-exchange protein for ADP-ribosylation factor (ARF) 1 and ARF3 that contains a Sec7-like domain. Proc Natl Acad Sci USA 93: 12856–12860, 1996
16. Franzusoff A, Schekman R: Functional compartments of the yeast Golgi apparatus are defined by the sec7 mutation. EMBO J 8: 2695–2702, 1989
17. Deitz SB, Wu C, Silve S, Howell KE, Melançon P, Kahn RA, Franzusoff A: Human ARF4 expression rescues sec7 mutant yeast cells. Mol Cell Biol 16: 3275–3284, 1996
18. Shevell DE, Leu WM, Gillmor CS, Xia G, Feldmann KA, Chua NH: EMB30 is essential for normal cell division, cell expansion, and cell adhesion in Arabidopsis and encodes a protein that has similarity to Sec7. Cell 77: 1051–1062, 1994

19. Meacci E, Tsai S-C, Adamik R, Moss J, Vaughan M: Cytohesin-1, a cytosolic guanine nucleotide-exchange protein for ADP-ribosylation factor. Proc Natl Acad Sci USA 94: 1745–1748, 1997
20. Chardin P, Paris S, Antonny B, Robineau S, Béraud-Dufour S, Jackson CL, Chabre M: A human exchange factor for ARF contains Sec7- and pleckstrin-homology domain. Nature (London) 384: 481–484, 1996
21. Peyroche A, Paris S, Jackson CL: Nucleotide exchange on ARF mediated by yeast Gea1 protein. Nature (London) 384: 479–481, 1996
22. Klarlund JK, Guilherme A, Holik JJ, Virbasius JV, Chawla A, Czech MP: Signalling by phosphoinositide-3,4,5-trisphosphate through proteins containing pleckstrin and Sec7 homology domains. Science 275: 1927–1930, 1997
23. Brown HA, Gutowski S, Moomaw CR, Slaughter C, Sternweis PC: ADP-ribosylation factor, a small GTP-dependent regulatory protein, stimulates phospholipase D activity. Cell 75: 1137–1144, 1993
24. Cockcroft S, Thomas GMH, Fensome A, Geny B, Cunningham E, Gout I, Hiles I, Totty NF, Truong O, Hsuan JJ: Phospholipase D: A downstream effector of ARF in granulocytes. Science 263: 523–526, 1994
25. Massenburg D, Han JS, Liyanage M, Patton WA, Rhee SG, Moss J, Vaughan M: Activation of rat brain phospholipase D by ADP-ribosylation factors 1, 5, and 6: Separation of ADP-ribosylation factor-dependent and oleate-dependent enzymes. Proc Natl Acad Sci USA 91: 11718–11722, 1994
26. Makler V, Cukierman E, Rotman M, Admon A, Cassel D: ADP-ribosylation factor-directed GTPase-activating protein. Purification and partial characterization. J Biol Chem 270: 5232–5237, 1995
27. Randazzo PA, Kahn RA: GTP hydrolysis by ADP-ribosylation factor is dependent on both an ADP-ribosylation factor GTPase-activating protein and acid phospholipids. J Biol Chem 269: 10758–10763, 1994
28. Kanoh H, Williger B-T, Exton JH: Arfaptin 1, a putative cytosolic target protein of ADP-ribosylation factor, is recruited to Golgi membranes. J Biol Chem 272: 5421–5429, 1997
29. Colombo MI, Inglese J, D'Souza-Schorey C, Beron W, Stahl PD: Heterotrimeric G proteins interact with the small GTPase ARF. Possibilities for the regulation of vesicular traffic. J Biol Chem 270: 24564–24571, 1995
30. Franco M, Paris S, Chabre M: The small G-protein ARF1 GDP binds to the Gt beta gamma subunit of transducin, but not Gt alpha GDP-Gt beta gamma. FEBS Lett 362: 286–290, 1995

Molecular and Cellular Biochemistry **193:** 159–161, 1999.

Index to Volume 193

Affar EB, Shah RG, Poirier GG: Poly(ADP-ribose) turnover in quail myoblast cells: Relation between the polymer level and its catabolism by glycohydrolase 127–135

Althaus FR, Kleczkowska HE, Malanga M, Müntener CR, Pleschke JM, Ebner M, Auer B: Poly ADP-ribosylation: A DNA break signal mechanism 5–11

Alvarez-Gonzalez R, *see* Pacheco-Rodriguez G

Alvarez-Gonzalez R, Watkins TA, Gill PK, Reed JL, Mendoza-Alvarez H: Regulatory mechanisms of poly(ADP-ribose) polymerase 19–22

Amé J-C, Jacobson EL, Jacobson MK: Molecular heterogeneity and regulation of poly(ADP-ribose) glycohydrolase 75–81

Amici A, *see* Raffaelli N *et al.*

ApSimon MM, *see* Boyonoski AC *et al.*

Auer B, *see* Althaus FR *et al.*

Axelsson B, *see* Pero RW *et al.*

Balducci E, *see* Moss J *et al.*

Berger NA, *see* Chatterjee S *et al.*

Berger SJ, *see* Chatterjee S *et al.*

Bobak DA: Clostridial toxins: Molecular probes of Rho-dependent signaling and apoptosis 37–42

Bortell R, Kanaitsuka T, Stevens LA, Moss J, Mordes JP, Rossini AA, Greiner DL: The RT6 (Art2) family of ADP-ribosyltransferases in rat and mouse 61–68

Boulares H, *see* Simbulan-Rosenthal CM *et al.*

Boyonoski AC, Gallacher LM, ApSimon MM, Jacobs RM, Shah GM, Piorier GG, Kirkland JB: Niacin defiency increases the sensitivity of rats to the short and long term effects of ethylnitrosourea treatment 83–87

Bürkle A, *see* Tatsumi-Miyajima J *et al.*

Cavanaugh E, *see* Moss J *et al.*

Chaplin D, *see* Pero RW *et al.*

Chatterjee S, Berger SJ, Berger NA: Poly(ADP-ribose) polymerase: A guardian of the genome that facilitates DNA repair by protecting against DNA recombination 23–30

Colanzi A, *see* Silletta MG *et al.*

Corda D, *see* Silletta MG *et al.*

Dantzer F, *see* Trucco C *et al.*

De Matteis MA, *see* Silletta MG *et al.*

de Murcia G, *see* Trucco C *et al.*

Di Girolamo M, *see* Silletta MG *et al.*

Dougherty G, *see* Pero RW *et al.*

Dutrillaux B, *see* Trucco C *et al.*

Ebner M, *see* Althaus FR *et al.*

Emanuelli M, *see* Raffaelli N *et al.*

Fiucci G, *see* Silletta MG *et al.*

Flatter E, *see* Trucco C *et al.*

Gallacher LM, *see* Boyonoski AC *et al.*

Gill PK, *see* Alvarez-Gonzalez R *et al.*

Graeff R, *see* Lee HC *et al.*

Greiner DL, *see* Bortell R *et al.*

Hanai S, *see* Miwa M *et al.*
Huang AC, *see* Jacobson EL *et al.*

Iyer S, *see* Simbulan-Rosenthal CM *et al.*

Jacobs RM, *see* Boyonoski AC *et al.*
Jacobson EL, *see* Amé J-C *et al.*
Jacobson EL, Shieh WM, Huang AC: Mapping the role of NAD metabolism in prevention and treatment of carcinogenesis 69–74
Jacobson MK, *see* Amé J-C *et al.*
Javier Oliver F, *see* Trucco C *et al.*

Kanaitsuka T, *see* Bortell R *et al.*
Kim HJ, *see* Moss J *et al.*
Kirkland JB, *see* Boyonoski AC *et al.*
Kleczkowska HE, *see* Althaus FR *et al.*
Konczalik P, *see* Moss J *et al.*
Küpper J-H, *see* Tatsumi-Miyajima J *et al.*

Lee HC, Munshi C, Graeff R: Structures and activities of cyclic ADP-ribose, NAADP and their metabolic enzymes 89–98
Lesma EA, *see* Moss J *et al.*
Lorenzi T, *see* Raffaelli N *et al.*
Luini A, *see* Silletta MG *et al.*

Magni G, *see* Raffaelli N *et al.*
Malanga M, *see* Althaus FR *et al.*
Masson M, *see* Trucco C *et al.*
Masutani M, Nozaki T, Nishiyama E, Shimokawa T, Tachi Y, Suzuki H, Nakagama H, Wakabayashi K, Sugimura T: Function of poly(ADP-ribose) polymerase in response to DNA damage: Gene-disruption study in mice 149–152
Mendoza-Alvarez H, *see* Alvarez-Gonzalez R *et al.*
Ménissier-de Murcia J, *see* Trucco C *et al.*
Mironov A, *see* Silletta MG *et al.*
Miwa M, Hanai S, Poltronieri P, Uchida M, Uchida K: Functional analysis of poly(ADP-ribose) polymerase in *Drosophila melanogaster* 103–108
Mordes JP, *see* Bortell R *et al.*
Moss J, Balducci E, Cavanaugh E, Kim HJ, Konczalik P, Lesma EA, Okazaki IJ, Park M, Shoemaker M, Stevens LA, Zolkiewska A: Characterization of NAD:arginine ADP-ribosyltransferases 109–113
Moss J, *see* Bortell R *et al.*
Moss J, Vaughan M: Activation of toxin ADP-ribosyltransferases by eukaryotic ADP-ribosylation factors 153–157
Munshi C, *see* Lee HC *et al.*
Müntener CR, *see* Althaus FR *et al.*

Nakagama H, *see* Masutani M *et al.*
Niedergang C, *see* Trucco C *et al.*
Nishiyama E, *see* Masutani M *et al.*
Nozaki T, *see* Masutani M *et al.*

Okamoto H: The CD38-cyclic ADP-ribose signaling system in insulin secretion
Okazaki IJ, *see* Moss J *et al.*

Pacheco-Rodriguez G, Alvarez-Gonzalez R: Measurement of poly(ADP-ribose) glycohydrolase activity by high resolution polyacrylamide gel electrophoresis: Specific inhibition by histones and nuclear matrix proteins 13–18

Park M, *see* Moss J *et al.*
Pero RW, Axelsson B, Siemann D, Chaplin D, Dougherty G: Newly discovered anti-inflammatory properties of the benzamides and nicotinamides 119–125
Piorier GG, *see* Boyonoski AC *et al.*
Poirier GG, *see* Affar EB *et al.*
Pisani FM, *see* Raffaelli N *et al.*
Pleschke JM, *see* Althaus FR *et al.*
Poltronieri P, *see* Miwa M *et al.*

Raffaelli N, Emanuelli M, Pisani FM, Amici A, Lorenzi T, Ruggieri S, Magni G: Identification of the archaeal NMN adenylyltransferase gene 99–102
Reed JL, *see* Alvarez-Gonzalez R *et al.*
Rolli V, *see* Trucco C *et al.*
Rosenthal DS, *see* Simbulan-Rosenthal CM *et al.*
Rossini AA, *see* Bortell R *et al.*
Ruggieri S, *see* Raffaelli N *et al.*

Santone I, *see* Silletta MG *et al.*
Shah GM, *see* Boyonoski AC *et al.*
Shah RG, *see* Affar EB *et al.*
Shieh WM, *see* Jacobson EL *et al.*
Shimokawa T, *see* Masutani M *et al.*
Shoemaker M, *see* Moss J *et al.*
Siemann D, *see* Pero RW *et al.*
Silletta MG, Colanzi A, Weigert R, Di Girolamo M, Santone I, Fiucci G, Mironov A, De Matteis MA, Luini A, Corda D: Role of brefeldin A-dependent ADP-ribosylation in the control of intracellular membrane transport 43–51
Simbulan-Rosenthal CM, Rosenthal DS, Iyer S, Boulares H, Smulson ME: Involvement of PARP and poly(ADP-ribosyl)ation in the early stages of apoptosis and DNA replication 137–148
Smulson ME, *see* Simbulan-Rosenthal CM *et al.*
Stevens LA, *see* Bortell R *et al.*
Stevens LA, *see* Moss J *et al.*
Sugimura T, *see* Masutani M *et al.*
Suzuki H, *see* Masutani M *et al.*

Tachi Y, *see* Masutani M *et al.*
Takebe H, *see* Tatsumi-Miyajima J *et al.*
Tatsumi-Miyajima J, Küpper J-H, Takebe H, Bürkle A: *Trans*-dominant inhibition of poly(ADP-ribosyl)ation potentiates alkylation-induced shuttle-vector mutagenesis in Chinese hamster cells 31–35
Trucco C, Rolli V, Javier Oliver F, Flatter E, Masson M, Dantzer F, Niedergang C, Dutrillaux B, Ménissier-de Murcia J, de Murcia G: A dual approach in the study of poly (ADP-ribose) polymerase: *In vitro* random mutagenesis and generation of deficient mice 53–60

Uchida K, *see* Miwa M *et al.*
Uchida M, *see* Miwa M *et al.*

Vaughan M, *see* Moss J

Wakabayashi K, *see* Masutani M *et al.*
Watkins TA, *see* Alvarez-Gonzalez R *et al.*
Weigert R, *see* Silletta MG *et al.*

Zolkiewska A, *see* Moss J *et al.*

Developments in Molecular and Cellular Biochemistry

Series Editor: Naranjan S. Dhalla, Ph.D., M.D. (Hon.), FACC

1. V.A. Najjar (ed.): *Biological Effects of Glutamic Acid and Its Derivatives.* 1981 ISBN 90-6193-841-4
2. V.A. Najjar (ed.): *Immunologically Active Peptides.* 1981 ISBN 90-6193-842-2
3. V.A. Najjar (ed.): *Enzyme Induction and Modulation.* 1983 ISBN 0-89838-583-0
4. V.A. Najjar and L. Lorand (eds.): *Transglutaminase.* 1984 ISBN 0-89838-593-8
5. G.J. van der Vusse (ed.): *Lipid Metabolism in Normoxic and Ischemic Heart.* 1989 ISBN 0-7923-0479-9
6. J.F.C. Glatz and G.J. van der Vusse (eds.): *Cellular Fatty Acid-Binding Proteins.* 1990 ISBN 0-7923-0896-4
7. H.E. Morgan (ed.): *Molecular Mechanisms of Cellular Growth.* 1991 ISBN 0-7923-1183-3
8. G.J. van der Vusse and H. Stam (eds.): *Lipid Metabolism in the Healthy and Diseased Heart.* 1992 ISBN 0-7923-1850-1
9. Y. Yazaki and S. Mochizuki (eds.): *Cellular Function and Metabolism.* 1993 ISBN 0-7923-2158-8
10. J.F.C. Glatz and G.J. van der Vusse (eds.): *Cellular Fatty-Acid-Binding Proteins, II.* 1993 ISBN 0-7923-2395-5
11. R.L. Khandelwal and J.H. Wang (eds.): *Reversible Protein Phosphorylation in Cell Regulation.* 1993 ISBN 0-7923-2637-7
12. J. Moss and P. Zahradka (eds.): *ADP-Ribosylation: Metabolic Effects and Regulatory Functions.* 1994 ISBN 0-7923-2951-1
13. V.A. Saks and R. Ventura-Clapier (eds.): *Cellular Bioenergetics: Role of Coupled Creatine Kinases.* 1994 ISBN 0-7923-2952-X
14. J. Slezák and A. Ziegelhöffer (eds.): *Cellular Interactions in Cardiac Pathophysiology.* 1995 ISBN 0-7923-3573-2
15. J.A. Barnes, H.G. Coore, A.H. Mohammed and R.K. Sharma (eds.): *Signal Transduction Mechanisms.* 1995 ISBN 0-7923-3663-1
16. A.K. Srivastava and J.-L. Chiasson (eds.): *Vanadium Compounds: Biochemical and Therapeutic Applications.* 1995 ISBN 0-7923-3763-8
17. J.M.J. Lamers and P.D. Verdouw (eds.): *Biochemistry of Signal Transduction in Myocardium.* 1996 ISBN 0-7923-4067-1
18. E.-G. Krause and R. Vetter (eds.): *Biochemical Mechanisms in Heart Function.* 1996 ISBN 0-7923-4118-X
19. R. Vetter and E.-G. Krause (eds.): *Biochemical Regulation of Myocardium.* 1996 ISBN 0-7923-4259-3
20. G.N. Pierce and W.C. Claycomb (eds.): *Novel Methods in Molecular and Cellular Biochemistry of Muscle.* 1997 ISBN 0-7923-4387-5
21. F.N. Gellerich and S. Zierz (eds.): *Detection of Mitochondrial Diseases.* 1997 ISBN 0-7923-9925-0
22. P.K. Singal, V. Panagia and G.N. Pierce (eds.): *The Cellular Basis of Cardiovascular Function in Health and Disease.* 1997 ISBN 0-7923-9974-9
23. S. Abdel-aleem and J.E. Lowe (eds.): *Cardiac Metabolism in Health and Disease.* 1998 ISBN 0-7923-8104-1
24. A.K. Srivastava and B. Posner (eds.): *Insulin Action.* 1998 ISBN 0-7923-8113-0

Developments in Molecular and Cellular Biochemistry

25. V.A. Saks, R. Ventura-Clapier, X. Leverve, A. Rossi and M. Rigoulet (eds.): *Bioenergetics of the Cell: Quantitative Aspects*. 1998 ISBN 0-7923-8118-1
26. G.N. Pierce, H. Rupp, T. Izumi and A. Grynberg (eds.): *Molecular and Cellular Effects of Nutrition on Disease Processes*. 1998 ISBN 0-7923-8171-8
27. K. Ahmed, E. Chambaz and O.G. Issinger (eds.): *Molecular and Cellular View of Protein Kinase CK2*. 1999 ISBN 0-7923-8208-0
28. M.V. Cohen, J.M. Downey, R.J. Gelpi and J. Slezák (eds.): *Myocardial Ischemia and Reperfusion*. 1998 ISBN 0-7923-8173-4
29. D.A. Bernlohr and L. Banaszak (eds.): *Lipid Binding Proteins within Molecular and Cellular Biochemistry*. 1999 ISBN 0-7923-8223-4
30. R. Alvarez-Gonzalez (ed.): *ADP-Ribosylation Reactions: From Bacterial Pathogenesis to Cancer*. 1999 ISBN 0-7923-8235-8
31. S. Imai and M. Endo (eds.): *Muscle Physiology and Biochemistry*. 1999 ISBN 0-7923-8264-1

KLUWER ACADEMIC PUBLISHERS – DORDRECHT / BOSTON / LONDON